T0271255

Contingency and Convergence

Vienna Series in Theoretical Biology

Gerd B. Müller, editor-in-chief

Johannes Jäger, Thomas Pradeu, Katrin Schäfer, associate editors

The Evolution of Cognition, edited by Cecilia Heyes and Ludwig Huber, 2000

Origination of Organismal Form, edited by Gerd B. Müller and Stuart A. Newman, 2003

Environment, Development, and Evolution, edited by Brian K. Hall, Roy D. Pearson, and Gerd B. Müller, 2004

Evolution of Communication Systems, edited by D. Kimbrough Oller and Ulrike Griebel, 2004

Modularity: Understanding the Development and Evolution of Natural Complex Systems, edited by Werner Callebaut and Diego Rasskin-Gutman, 2005

Compositional Evolution: The Impact of Sex, Symbiosis, and Modularity on the Gradualist Framework of Evolution, by Richard A. Watson, 2006

Biological Emergences: Evolution by Natural Experiment, by Robert G. B. Reid, 2007

Modeling Biology: Structure, Behaviors, Evolution, edited by Manfred D. Laubichler and Gerd B. Müller, 2007

Evolution of Communicative Flexibility, edited by Kimbrough D. Oller and Ulrike Griebel, 2008

Functions in Biological and Artificial Worlds, edited by Ulrich Krohs and Peter Kroes, 2009

Cognitive Biology, edited by Luca Tommasi, Mary A. Peterson and Lynn Nadel, 2009

Innovation in Cultural Systems, edited by Michael J. O'Brien and Stephen J. Shennan, 2010

The Major Transitions in Evolution Revisited, edited by Brett Calcott and Kim Sterelny, 2011

Transformations of Lamarckism, edited by Snait B. Gissis and Eva Jablonka, 2011

Convergent Evolution: Limited Forms Most Beautiful, by George McGhee, 2011

From Groups to Individuals, edited by Frédéric Bouchard and Philippe Huneman, 2013

Developing Scaffolds in Evolution, Culture, and Cognition, edited by Linnda R. Caporael, James Griesemer, and William C. Wimsatt, 2014

Multicellularity: Origins and Evolution, edited by Karl J. Niklas and Stuart A. Newman, 2015

Vivarium: Experimental, Quantitative, and Theoretical Biology at Vienna's Biologische Versuchsanstalt, edited by Gerd B. Müller, 2017

Landscapes of Collectivity in the Life Sciences, edited by Snait B. Gissis, Ehud Lamm, and Ayelet Shavit, 2017

Rethinking Human Evolution, edited by Jeffrey H. Schwartz, 2017

Convergent Evolution in Stone-Tool Technology, edited by Michael J. O'Brien, Briggs Buchanan, and Metin I. Erin, 2018

Evolutionary Causation: Biological and Philosophical Reflections, edited by Tobias Uller and Kevin N. LaLand, 2019

Convergent Evolution on Earth: Lessons for the Search for Extraterrestrial Life, by George McGhee, 2019

Contingency and Convergence

Toward a Cosmic Biology of Body and Mind

Russell Powell

The MIT Press
Cambridge, Massachusetts
London, England

This book was set in Times Roman by Westchester Publishing Services, Danbury, CT.

Library of Congress Cataloging-in-Publication Data

Names: Powell, Russell, author.
Title: Contingency and convergence : toward a cosmic biology of body and mind / Russell Powell.
Description: Cambridge, MA : MIT Press, [2019] | Series: Vienna series in theoretical biology | Includes bibliographical references and index.
Identifiers: LCCN 2019009760 | ISBN 9780262043397 (hardcover : alk. paper)
Subjects: LCSH: Evolution (Biology)--Philosophy. | Convergence (Biology)--Philosophy.
Classification: LCC QH360.5 .P68 2019 | DDC 576.8--dc23
LC record available at https://lccn.loc.gov/2019009760

For Alexander Teoman Powell

Contents

Series Foreword

Biology is a leading science in this century. As in all other sciences, progress in biology depends on the interrelations between empirical research, theory building, modeling, and societal context. But whereas molecular and experimental biology have evolved dramatically in recent years, generating a flood of highly detailed data, the integration of these results into useful theoretical frameworks has lagged behind. Driven largely by pragmatic and technical considerations, research in biology continues to be less guided by theory than seems indicated. By promoting the formulation and discussion of new theoretical concepts in the biosciences, this series intends to help fill important gaps in our understanding of some of the major open questions of biology, such as the origin and organization of organismal form, the relationship between development and evolution, and the biological bases of cognition and mind.

Theoretical biology has important roots in the experimental tradition of early twentieth-century Vienna. Paul Weiss and Ludwig von Bertalanffy were among the first to use the term *theoretical biology* in its modern sense. In their understanding the subject was not limited to mathematical formalization, as is often the case today, but extended to the conceptual foundations of biology. It is this commitment to a comprehensive and cross-disciplinary integration of theoretical concepts that the Vienna Series intends to emphasize. Today, theoretical biology has genetic, developmental, and evolutionary components, the central connective themes in modern biology, but it also includes relevant aspects of computational or systems biology and extends to the naturalistic philosophy of sciences. The Vienna Series grew out of theory-oriented workshops organized by the KLI, an international institute for the advanced study of natural complex systems. The KLI fosters research projects, workshops, book projects, and the journal *Biological Theory*, all devoted to aspects of theoretical biology, with an emphasis on—but not a restriction to—integrating the developmental, evolutionary, and cognitive sciences. The series editors welcome suggestions for book projects in these domains.

Gerd B. Müller, Johannes Jäger, Thomas Pradeu, and Katrin Schäfer

Acknowledgments

I could never have written a book like this without the world-class training in philosophy of biology that I received during my graduate work at Duke University under the guidance of Robert Brandon and Alex Rosenberg, as well as Dan McShea, Karen Neander, Louise Roth, and other brilliant faculty members and graduate students who have crafted a seamless interface between the philosophy and biology departments at Duke. I am beyond lucky to have spent my formative academic years in such an intellectually rich and constructive environment, which has never ceased to support me, both personally and professionally, since I left the nest nearly a decade ago.

The work that led to the writing of this book also benefited greatly from conversations with John Beatty, Michael Benton, Matt Cartmill, Milan Ćirković, Simon Conway Morris, Adrian Currie, Ford Doolittle, Doug Erwin, Chris Haufe, Carlos Mariscal, George McGhee, Maureen O'Malley, Dan McShea, Treavor Pearce, Grant Ramsey, Nick Shea, Elliott Sober, Kim Sterelny, Derek Turner, and Günter Wagner. Many thanks to Gerhart von der Emde for showing me around his extraordinary electrolocation lab in Bonne, and to the Konrad Lorenz Institute and National Evolutionary Synthesis Center for supporting my research as a junior scholar. I am especially indebted to Irina Mikhalevich for not only providing detailed comments on an earlier draft of this manuscript but also for sketching the beautiful artwork contained in these pages. I owe an incalculable debt of intellectual gratitude to Stephen Jay Gould, whose popular work inspired me to leave the practice of law in order to explore philosophical problems in evolutionary theory. I am truly grateful to the John Templeton Foundation (grant #43160) for supporting ambitious and risky research on some of the biggest questions that human beings can presently ask. I hope that this book takes one small step toward answering some of those questions, or at least, toward developing a better understanding of what it would take to answer them. And many thanks as well to the editors at MIT Press, including Bob Prior, Anne-Marie Bono, and Chris Eyer, for their generous support of this project, as well as to John Donohue at Westchester Publishing Services for his boundless patience and unparalleled copyediting expertise.

Lastly, I am forever thankful to my family: to Sanem, for encouraging me to apply for the grant that led to this book and for supporting me through the ups and downs of our Take Five life; to Irina, for all the enchanted creatures we have found and nurtured together, and for all the new ones yet to arrive; to my father, Jeffrey, for writing science fiction stories that he would read to me at bedtime as a child, leaving me to dream of intelligent life and its evolutionary possibilities; and finally, to my son, Alex Teo, for helping me reexperience the magic of the living world.

Introduction: The Big Picture

> The nature of life on earth and the quest for life elsewhere are two sides of the same question: the search for who we are.
> —Carl Sagan

In his book *Wonderful Life: The Burgess Shale and the Nature of History*, the paleontologist Stephen Jay Gould poses the following evolutionary thought experiment: Imagine replaying the "tape of life" from an early period in the history of animal evolution, and consider how its story would again unfurl. Would the replay result in a quintessentially alien biological world—a parade of forms that bear little resemblance to life as we know it? Or would a strikingly familiar set of anatomical designs, functional properties, and organizational patterns reemerge? The answer to this question has profound implications not only for the prospect and nature of complex life on other worlds, but also for the epistemic status of biological science itself and its potential to uncover universal laws of life.

Gould argued in *Wonderful Life*, and throughout his remaining career, that replaying the tape of life would result in radically different evolutionary outcomes. Play the tape again, and not only would no primates, mammals, and tetrapods evolve, but vertebrates and other major animal groups like mollusks, arthropods, echinoderms, annelids, and brachiopods would be relegated to the evolutionary dust heap of morphological *possibilia*, replaced by some other fortuitous set of equally workable animal designs. On this view, the familiar spate of animal body plans arose, along with many fantastical forms, in the initial explosive phases of animal evolution when developmental parameters were still largely unconstrained. These familiar forms survived the early major extinctions of animals, whereas other forms perished, for reasons unrelated to their relative functional merit. The body plans of these fortuitous survivors then solidified for the remainder of life's history due to a causal structure of embryonic development that came into effect once a certain degree of morphological complexity had been laid down. Existing animals thus represent but a small fraction of the set of functionally equivalent forms that either went extinct or never arose due to an unrepeatable sequence of events. If Gould is right, then history reigns supreme in macroevolution, and there is no law-like necessity to the overarching shape of animal life on Earth.

Gould's "radical contingency thesis" (RCT) and the critical attention it has received have generally focused on the replicability of complex morphology. However, the RCT is particularly provocative for the implications that it might have for the emergence of complex *cognitive* properties. Astrophysicists, especially those who have advocated the search for extraterrestrial intelligence, have long been optimistic about the frequency of intelligent life forms in the universe. Thus far, however, the cosmologist's optimism has relied mainly on general statistical principles, the ubiquity of geochemical conditions that are amenable to life, and certain (often implicit) progressivist assumptions about macroevolution, rather than from any specifically biological considerations. Gould's arguments for the primacy of contingency in macroevolution pose a formidable challenge to the view that galaxies are teaming with intelligent life, though the nature and force of this challenge has yet to be analyzed in any depth.

Is the evolution of the mind a historical accident, unlikely to be repeated in hypothetical replays of the tape of life on Earth and actual replays on Earth-like planets elsewhere? Or is mind, in any or all of its variegated forms, likely to be an evolutionarily important feature of any living world? How is the evolution of complex minds related to the evolution of complex bodies? And what does all this say about the nature of our own existence and humanity's place in the cosmos? These are big questions with correspondingly big methodological problems. Most glaring among these is the so-called $N=1$ problem: given that we are privy to but a single history of life, how can we even begin to approach let alone adjudicate questions about the contingency of bodies and minds?

Despite the seemingly intractable nature of the problem, and notwithstanding the lack of an extraterrestrial data set, biotheoretic support for the cosmologist's optimism has started to mount, and the pendulum has begun to swing away from contingency. This is due in large part to studies of "convergent evolution"—the repeated origination of similar biological forms and functions—which offers a promising avenue for empirically adjudicating the contingency question. Countless cases of sometimes striking convergence have been documented at all hierarchical levels of the biological world, from molecules and morphology to mechanisms of mind. Textbook examples of convergence include the evolution of camera-type eyes in fish (vertebrates), cephalopods (mollusks), jumping spiders (arthropods), and cubozoan jellyfish (cnidarians), lineages that diverged from a common ancestor close to 1 billion years ago in the Precambrian;

the evolution of dolphinoid forms in porpoises (mammals) and ichthyosaurs (Mesozoic marine reptiles), whose last common ancestor was a terrestrial tetrapod that lived around 300 million years ago in the Carboniferous; and the suite of saber-toothed lion morphology that evolved separately in placental and marsupial mammals, lineages that split in the mid-Jurassic. We will encounter many such instances of evolutionary iteration over the course of this book, some of which will be analyzed in depth either to illustrate a general point or because they serve as a crucial premise in the arguments being considered.

On its face, the logic of the evidence is straightforward and compelling: convergent evolution is tantamount to natural experimental replication in the history of life; and to the extent that evolutionary replication is ubiquitous, this would seem to cast doubt on the RCT. Indeed, this is precisely how patterns of convergent evolution have been interpreted by some biologists, philosophers of science, and cosmologists. But is this reading of convergent evolution correct? Do episodes of convergence provide evidence against the RCT, and if so, under what conditions might they do so? Are certain cases of convergence more theoretically significant than others, and how might such cases be delineated in concept and identified in practice? More broadly, what mechanisms could explain these convergent regularities on Earth, and what might these patterns and mechanisms tell us about the prospect of complex bodies and minds on other worlds? This book will sharpen some critical points of the "critique from convergence" while blunting others, with the aim of developing a clearer picture of the interplay between contingency and convergence in the great dialectic of macroevolution.

To begin to work out answers to these questions, we must first have a clearly articulated account of evolutionary contingency in hand. Roughly speaking, evolutionary outcomes are radically contingent if they are historical accidents that are unlikely to be repeated across replays of the tape of life. Although the RCT has received much attention from biologists and (to a lesser extent) from philosophers of science, it remains decidedly underspecified, liable to misinterpretation, and riddled with conceptual and methodological difficulties that have only barely been explored. Furthermore, while convergent evolution constitutes a potentially fruitful source of evidence against the RCT, the logic of this evidential relationship has been underexamined. The lack of an extended analytical treatment of the theoretical relationship between contingency and convergence has hampered the design and interpretation of evolutionary research, resulting in critiques of the RCT that are less effective than they could have been. Challenges to Gould's thesis that appeal to the general frequency of convergent evolution or to specific compelling episodes of convergence have generally failed to engage with the core claims and theoretical commitments that underpin the RCT. By the same token, contingency theorists have yet to situate the phenomenon of convergent evolution within a broader, historically contingent view of life.

This book represents the first attempt to weave these disparate philosophical and empirical threads together into a critical discussion of the interaction between contingency and convergence in macroevolution, as it relates to both complex life in general and cognitively complex life in particular. It would clearly be impossible for a single book to treat these matters in any depth without an appropriate focus to rein it in. The present analysis is accordingly narrow but at the same time ambitious: its goal is to shed light on what the phenomenon of convergent evolution tells us about constraints on the history of life as it has unfolded on Earth, and how it might do in other nooks and crannies of the habitable universe.

This, in turn, requires that we engage with philosophical discussions not only of the central organizing concepts of contingency and convergence, but also of laws, chance, causation, constraint, explanation, evidence, and experiment as these concepts relate to the evolutionary process. As applied to mind, the contingency question compels us to wrestle with problems surrounding the nature of cognition, its phylogenetic distribution, and the circumstances and consequences of its evolution. The overarching aim of this analysis is to foreground the conceptual, methodological, and empirical problems that must be resolved if we are to assess the significance of contingency and convergence in this, or any, history of life.

Part I, "Convergent Bodies," considers what we can infer about the shape of life as we do not know it by studying patterns of convergent evolution in the history of life on Earth—and in particular, and by examining the causes that underlie these patterns. The evolutionary replicability of complex morphological outcomes hinges, of course, on the replicability of organizational stepping stones along the way, including first and foremost the evolution of life itself. We thus begin in chapter 1, "Visions of the Living Cosmos," by contrasting the emerging consensus on the cosmic replicability of basic microbial life from the greater controversy that surrounds the evolution of more complex organisms. An apparent theoretical tension between the rapid timing and singular nature of life's origin is highlighted and resolved, and the limits of coarse-grained statistical approaches to contingency questions are made clear. In particular, observer selection biases prevent us from projecting aspects of our own history of life onto other life-amenable worlds, underscoring the crucial need for biotheoretical input.

From this launching point, we go on to reconstruct the conceptual and empirical foundations of the RCT, consider several alternative ways of understanding contingency in macroevolution, and examine the use of convergence as evidence against the most plausible interpretation of Gould's thesis. Chapter 2, "The Radical Contingency Thesis," discusses the epistemic role of counterfactual thinking in the historical sciences writ large before turning to Gould's "rewinding the tape" thought experiments and distilling the theoretical pillars that undergird his thesis. These premises include particular taxonomic and paleoecological interpretations of the earliest animal fauna, the role of early stochastic extinctions in the culling of higher taxa, and the developmental canalization of fortuitous surviving lineages resulting in the deep conservation of body plans at high and moderate taxonomic levels. Critiques of Gould's taxonomic and paleoecological presuppositions are considered, but deemed nonfatal and to some extent orthogonal to the core contentions of the RCT.

Chapter 3, "A Philosophical Theory of Evolutionary Contingency," offers a more sustained analysis of the concept of contingency in macroevolution. It argues that the RCT is best understood not as a series of assertions about unpredictability, chance, stochasticity, path-dependency, or repeatability per se in macroevolution, but rather as a "modal" thesis about the stability of evolutionary outcomes across possible evolutionary worlds. Contingency, it is shown, is distinct not only from the metaphysics of chance and epistemic issues of predictability, but also from questions about the existence or absence of laws in biology. Although each of these matters are related in various ways to contingency, conflating them with contingent dynamics is liable to confuse more than to illuminate. This enables us to more clearly frame the antithesis—the "robust replicability" view—whose theoretical commitments are then laid bare and distinguished from other related but distinct theses such as adaptationism.

With this account of contingency and its antithesis in hand, chapter 4, "The Critique from Convergent Evolution," canvasses a challenge to the RCT that appeals to the ubiquity of convergent evolution. It identifies several misconceptions that have prevented this critique from making contact with Gould's thesis. For instance, contingency is frequently but incorrectly equated with unpredictability or nonrepeatability, due in part to Gould's own inconsistent writings on the topic. This chapter shows that such readings do not go to heart of the RCT and have resulted in much talking at cross-purposes. Once these exegetical shortcomings are recognized, a more effective case against macroevolutionary contingency can be made.

Chapter 5, "Convergent Evolution as Natural Experiment," shows that convergence can, under certain conditions, constitute natural experimental replications that undercut the RCT. Yet not only do many cases of convergence fail to meet these experimental validity conditions, but some evolutionary iterations actually *bolster* the RCT insofar as they are caused by shared developmental constraints that make certain adaptive innovations more accessible to selection. Shared internal constraints undermine the independence of natural experimental replications by constricting the space of evolutionary possibility for converging lineages, which in turn limits the evolutionary robustness of the iterations observed. Such internally constrained iterations are referred to as "Gouldian repetitions," and they are distinguished from related phenomena such as "parallelisms." The key contrast, however, is between Gouldian repetitions and "true convergence"—or iteration that transcends the developmental history of the lineages involved. This chapter shows that there are significant challenges to both conceptualizing Gouldian repetitions and detecting them in practice, but that meeting these challenges is necessary if we are to establish the evidentiary relevance of convergence to the contingency debate. Because the existing body of convergence data does not discriminate between these different types of convergence, the case against the RCT remains inconclusive.

Chapter 6, "The Entanglement Problem," shows that the problem of evidential underdetermination that beleaguers the critique from convergence is simply one aspect of the broader challenge of disentangling radically contingent features of the living world from the robustly replicable features with which they are intertwined. The entanglement problem is exacerbated by the human cognitive penchant to treat familiar trait bundles as law-like clusters that can be projected outward into the cosmos. The bundling bias is counteracted, and the underdetermination problem remedied, by proposing criteria for differentiating convergent regularities based on the modal robustness of the generalizations they support. These criteria are then applied to a smattering of better and lesser-known cases of convergence, including some that prove to be particularly important for the remainder of the book. In the coda to part I, "Convergence at the Grandest Scales," special attention is given to the so-called major transitions in evolution, or key organizational shifts in the history of life that have served as structural, informational, and energetic scaffolding for the subsequent step-ups in hierarchical complexity that would ultimately lead to animal bodies and brains.

The upshot of part I is that convergent regularities vary in the temporal and phylogenetic depth of the evolutionary "rewinds" across which they remain modally stable. For instance, some iterations only hold up across the evolution of mammals, tetrapods, or vertebrates, while others appear to be stable across "deeper" rewinds such as the evolution of animals, eukaryotes, and perhaps even life itself (with caveats for some known unknowns). The origins of eukaryote-grade life (complex cells) as well as animals with guts and neurons remain two nagging pressure points in the case for contingency with respect to which the convergence data are presently inconclusive.

Does this analysis ultimately vindicate or rebuff Gould's thesis? The short answer is that it does both. Important elements of the RCT survive, including its claims about the radical contingency of animal body plans and the evolutionary iterations that causally hinge on them. However, other evolutionary outcomes look to be robustly replicable, and perhaps even law-like, at a level of detail that begins to put significant pressure on the RCT. This is true for some of the increases in hierarchical complexity upon which sophisticated sensory, neural, cognitive, behavioral, and social systems have been built, as well as for some of these systems themselves.

This partial defense of the RCT in part I paves the way for a substantive critique to follow in part II, "Convergent Minds," where the focus shifts from complex life in general to complex *cognitive* life in particular. Among those evolutionary outcomes that might defy the RCT, cognition is particularly important, not only for human-specific inquiries into the evolution of intelligence and consciousness but also for the large-scale ecological organization of living worlds. What do patterns of cognitive convergence, and the lack thereof, tell us about the nature of mind, its evolution, and its place in the universe?

The evolution of mind starts with the senses, and in particular, with image-forming sensory modalities. The investigation of convergent minds thus begins in chapter 7, "Convergent Ways of Seeing," by exploring the convergent evolution of three image-forming sensory modalities: vision, echolocation, and electrolocation. These three modes of "seeing," despite their different energetic bases, representational capacities, and phenomenological qualities, allow organisms to construct rich three-dimensional scenes of their surrounding world. From these diverse modes of seeing we can glean some general lessons about the universal constraints on image-forming organs as well as their evolvability and ecological utility. This chapter argues that due to the laws of physics, there are a limited number of waveform energies that organisms can use to produce informationally rich, real-time, ecologically relevant images of the kind that can support active, goal-directed behavior in a three-dimensional world. These energies include light, sound, and electromagnetism—all potential reservoirs of detailed spatial information that evolution has tapped repeatedly in distant groups. This suggests that image formation is widely accessible, multiply realizable, and confined to a small number of possible solutions—properties that smack of law-like-ness. Furthermore, the evolution of object perception in different modalities requires convergent solutions to a similar set of cognitive problems, such as how to bind properties, valences, and identities to objects, and how to produce a stable, unified model of the surrounding world.

Chapter 8, "Convergent Evolution of the Umwelt," argues that image-forming sensory systems serve as the gateway to the emergence of intelligent life. As their perceptual acuity was honed, image-forming sensory organs—in particular, convergent visual systems—would have generated selection pressures for increasingly sophisticated cognitive mechanisms and their associated neural architectures, enabling organisms to make adaptive use of the valuable information that such sensory systems afford. These neural-cognitive mechanisms and honed optical functions would have then fed back into the evolution of sensory, motor, and proprioceptive systems in a complexifying feedback loop that resulted in the convergent evolution of the Umwelt. Adapting what was originally a nonevolutionary concept in German biology, the "Umwelt" is understood here as a unified phenomenal world of spatially distributed, bound, and meaningful (valenced) objects with the body and/or subject of experience at its center. Although the Umwelt can be conceived in wholly cognitive (e.g., representational) terms, the subjective quality of the Umwelt is also taken seriously as an evolutionary explanandum.

Does Umweltian cognition and consciousness reflect a deep structure of mind that is likely to be found elsewhere in the living cosmos? This is the central question taken up over the next two chapters, which explore the distribution of mind in nature. Although mind cannot be observed directly, its causal signature can be inferred from two mutually informing lines of evidence: brains and behavior, which, when taken in the light of evolution, permit us to draw inferences about the replicability of mind on Earth and beyond.

Chapter 9, "Finding Minds: Evidence from Neuroanatomy," reviews evidence for the convergent evolution of brains in animals, as well as the muddled evolutionary history of the neuron. It first wrestles with an apparent evidential conflict: on the one hand, phylogenetic and fossil data point to at least three separate origins of brains as the most parsimonious explanation of observed patterns of neuroanatomy; on the other hand, there are genetic, structural, and developmental data that seem to indicate one brain origin and numerous brain losses. If the latter "brain homology" hypothesis were true, it would entail more than a dozen losses of head/brain/eye complexes and the corresponding reorganization of body plans and lifeways in many animal groups. This chapter concludes that, all things considered, the brain homology hypothesis is less plausible than the brain convergence scenario, which suggests that vertebrates, cephalopod mollusks, and arthropods (the group that includes insects, arachnids, myriapods, and crustaceans) independently evolved central nervous systems. The chapter goes on to examine the even harder-to-parse phylogeny and evolvability of the neuron cell type, as well as the signaling systems and action potentials that allow for sophisticated information transfer in organisms that lack proper nervous systems, such as in green plants.

Even if brains originated multiple times, this does not necessarily imply that minds arose multiple times, because the connection between brains and minds is not clear-cut. Central nervous systems may or may not generate cognitive mechanisms that rise to the level of what we would want to call "mind." To better interpret the neuroanatomical data, therefore, we need to look for the behavioral signatures of cognitive functions that are linked to intelligence, thinking, planning, concept formation, spatial cognition, and other mind-like properties. Chapter 10, "Finding Minds: Evidence from Behavior," reviews the behavioral evidence for sophisticated forms of perception and cognition in invertebrates. The chapter focuses on invertebrates because it is only across the great phylogenetic chasms of animal evolution—in lineages in which brains arose separately and independently—that the robust replicability of the Umweltian minds can be established. The chapter begins with a review of the comparatively limited but better-known research on cephalopods, and then moves into a lengthy exploration of the stunning body of research on arthropod perception, cognition, and behavior.

Work with bees has replicated many of the complex learning tasks that have been demonstrated for "advanced" vertebrates like mammals and birds. The bulk of chapter 10 is dedicated to making the case for holistic perception and object recognition in insects as well as examining their ability to categorize objects, learn abstract concepts, perform complex unnatural motor sequences, reason about causes, plan routes, perceive time, count, and learn through observations of conspecifics. Evidence for attentional processes, emotional states, and even consciousness in arthropods is also reviewed. Taken in conjunction with neuroanatomical and evolutionary data, this work suggests that the Umwelt has arisen at least three times in the history of animal life. The chapter concludes by considering how the Umwelt turbo-charged goal-directed behavior and, in so doing, ushered in the modern era of animal ecology.

Part II concludes that if complex bodies are common in the universe, then complex minds are probably common as well. This raises the prospect, to borrow Stuart Kauffman's felicitous phrase, that mind may one day yet find itself "at home in the universe." In the coda to part II, "Homage to *Homo Sapiens*," the pendulum swings back once again toward contingency, as the discussion moves away from the convergent and toward the singular. Despite convergence on complex cognition, social learning, and tool use, cumulative cultural species are likely to be extremely rare due to the peculiar suite of phylogenetic and ecological contingencies that make the evolution of robust technological taxa possible. The key to robust technology is the evolution of cumulative culture, which allows for the maintenance, incremental improvement, and reliable transmission of complex technical designs.

What accounts for the striking lack of convergence on cumulative culture in nonhuman animals? Part of the puzzle is that there is a sizable temporal gap between the origins in early humans of the cognitive and cultural capacities that are implicated in cumulative technology—such as language, morality, causal reasoning, foresight, coordinated intentions, and culturally constructed learning environments—and the origins of cumulative culture, which occurred only quite recently in human evolutionary history. Moreover, many of the traits that underwrite this complexly configured adaptation are uniquely human, even if some of their building blocks can be found scattered among nonhuman animals. Unlike the Umwelt, which is likely to have evolved in positive feedback with image-forming sensory modalities and motor systems, the evolution of cumulative technological capacities appears to occur only when a suite of largely orthogonal factors, each with a low probability of occurring, are simultaneously in place.

This investigation of contingency and convergence in the history of life leads, therefore, to an unexpected conclusion. Although there are distinctively biological reasons to believe that intelligent life is common in the universe wherever microbial life evolves and is sustained for billions of years, exceedingly few of these intelligent species will ever develop robust technological capacities with signatures that are detectable across the immensity of space and time. Wherefore an eerie silence above, a mindful cacophony below.

I CONVERGENT BODIES

1 Visions of the Living Cosmos

Deep sky surveys performed by the Hubble Space Telescope in 2016 revealed that the observable universe contains upward of 2 trillion galaxies, each containing anywhere from millions to hundreds of trillions of stars. These are truly staggering numbers that make even astronomically improbable events inevitable over the vastness of space and time. Even if life is a vanishingly rare phenomenon, in a universe that is brimming with trillions of galaxies and which may, in fact, be infinite, we can expect that familiar forms of life, and perhaps many forms we could never even dream of, will emerge in the cosmos from time to time. The notion that there is an infinite number of extraterrestrial civilizations in an infinite universe is irrelevant, of course, to the practical search for extraterrestrial intelligence (SETI), which is confined to areas of the cosmos that are amenable to empirical investigation. The key issue, therefore, is not whether we are alone in the universe, but how common we should expect our cosmic companions to be and whether there is anything biologically meaningful we can say about them from our blinkered vantage point on Earth.

The question "Are we alone in the universe?"—which motivates the SETI search—is reflexively ambiguous. It could refer to we as intelligent technological civilizations, we as morphological humanoids, we as conscious critters, we as animals with vertebrae, we as complex multicellular organisms, or simply we as living things. To the extent that the concept "we" presupposes higher cognitive properties like self-awareness and social identity, the question may be pondering the prevalence of other complexly cognitive "we's" in the universe. Of course, we cannot even begin to estimate the astrobiological frequency of mindful creatures like ourselves without first considering the ubiquity of life and its basic organizational forms in the cosmological horizon.

Thinking about the prospect and nature of life in the universe is important for reasons quite apart from our pursuit of cosmic companionship. It is only by understanding the cosmic diversity—and uniformity—of biological forms and functions that we can judge whether the project of a universal biology will succeed. And the success or failure of universal biology has implications for the epistemic status of biological science itself. To establish biology as a truly law-like, rather than fundamentally historical, discipline, we must identify contentful evolutionary outcomes that can reliably be projected onto other life worlds, and distinguish these law-like features of life from the quirky historical results of earthly evolution. A central aim of this book is to outline a conceptual framework for separating law-like from accidental features of the evolutionary process and to apply this framework to the various possible senses of "we" that are packed into the SETI-motivating refrain.

1. The Cosmic Imperative of Life

Let us begin this grand disambiguation project, as evolution does, with the small and simple. There is a consensus building among the ranks of astrobiologists that the evolution of simple microbial life is probably common in the universe, with some going so far as to call it a "cosmic imperative."[1] As microbiological theorists will be quick to point out, however, prokaryotic (bacteria-grade) life is anything but simple, and its relevance to the evolution of complex multicellular life goes far beyond merely providing a necessary stepping stone on the evolutionary road to an intelligent *we*. The metabolic machinery of the microbial world powers the biogeochemical cycles that sustain all life on the planet, and as we shall see later, made transitions to higher levels of organization possible at critical junctures in the history of life.

1.1 On Lacunae and Laws

Let us refer to the proposition that basic microbial life is ubiquitous in the universe as the "cosmic imperative of life thesis." This thesis hangs on the law-like nature of the mechanisms and processes that produced terrestrial prokaryotes. Unfortunately, there remain yawning gaps in our theoretical understanding of the origin of life and its drivers. There is, for instance, a great lacuna between our knowledge of the origins of nucleic acids—the chemical building blocks that comprise the informational backbone of life's replication machinery—on the one hand, and the configuration of the first proper cell equipped with biosynthetic membrane and metabolism, on the other. How we got from a postulated "RNA world" to encoded protein synthesis remains, in the words of one preeminent microbiologist, "the most challenging and important problem the field of Biology has ever faced."[2]

There is currently no consensus among origin of life researchers as to whether replicating molecules or metabolizing units arose first; whether prokaryote cells originated once or multiple times independently in Eubacteria and Archaea (the two prokaryote domains); or even, more strikingly, whether the last universal common ancestor of life on Earth (LUCA) was phenotypically bacteria-like or rather a complex eukaryote-like cell that repeatedly gave rise to the simpler, more streamlined prokaryote lineages that exist today.[3] If the notion that LUCA was more, rather than less, complex than extant prokaryotes seems deeply counterintuitive, it is probably because it flies in the face of the "simple to complex" narrative of macroevolution in which most students of biology have been steeped. It may turn out, however, that losses of biological complexity are just as important and pervasive as complexity gains,[4] and there could even be a rough "law of conservation of complexity" operating in the formation of more complex, higher-level evolutionary individuals (see the coda to part I).[5]

Another major unresolved issue concerns the place of Eukaryota—the group that includes animals, plants, fungi, algae, and protists—within the tree of life. The preponderance of evidence now suggests that eukaryotes are more closely related to Archaea bacteria than they are to Eubacteria, having arisen from within the Archaea domain after its split from Eubacteria. This picture is further complicated by the subsequent endosymbiotic acquisition by an archean of the ancestral eubacterium that would eventually become the mitochondrion organelle of eukaryotes. If eukaryotes are essentially a peculiar branch of Archaea, then the term "prokaryote" (like the far more problematic term "microbe") is essentially a phenetic (or property-based) classification, rather than a genealogical one. That is, it picks out a morphological "grade" rather than a proper "clade" (the latter is the central unit of modern taxonomic classification, which includes a single ancestor and all of its descendants). Grade-based classifications are controversial, as they are vulnerable to the charge of subjectivity and could obscure evolutionary relationships.[6]

A further impediment to resolving the largest branches of the tree of life stems from the ubiquity of horizontal gene exchange. Prokaryotes do not have sex, but they can exchange genes through several ancient mechanisms, and this exchange can take place between very distant lineages of life. There has been far more horizontal gene transfer between bacteria, archaea, and even eukaryotes than had initially been imagined, blurring tidy Darwinian lines of descent with modification. The horizontal exchange of genetic elements across distant lineages is accomplished through viruses, plasmids, transposons, and other mobile genetic elements, and is now thought to be a central mode of evolutionary innovation in prokaryotes (and to far lesser extent in eukaryotes). As we shall see, the ubiquity of horizontal gene exchange, particularly in the earliest phases of life on Earth, significantly clouds our picture of life's origin(s).

Despite these lingering lacunae in our understanding of the origin of life on Earth, the prevailing scientific view (or hunch) is that the emergence of prokaryote-grade life is nomically expectable, or the reliable outcome of natural laws and mechanisms, even if our understanding of these laws and mechanisms is presently incomplete. The commonly accepted idea is that there is a high probability of life emerging when particular generic conditions obtain, due to certain lawful connections. How might the existence of such lawful connections be inferred?

The now flourishing field of astrobiology was long pejoratively characterized as a science without a subject matter. It is true that there are as of yet no observations of extraterrestrials, not even of the basic microbial variety, from which we can draw generalizations about life writ cosmically large. But this does not mean that we are completely in the dark about the prospects of life on other worlds. The cosmic imperative of life is supported by several mutually informing lines of evidence. One is the observation that the basic chemical building blocks of life are readily generated under simple conditions, from which it is inferred that the molecular basis of life as we know it is pervasive in the universe. The biochemical molecules that are critical to life—such as the amino acids that make up proteins and the nucleobases that comprise DNA and RNA—can be created in the laboratory under fairly generic conditions. More recent analyses of carbon-rich meteorites, as well as spectroscopic studies of the solar system, confirm that amino acids and nucleobases are pervasive in the universe and likely deliverable by asteroid impact.[7]

Likewise, aqueous chemistry, which is often thought to be necessary for metabolism and hence the evolution of life, has also been shown to exist throughout the solar system in water, ammonia–water combination, and hydrocarbon forms. Water ice is ubiquitous, and liquid water—including oceans—have been found to have existed, or to currently exist, on many planets, minor planets, and moons in our solar system, where they are prevented from freezing by solar or geothermal energy (such as tidal flexing, radiogenic heating, or other forms of hydrothermal activity that drive plate tectonics and similar forces). This indicates that, at least in terms of its basic molecular substrate, the evolution of life on Earth did not depend on an exotic cocktail of chemical conditions that are unlikely to be replicated in the cosmos.

Nevertheless, as noted previously, there is a major explanatory gap between the origins of the basic building blocks of life, and the emergence of replicating, membrane-bound, homeostatic systems of the sort that configured the first proper organisms on Earth. Absent a thorough understanding of this transition, how can we be confident that the emergence of life is not a profoundly unlikely accident or even a one-off event in our galaxy or cosmic horizon? The strongest evidence for the cosmic imperative of life thesis can be found not out in the cosmos, but right here at home in the fossil record of life on Earth.

1.2 The Timing and Likelihood of Life's Origin

As with comedy, the key to the cosmic imperative of life thesis is timing. Bacterial life forms appear in the fossil record immediately after the Earth cooled from numerous world-sterilizing impacts that took place during the formation of the solar system. Layered structures called "stromatolites," which are produced by communities of cyanobacteria (traditionally referred to as "blue-green algae"), have been found in the oldest sedimentary rocks on Earth that retain fossil signatures.[8] These accretionary structures were left by sophisticated organisms that already possessed exquisite photosynthetic adaptations, which indicates that LUCA emerged much earlier than the observed fossil signatures. Filamentous bacteria-like fossils have been found in rocks associated with submarine hydrothermal vents that may be as old as 4.28 billion years, which suggests that life emerged as soon as the oceans formed.[9] Life is increasingly thought to have originated around hydrothermal vents, which offer pockets of far-from-equilibrium chemistry that are protected from surface impacts and are probably ubiquitous in the universe.[10] If the earliest life-forms on Earth are found around deep-sea vents, this would seem to rule out "panspermia"—the thesis that life originated elsewhere (perhaps on a much longer timescale) and was transported to the nascent Earth via asteroid or comet.

These findings are consistent with "molecular clock" analyses dating LUCA to the Hadean Eon, more than 4 billion years ago.[11] This pushes back the origin of life to just before a period in the Earth's history known as the "late heavy bombardment" (approximately 3.9 billion years ago). During this spike in Earth-crossing asteroid impacts, the Earth's crust is thought to have been repeatedly sterilized several kilometers deep by gigantic asteroid collisions and the fiery rain of their impact ejecta. Models indicate that life could have survived during the late heavy bombardment in protected pockets (refugia) of the geophysical habitable zone, especially in deep-sea hydrothermal vents.[12] In sum, we can say with some confidence that life arose as soon as minimally suitable conditions obtained on Earth, and that it did so despite the presence of literally hellish conditions long thought to preclude its emergence. Life appears to be a robust, not precarious, phenomenon.

Carl Sagan exclaimed that "the origin of life must be a highly probable circumstance; as soon as conditions permit, up it pops."[13] The rapid timing of life's origin does seem, at least intuitively, to support the cosmic imperative of life thesis over what we might call the "cosmic accident hypothesis." One way to gloss the logic of this intuition is in terms of a "likelihood" argument. According to likelihood arguments, some observation O confirms hypothesis H1 over rival and mutually exclusive hypothesis H2 only if the probability of O is greater given that H1 is true than it is given that H2 is true. Likelihood analyses view evidence as inherently contrastive: observations do not support hypotheses *simplicter*; rather, they favor one hypothesis as against another. From this perspective, confirmation is a three-place favoring relation among an observation and two rival hypotheses.

A likelihood argument for the cosmic imperative of life thesis might go something like this: the probability that we would observe basic microbial life emerge just when the Earth had cooled sufficiently to support liquid water and by implication biological functions that rely on aqueous chemistry (O_{TIMING}) is higher given that the cosmic imperative thesis (H1) is true than it is given that the cosmic accident hypothesis is true (H2). According to H1, life emerges reliably wherever certain generic conditions obtain. Although H1 does not provide a specific timeline for the origin of life, we might expect life to emerge with relative geological rapidity wherever suitable conditions arise, just as stars and planets reliably form (typically on the order of 1 million to 10 million years) given certain concentrations of matter and the absence of disrupting forces. And this is precisely what we observe: life's origin coincides with the emergence of life-friendly environments on Earth, as determined by independently supported observations and auxiliary hypotheses that allow us to pinpoint the timing of life's origin and the prevailing conditions on Earth at the time.

By contrast, the cosmic accident hypothesis (H2) does not make any particular predictions about when or under what conditions we should expect life to arise. According to this view, life is the astronomically unlikely product of a series of complexly configured causes that are unlikely to be replicated on Earth or elsewhere. Although O_{TIMING} is nomically consistent with H2, it does not relationally confirm H2. To the contrary, O_{TIMING} makes H2 less likely to be true than its mutually exclusive rival H1. For either H2 is true and O_{TIMING} is a remarkable coincidence, given all the potential values that it could have taken; or H1 is true and O_{TIMING} is a nomically constrained expectation. Our intuitions are therefore vindicated: O_{TIMING} favors H1 over H2.

On the other hand, had we observed a substantial delay between the existence of hospitable conditions and the emergence of the first microbes on Earth—such as a temporal lag on the order of billions of years—this might raise serious questions about the cosmic imperative of life thesis, particularly given that the habitability window for Earth-like planets orbiting Sun-like stars may total around 5 billion years.[14] This last point is important because it brings to the fore an implicit auxiliary hypothesis (A1) that is crucial to adjudicating the cosmic imperative of life thesis: namely, that the life cycles of Sun-like stars, and the ability of planets to sustain atmospheres via geothermal activity (even when they are located round a long-lived, red-dwarf star), pose outer boundaries with respect to how long the emergence of life can (on average) take and yet still be cosmically ubiquitous.

To illustrate this point, let us distinguish a further hypothesis (H3), which holds that life emerges reliably due to certain law-like connections but requires many billions of years on average to do so. The conjunction of H3 and A1 does not support the cosmic ubiquity of life thesis, due to solar and geophysical constraints on the window of habitability. However, once again, it would be an incredible coincidence for life to emerge right when favorable conditions prevailed, as predicted by H1, as opposed to anytime later in the many-billion-year time frame for its initiation as predicted by H3. Although H3 is consistent with O_{TIMING}, it is, like H2, comparatively disfavored by O_{TIMING} in relation to H1; and so O_{TIMING} relationally confirms H1 over H3, as well.

Some important complications are introduced by observer selection effects. For example, say it were the case that the only histories of life that produce beings who can reflect on these heady questions are those histories that initiate rapidly, because moderate to late initiations do not provide enough time for robustly cognitive and cultural species to emerge. In that case, O_{TIMING} would not support H1 over H2 or H3, but neither would it support H2 or H3 over H1. But let us bracket observer selection effects until later in this chapter.

1.3 The Single Origin Anomaly

There is one glaring observation that sits rather uneasily with the cosmic imperative of life thesis, however. And that is the well-supported inference, based on comparative genomic data, that all known life on Earth shares a single common ancestor, which possessed the specific genetic code that is now universal to all life on Earth. Independent origins of life are unlikely to resolve in the morphological fossil record, so we must look to comparative molecular data—the molecular fossil record—to evaluate this possibility. The LUCA inference is based not only on the universality of the genetic code itself, but also on the finding that there is a handful of particular genes that are present in all known life on Earth. Most of these universal genes code for conserved proteins that are involved in transcription and translation machinery.[15]

The origin of cellular membranes, which enclose replicating machinery into compartments and contain energy-gathering redox reactions, is somewhat hazier. Unlike the universal genetic code, it is plausible and perhaps even likely that membranes have multiple origins. The basis of this conclusion is that membrane lipid structure and biogenesis is radically different between Eubacteria and Archaea—so different, in fact, that archean membranes are unlikely to be derived from eubacterial ones. This offers strong evidence for the independent evolutionary origins of true cells. And if it is right, it implies that LUCA was not a free-living cell with a biogenic membrane—rather, it must have existed within inorganic, geologically created compartments that permitted replication and metabolism by preventing the diffusion of chemical reactions into the ocean.[16] So if by the "origin of life" one means the origin of free-living protocells (i.e., cells with biogenic cell membranes), there is direct evidence that life arose more than once on Earth, which lends further support to the cosmic imperative thesis.

This may not be quite enough to dispel the LUCA anomaly, however, for it still leaves us with the single origin of complex replication machinery. Recall that if the cosmic imperative thesis is true, then we should expect life to originate with some reliability and rapidity wherever suitable conditions obtain and defeasors are absent. Even the independent origins of membranes appear limited to the earliest phases of life. Why has replication and membrane machinery not appeared over and over again during the long history of life on Earth? Because suitable conditions have been present on Earth for nearly 4 billion years, with no obvious disturbing conditions that would otherwise impede subsequent origins of life, the LUCA observation (O_{LUCA}) appears to disconfirm the cosmic imperative hypothesis (H1) relative to the cosmic accident thesis (H2). It is as though we had examined a large sample of habitable worlds with identical conditions to those on Earth and found that life never originated on any of them.

There are several problems with this line of reasoning, however. Simply put, the fact that all known life on Earth shares a single common ancestor (LUCA) does not imply that life only originated once on Earth. Life may have originated numerous times, but the descendants of those independent origins may (1) fail to be recognized as independent trees of life because they are empirically indistinguishable from the dominant tree, (2) emerge at some regular frequency but be quickly metabolically dispatched (eaten) or outcompeted by members of the dominant tree—or else go extinct for stochastic reasons that prevent them from gaining a phylogenetic foothold to the point of detection, and/or (3) exist in substantial numbers and in substantially different genetic forms but, due to their microscopic nature, have yet to be discovered. If there are good reasons to think that any of these possibilities is true, then O_{LUCA} does not favor either hypothesis over the other.

The comparative genomic data on which the O_{LUCA} inference is predicated relies on the highly plausible—but nonetheless contested—assumption that the universal genetic code is only one among a very large number of functionally equivalent codes that could have evolved but, for quirky historical reasons, did not actualize. In other words, the O_{LUCA} inference is only permissible against a certain background theory, namely that there are weak functional constraints on the shape that genetic codes may take—an assumption that underpins Francis Crick's "frozen accident" theory of the genetic code.[17] If, in fact, codes vary widely in their functionality—such as with respect to their mutational robustness or ability to carry information—and if a wide range of code space is accessible to selection in any given history of life (especially early on), then the same functional code is likely to evolve independently, whereas functionally suboptimal codes will tend to be driven to extinction. On this view, the known code is universal not because it is a frozen accident but because it reflects an optimal attractor in "code space."[18] If this is so, then numerous origins of genetic codes will be obscured by molecular convergence and early competitive replacements, which have left no trace of countless failed evolutionary experiments in the transmission of biological information.

Even if this view is wrong, and functional constraints on genetic codes are weak (as they are widely believed to be), the first actual code among the early codes to evolve may gain an "incumbent advantage" that precludes the evolution of alternative codes by packing the available niches. In other words, once life on Earth was pervasive, any subsequent origins of replicating molecules, as well as the resources any new life-forms would have had to compete for, would have been quickly gobbled up by the plethora of existing organisms. These voracious incumbents—descendants of the first functionally adept origin of life—would have already been well-suited to their lifeways, preventing any new functional codes from becoming established in the billions of years after LUCA. These first true cells would have had sufficient time to populate and adaptively radiate into the empty ecological landscape before any major alternative trees of life could be established.

In their seminal work on incumbent replacement, Rosenzweig and McCord noted that challengers are faced with an evolutionary catch-22: in order for an invading lineage to displace an incumbent, it must evolve the suite of fine-scale local adaptations that give it a selective edge over the incumbent.[19] The catch, however, is that selection can only produce this suite of adaptations in the local environment that is currently dominated by the incumbent. Thus, selection cannot produce superior adaptations until the invader displaces the incumbent; but it cannot displace the incumbent without the superior suite of traits that it needs to succeed in the incumbent's niche. It stands to reason, therefore, that absent some separate extinction event that weakens or knocks out the incumbent, incumbent species are safe in their niches.

Any incumbent advantage that did belong to the first prokaryote tree of life would have been massively accentuated by the exchange of "public biological goods" through lateral gene transfer, which appears to have dominated the initial evolutionary phases of life on Earth.[20] Recent modeling work by microbiologist Carl Woese and his collaborators suggests that lateral gene transfer within communal innovation pools will tend to drive a single genetic code to fixation. Their argument runs as follows. Lateral gene transfer, a major source of innovation and evolvability in prokaryotes, can only occur if there are sufficient levels of code similarity between sharing lineages. As a result, lineages with "alien" codes that were very different from the incumbent code would have been at a severe evolutionary disadvantage because they would have been unable to avail themselves of the enormous communal pool of genetic goods that was available through lateral gene exchange. This, in turn, would have denied them access to the very substantial reservoir of adaptive variation—and hence evolvability—that lateral gene-sharing affords.[21]

Their argument goes further, however. Not only would sufficiently alien codes have been driven to extinction, but in addition we should expect natural selection to bring existing alternative codes closer to one another in an evolutionary process they call "code attraction." Together, the selective purging of alien codes and the process of code attraction can be expected to result in the elimination of alternative trees of life and the fixation of a single, universal code: ergo, O_{LUCA}. It follows that even if alternative codes arose from time to time in the history of life, and even if some of these alternative codes were functionally superior to the incumbent code, they would have been precluded from gaining a foothold due to incumbent advantage and would be eliminated or convergently modified via code attraction in the rigorous competition between innovation pools. Once the incumbent code solidified and rose to dominance, an increased premium would have then been placed on fidelity of transmission, and developmental constraints would prevent any further exploration of code space. We would then be left, for the remainder of life's history, with a single code of life.

Woese's picture of communal evolution as an explanation of code universality is striking for many reasons, not the least of which is that it implies that there was no single, discrete common ancestor of all extant life. Instead, there was a single ancestral *communal population.* On this view, proper "tree-like" evolution, with clean lines of vertical descent, would not come until later in life's story. The communal ancestor model of LUCA also contrasts sharply with Crick's frozen accident theory because it offers a window for the selective optimization of the code through lateral gene exchange. But most importantly for present purposes, it explains away the anomaly of single origin by rendering O_{LUCA} as the nomically expectable result of prolific lateral gene exchange in the early phases of life on Earth.

In short, even if chemical conditions on Earth have been amenable to the emergence of life for billions of years, *biotic* conditions may have prevented independent life lineages from becoming established to the point of scientific detection. Even if separate prokaryote lineages have been established and remain with us, there is a good chance they will not be detected. Macroscopic life forms are far more likely to have their genomes analyzed; yet all macroscopic organisms are eukaryotes, a relatively young evolutionary branch of the incumbent tree, and thus not where we would expect to find alternative codes. Until a far greater proportion of microbial diversity is analyzed, we cannot be sure that there are no tiny aliens living among us.

Establishing a new tree of life will always be a Quixotic endeavor, given the high probability of extinction as a lineage battles away from the absorbing boundary of extinction. Stochastic (or "random walk") models of extinction and diversification suggest that only one of numerous independent origins of life can be expected to survive, if any life survives.[22] If this is right, then any history of life that is successfully established is likely to have emanated from a large number of beginnings rather than from one, even though all organisms at nearly every given time slice in that successful history of life can be traced to a single origin. In other words, any tree of life that is around long enough to permit one of its tiny twigs to read these pages will, in all likelihood, have emanated from a multiple origins scenario.

For all these reasons, the observation of code universality does not compellingly speak to a single origin of life's replicating machinery. When we take into account the evidence for an extremely early origin of life, the repeated evolution of free-living cells, the cosmic ubiquity of the chemical building blocks of life, and models of code convergence under heavy horizontal gene transfer, a strong case can be made for the cosmic imperative of life thesis, even if a single alien microbe has yet to be found. If this cosmic imperative thesis is eventually confirmed beyond a reasonable doubt, it will most likely be through sophisticated next-generation telescopes that analyze the gaseous compositions of exoplanets for the biochemical signatures of life, such as gases produced by metabolic redox reactions and other living activities. Yet even this endeavor is fraught with the risk of false positives (because geological processes can mimic living ones) and false negatives (because we may fail to recognize the diversity of metabolic forms that life can take).

2. Extending the Cosmic Imperative: Nonbiological Approaches

If the cosmic imperative thesis for basic microbial life is on reasonably firm foundations, the case for the cosmic ubiquity of more complex forms of life is on far shakier ground. It is here that scientific views begin to diverge rather sharply along disciplinary lines, resulting in two contrasting visions of the living cosmos. Astrophysicists, especially advocates of SETI, have historically been optimistic about the possibility of complex—and complexly cognitive—lifeforms emerging in a sizable fraction of cases where basic microbial life evolves. Even if this fraction is small and complex life is rare among life worlds, given that nearly all stars appear to have planetary systems, there is likely to be a staggering number of habitable planets in our galaxy.[23] Thus, the Milky Way should be teaming with complex life, as is often depicted in science fiction. Evolutionary biologists, meanwhile, for reasons that will soon be clear, have been less sanguine about such prospects.

The astrophysicist's optimism about macroscopic life in the universe has reposed on general statistical considerations, assumptions about the uniformity of nature, and extrapolation from the familiar modes of inference and explanation that are endemic to the physical sciences, rather than from any specifically biological justifications. As we shall now see, these approaches cannot take us very far toward gauging the cosmic frequency of complex life.

2.1 The Copernican Principle and Observer Selection Effects

The Copernican principle holds, in essence, that we should be suspicious of scientific theories that take humans to be special or privileged observers. The sixteenth-century Polish astronomer Nicolaus Copernicus famously showed, contrary to long-standing wisdom, that the Earth does not lie at the center of the solar system but rather revolves with numerous other planets around the Sun. Centuries later, we now know that although the Sun lies at the center of our solar system, it is in fact a mundane star located in a low-key suburb of an inconspicuous galaxy that is buried in a field of perhaps trillions of galaxies. Nothing in this picture smacks of a universe centered, literally or figuratively, around humans. As cosmologist and SETI theorist Milan Ćirković remarks, "the relentless march of Copernicanism has repeatedly threatened and destroyed our cherished myths and prejudices in the course of the last … five centuries."[24] It is only natural, therefore, to think that the notion we are the cosmic exception is merely the next mythic construction of anthropocentricism teed up for Copernican demolishment.

The Copernican principle can be read in either of two ways: one psychological, the other statistical. Read psychologically, the Copernican principle is less of a "principle" and more a cognitive corrective for the human predilection to interpret data in a self-centered, ad hoc, or politically motivated way so as to preserve theories or religious doctrines that secure a privileged position for humanity in the universe (such as the geocentric model of the solar system or the biological theory of special creation). The reason why this penchant for privilege is properly considered a bias is that it has led us systematically astray in the past—and thus, so goes the induction, it is likely to lead us epistemically astray in the future.

Statistical readings of the Copernican principle, such as the "mediocrity principle"[25] or "typicality principle"[26] formulations, attempt to provide a more formal justification for the directive that we ought to be wary of privilege-preserving theories. If we assume that the shape of life on Earth falls squarely within the bell curve of living worlds, then we can predict with high confidence that life on other habitable worlds will not only exist but in some cases will be highly similar to our own. This reasoning is surely flawed. Even Carl Sagan, probably the twentieth century's most prominent advocate for SETI, conceded that applying the Copernican and mediocrity principles to the question of life elsewhere in the universe is little more than an act of faith. The problem with using these principles to draw inferences about extraterrestrial life is not that they are faith based, but rather that they are distorted by observer selection effects.

No matter how infinitesimally rare the evolution of higher cognitive life may be, any intelligent observer contemplating the frequency of other we's in the universe will by necessity occupy a planet where intelligent life *did evolve*, and where it evolved *in a particular way* (e.g., where it was instantiated in a morphological humanoid form). So long as the only example of life we have is our own, it would be a grave mistake to infer on the basis of the Copernican or mediocrity principles that intelligent organisms—let alone intelligent humanoids—are likely to evolve whenever or even in a significant fraction of cases wherever life arises. We cannot infer from what did happen to what had to happen purely on the basis of statistics.

Consider the sizable,nearly 4-billion-year time lag between the origin of basic microbes and the evolution of intelligent multicellular life on Earth, a period that spans nearly one-half of the entire main sequence of the Sun. One might conclude from this observation, drawing on the mediocrity principle, that the evolution of intelligent life is a run-of-the-mill occurrence, taking place, as it did, smack dab in the middle of the Sun's main sequence. Theoretical physicist Brandon Carter argued, however, that this intuitive conclusion is misguided because it fails to take into account observer selection effects.[27] Carter reasoned this way: any being computing the average time to intelligent life on a planet must itself have emerged within the main sequence of its star; in other words, observer selection effects prevent any observations from occurring beyond the habitable sequence of the local star. Taking this into account, the fact that the time it took for intelligent life to emerge on Earth is close to the total duration of the habitable sequence of the Sun (within a factor of 2) suggests that the average time to intelligent life in any history of life is in fact much longer than the main sequence of Sun-like stars.

In essence, Carter argues that it is astronomically unlikely that the time it took to produce intelligent civilization on Earth just happens to be comparable to the duration of the main sequence of the Sun, because the astronomical and biointelligence timescales are not causally correlated. This observation is precisely what we should expect, taking into account observer selection effects, if the average time to intelligent civilization is much longer than the main sequence of Sun-like stars. Thus, Carter concludes that our own case is likely to represent one in which intelligent life proceeded much more rapidly than is cosmically typical.

Recent studies on the effects of solar output on the Earth's climate strengthen this conclusion by narrowing the habitable portion of the Sun's life cycle. Modeling work suggests that the Earth will become inhabitable to all but basic microbial life in less than 1 billion years due to the effects of increasing solar output on the biosphere.[28] If this is right, then the emergence of intelligent life on the very tail end of the habitable portion of the Sun's life, taking into account observer selection effects, could indicate that the average time to the emergence of intelligent life on Sun-like stars may be significantly greater than the habitable portion of their main sequence.

Carter's argument has been subject to its fair share of criticisms, and I will not say more about it here.[29] However, the SETI skepticism that flows from Carter's argument could be tempered by the recent finding that red dwarfs—very low mass stars that burn stably for trillions of years—are by far the most common type of star in our galaxy and could present a substantially greater window of habitability than main sequence stars. Yet if red dwarf stars are a hundred times more numerous than Sun-sized stars and equally habitable, then why is it that we do not hail from a red dwarf star system? These mysteries would be dissolved if it turns out, as some recent evidence suggests, that red dwarf star systems are inhospitable to life or at least to complex life. Even if red dwarf systems are habitable, their window of habitability may be comparable to that of main sequence stars due to the outer limits of geothermal processes that support aqueous oceans and molten metal cores (and hence magnetospheres, which deflect solar winds that would otherwise deplete atmospheres, as appears to have doomed the early Mars).[30]

If we zoom out to cosmic timescales, then what looks like a late origin of intelligent life in the evolution of our *solar system* is revealed to be an incredibly early origin of intelligence in the evolution of the *universe*. Star formation is not expected to end in galaxies for at least a trillion years: why is it, then, that we hail from such an early phase in the history of the universe, rather than from a time slice in the distant future, which would appear to be millions of times more likely? As with the rapid origin of life on Earth, the timing of the origin of intelligence in the universe is not affected by observer selection biases. The fact that we hail from the first one billionth of the life of the universe as we orbit a second-generation main sequence star seems to speak in favor of the cosmic ubiquity of intelligence.

Let us return, however, to the rapid origins of life itself. On its face, the evidential value of O_{TIMING} is not undermined by observer selection effects; for while it is true that the *origin* of basic microbial life is subject to observer biases—in that any macroscopic observer will necessarily come from a planet where microscopic life has arisen (and probably many times)—the *timing* of life's origin is not. An observer could just as well hail from a history of life that took several billions of years to commence after suitable conditions arose, as it could from a history of life that got rolling right when suitable conditions obtained. This is why the rapid timing of life's origin on Earth supports the cosmic imperative thesis about the evolution of basic microbial life, whereas the enormous timespan that was required to develop complex multicellular life (about 3.5 billion years) and then, on top of that, intelligent/technological species (another 500 million years) bespeaks the comparative cosmic rarity of such outcomes.

Having said this, one could argue that there is, in fact, an observer selection effect in relation to the timing of life's origin: if it takes an exceedingly long time on average for intelligent life to arise, then intelligent observers will tend to hail from histories of life that get off to an early start, just as ours did, even if early starts are vanishingly rare. The takeaway lesson is that we must steer a careful course between deluded anthropocentrism, on the one hand, and anthropomorphic projections that neglect observer selection biases, on the other. The Copernican principle is simply too coarse-grained, too inductively unsupported, and too subject to observation selection effects to navigate these deeply biological waters. Unfortunately, as historian of science George Basalla noted, this has not stopped "virtually every commentator on the subject of extraterrestrial life, and every scientific research program seeking intelligent alien life, [from] operat[ing] upon this principle."[31]

These problems demonstrate why it so important to search for biologically specific theory and data to inform our prognostications about the prospects of complex life in the universe. One reason why this book is addressed to the phenomenon of convergent evolution is because, as we shall see, evolutionary iteration offers a distinctively biological source of evidence that gets around pesky observer selection effects.

2.2 The Uniformity of Nature

Physical science has generally operated on the assumption that the basic laws of nature are spatiotemporally invariant—that they hold at all times and in all places. If some sequence of events occurred one way in the past or in one region of the universe, we can presume that it will unfold in the same way in the future or in another region of the universe. The eighteenth-century titan of philosophy David Hume famously showed that the "uniformity of nature" assumption on which the method of scientific induction rests cannot be grounded in logic because it does not follow deductively and cannot be induced from past experience without vicious circularity.

Many are tempted to argue that the uniformity of nature assumption is justified because it has been borne out by the remarkable predictive and explanatory successes of science. Hume argued, however, that we cannot conclude from the past successes of the inductive method that induction will be successful in the future because there is no guarantee that the future will be governed by the same laws as the past or that one region of the universe will be governed by the same laws as another region—even though his has, thus far, been the case. Induction, moreover, cannot ground itself. We must therefore look to other, more pragmatic sources of justification for the scientific method. For instance, the philosopher of science Hans Reichenbach argued that if any method will be capable of making successful predictions, then induction will succeed; in other words, if the future is amenable to prediction, then induction will work. If the future is not amenable to prediction, then no method will work. So we have everything to gain and nothing to lose by employing the rules of induction.[32]

Several centuries of scientific observation give us good reason to believe that many physical, chemical, and geological regularities are spatiotemporally invariant—or at least we may proceed, methodologically, as if they were. So why not operate on the assumption that the same regularities that characterize Earthly biology will apply to living systems everywhere in the universe? The trouble with uncritically extending the uniformity assumption into the life sciences is twofold. First, biologists have no similar methodological or inductive base from which to justify the presupposition of uniformity. We learned from Newton that the same physical laws that govern the interaction of terrestrial objects also govern the interaction of celestial ones. In biology, however, there are no observations of extraterrestrial life against which to test the uniformity hypothesis. Whether terrestrial biology can be projected onto other life worlds in the way that terrestrial physics can be projected onto the heavens remains unknown.

Second, even if the assumption of uniformity is methodologically warranted, this does not preclude many regularities in the universe from being fundamentally accidental rather than the necessary outcomes of laws. Some generic biological laws, such as the principle of natural selection, may apply to life everywhere and every-when because they amount to a priori mathematical models[33] or applications of probability theory.[34] Yet for all we know the specific biological regularities observed on Earth—from particular metabolic innovations, to the structure of the genetic code, to the morphological shape of life—are contingent accidents that result from quirky chains of events that are unlikely to be replicated on other living worlds.

In an epic illustration of the dangers of prognosticating on the limits of science, towering German philosopher Immanuel Kant infamously proclaimed from the comfort of his eighteenth-century armchair that there would never be a Newton for the blade of grass. Kant was skeptical that humans could ever explain the origins of natural ends (biological functions) without recourse to an intelligent designer. Many authors have been quick to tout Charles Darwin as precisely such a "Newton." Darwin's theory of natural selection provided an elegant mechanistic explanation of the exquisite ecological match between organisms and their environment, which Kant, in his penchant for categoricity, had proclaimed was in principle unsolvable.

Yet in an important sense Darwin was not another Newton. Newton's theory of gravitation causally unified terrestrial and celestial events that were previously regarded as disparate kinds of phenomena. Similarly, the great achievement of the Darwinian revolution—which combined the theory of natural selection with the theory of common descent—is that it unified an astounding range of observations.[35] However, unlike Newton's physical theory and its modern counterparts, the theory of natural selection does not make any specific predictions about how histories of life will unfold elsewhere in the universe, or even whether they will. The reason for this relates to the "schematic" nature of natural selection.

As philosopher of biology Robert Brandon shows, when the principle of natural selection is articulated in general form, it applies to all evolving systems in the universe; but framed in this generic manner, it makes no specific predictions about what sorts of evolutionary outcomes we can expect to occur.[36] When the theory of natural selection is framed in specific terms, such as in relation to a particular population of organisms with particular traits, genetic variations, developmental interconnections, and ecological pressures, then it admits of specific predictions—but there is no reason to think that these specific predictions will project to life elsewhere in the universe. The developmental and ecological details on which such predictions depend may, for all we know, be entirely accidental and thus restricted to a narrow time slice of life on Earth. This has motivated skepticism among evolutionary biologists and philosophers of science about the feasibility of developing a truly law-like science of life.

3. The Cosmic Contingency Thesis

Evolutionary biologists have long complained that committees tasked with computing the probability of intelligent extraterrestrials have tended to focus on (bio)physics and (bio)chemistry while on the whole failing to engage with evolutionary theory and data regarding large-scale patterns in the history of life on Earth. For instance, when participants of the historic Green Bank Conference in 1961 attempted to assign probabilities to factors in the (in)famous Drake equation—which attempts to quantify the probability of extraterrestrial civilizations in the galaxy—not only did they assume that intelligence will inevitably evolve on any living world, but in addition they estimated that around one in five intelligent extraterrestrial lineages would develop powerful technological civilizations with capacities for radio communication. Although these conclusions were minimally informed by work on the evolution of intelligence—such as by New Age psychoanalyst John Lilly's research on dolphin communication—no one at Green Bank was operating with a truly broadscale picture of life on Earth or an in-depth understanding of the evolutionary process. The historical lack of biotheoretical input into SETI theorizing is important because macroevolutionary science may have a very different story to tell.

3.1 The Biologist as SETI Spoilsport

To many (perhaps most) biologists, the history of life is not a linear narrative of progression toward ever more complex forms of life culminating in a single, self-aware species of hominin. Rather, macroevolution presents as a series of historically contingent events with unpredictable consequences, a pattern that provides little assurance of the evolution of complex life, let alone complexly cognitive beings like ourselves.

As macroevolutionist G. G. Simpson noted more than half a century ago, theorists with a bird's-eye perspective of life on Earth have tended to play the role of SETI spoilsport. Simpson's own SETI skepticism was motivated in part by astrophysical assumptions that have been exploded in the decades following publication of his influential "anti-SETI" manifesto.[37] For instance, we now know that extrasolar planets are not a rare occurrence but rather a reliable outcome of stellar system evolution—yet another instance of successful Copernicanism in action. The argument with the most staying power is Simpson's appeal to the role of contingency in evolution. Simpson contended that "even slight changes in earlier parts of the history would have profound cumulative effects on all descendent organisms through the succeeding millions of generations."[38] Accordingly, the trajectory of life on Earth represents but a tiny fraction of the possible directions, organizations, and functional morphologies that life could take. If so, then there will be few meaningful properties of the actual history of life that can be projected onto other life worlds. Let us call this view the "cosmic contingency thesis."

Echoing Simpson's trenchant criticisms of exobiology three decades earlier, the evolutionist and SETI skeptic Ernst Mayr complained in a letter to the journal *Science* that the "expert" views on which SETI grants were based are almost entirely those of astronomers, physicists, and engineers, even though the factors that will determine the success or failure of the SETI project are distinctively biological and sociological.[39] Mayr argued that the prospect of making contact with advanced extraterrestrial civilizations is too astronomically miniscule to justify supporting SETI programs with taxpayer money. He also criticized physical scientists for thinking "deterministically" about the evolution of complex life—that is, for assuming that once basic microbial life emerges, the evolutionary process will drive inexorably toward intelligence.

Other prominent students of macroevolution have followed Simpson's spoilsport model in tending toward a less sanguine view of SETI success, stressing the role of contingency, unpredictability, and unrepeatability in the great twists and turns of the history of life on Earth.[40] As the paleobiologist David Raup notes, most evolutionary biologists have been "quite negative and their views have been trotted out by the anti-SETI forces to argue for the futility of search programs, or at least for the very small probability of success."[41] Evolutionary anthropologist Jared Diamond, for example, has reinforced Simpson and Mayr's anti-SETI arguments, arguing that although certain isolated traits might be evolutionarily repeatable, the overarching lesson of macroevolution is that complex outcomes, such as the bundle of traits that comprise specific taxa, are accidental regularities limited to life on Earth.[42]

This is not to say that all evolutionary biologists have been down on the SETI project. Even some evolutionists who take contingency seriously have been SETI supporters (David Raup and Stephen Jay Gould[43] are prominent examples). Nevertheless, there are sound theoretical reasons for thinking that the evolutionary processes that produce life—and hence the epistemological character of the science that studies it—are of a fundamentally different character from the processes that govern and sciences that explain the dynamics of purely physical and chemical systems. Put simply, the study of species may have little in common with the study of stars. If there is a fundamental difference between these disciplines, it lies in the irreducible historicity of biology. This historical view of life is most forcefully articulated in the work of Stephen Jay Gould, whose seminal argument for evolutionary contingency will serve as the launching point for the rest of this book.

2 The Radical Contingency Thesis

On June 28, 1914, Archduke Franz Ferdinand, heir to the Austro-Hungarian Empire, traveled with his pregnant wife Sophie by royal car down the streets of Sarajevo, a Bosnian city stoked with Slavic nationalism and stirring with rebellion. The archduke and his family had successfully evaded an assassination attempt in the form of a lobbed grenade only hours earlier during their trip to city hall, and the return journey took the royal procession along the same fateful route. In the "shot heard round the world," the archduke and his wife were murdered in their vehicle by an assassin who fired two single shots from his pistol.

The archduke had been an influential moderating force in attempts to diffuse the growing prospect of military conflict between the European powers, and his assassination greatly affected the psychology of the Austro-Hungarian leadership. The chain of events that ensued was complex, unpredictable, and formative of global events for generations to come. The assassination had been organized by a secret Serbian military society and was bankrolled by the government of Serbia, which led the leaders of Austro-Hungary and Germany to issue an ultimatum to Serbia—worded in a way that made Serbia likely to reject it.

The assassination bolstered the case for war that was being made by hawkish elements within the empire, and it served as the ideal pretext for an aggressive attack on Serbia. If Russia chose to intervene on Serbia's behalf, Kaiser Wilhelm II, the German emperor, would then be absolved of any blame for escalating the conflict. Upon Serbia's partial rejection of the ultimatum, Austro-Hungary declared war on Serbia, and under the terms of a secret pact, Russia and France were compelled to enter the war on Serbia's side. This, in turn, triggered the military mobilization of the Austro-Hungarian Empire and its invasion of neutral Belgium, with the aim of decisively defeating France to avoid a two-front war. This then brought Britain into the war on the side of France and Russia.

In a matter of weeks, nearly all major world powers (save for the United States, whose entry would come later) had entered into a war that would take the lives of 40 million people and culminate in the signing of the punitive Versailles Treaty. This draconian treaty would sow the seeds for a second and even more devastating world war and its unparalleled war crimes and genocides, which would follow only two decades later. The resolution of this second global catastrophe would be followed by the Cold War, the collapse of the Soviet Union, and solidification of U.S. geopolitical hegemony—a sequence of events that some historians take to be so causally intertwined that they constitute a single, protracted, historical episode: all emanating from a single assassin's bullet.

As the *New York Times* presciently reported just one year after the archduke's assassination,

> Those two shots brought the world to arms, and the war that followed … brought devastation upon three continents and profoundly affected two others.… Nation after nation has been drawn into the whirlpool, and more are drawing toward it, and the end is far off. What face the world will wear when it is all over no man can predict, but it will be greatly changed, and not geographically alone.[1]

The assassination of Archduke Ferdinand and the ensuing descent into global catastrophe is perhaps the most famous illustration of the role of contingency in human affairs. Had a breeze or tremor of the hand caused the assassin's bullet to miss the archduke's jugular vein, or had the conspiracy been foiled, then World War I could very well have been avoided—and had it been avoided, then the geopolitical landscape of the contemporary world would have assumed a dramatically different shape. There is nothing inevitable about the forces and pathways that led to our particular political world.

The same may be said of more recent world-shaping historical events, such as the 9/11 attacks by Al-Qaeda on New York City and Washington, DC, which garnered support for the 2003 U.S. invasion of Iraq. Postwar mismanagement of the Iraqi army opened up a power vacuum in the Levant in which the Islamic State could recruit and operate; this then created fertile grounds for a cascade of rebellions that comprised the Arab Spring, resulting in the catastrophic collapse of Syria into a multidimensional civil war. Syria then became the epicenter of a massive refugee crisis that stirred the extreme nationalism, nativism, and xenophobia that helped bring about the Russia-assisted Trump presidency and the ascendancy of similarly authoritarian political parties in Europe and around the world, as well as the destabilization of NATO and the European Union. Like the *New York Times* in 1915, contemporary policy makers, journalists, historians, political scientists, and international relations theorists can scarcely make out the shape of the global geopolitical landscape even just one decade out.

As counterfactual historian Richard Ned Lebow has put it, "a small and credible re-write of history has the potential over time to bring about a very different world."[2] One reason for this is that human historical trajectories are sensitive to "micro" events, such as the quirky beliefs and desires—or more reductively, brain states—of executive leadership, be they presidents, prime ministers, generals, directors, advisors, experts, or policymakers, who sit at or near the apex of organizational hierarchies and thus have a ramifying causal influence. If these executive levers of causal influence are pressed, they can drive the system into unlikely historical terrain. It is for this reason that we have difficulty predicting the human historical future, even while history can be forensically pieced together once it has been made.

In his book *1776*, historian David McCullough describes the outcome of the American Revolution, and the shape of history that unfolded from it, in beautifully contingent terms: "Especially for those who had been with Washington and who knew what a close call it was at the beginning—how often circumstance, storms, contrary winds, the oddities or strengths of individual character had made the difference—the outcome seemed little short of a miracle."[3]

Is the history of life on Earth similarly contingent? If so, what features of the evolutionary process are responsible for these dynamics? We will begin by reconstructing the most influential argument for contingency in macroevolution: Stephen Jay Gould's analysis of early animal evolution in the Cambrian period and the philosophical lessons he draws therefrom. Before taking a deep dive into counterfactual Cambrian seas, however, let us first unpack some of the challenges that face counterfactual reasoning in the historical sciences more broadly, as this will portend some of the difficulties that confront the adjudication of Gould's thesis.

1. Counterfactuals in the Historical Sciences

In his epic and ruminating tome *The Structure of Evolutionary Theory*, published around the time of his death, Stephen Jay Gould noted that it was his admiration for the study of history in the broadest sense that compelled him to investigate the role of contingency in evolution.[4] The counterfactual investigation of human history—studies of how history would have unfolded had certain events not occurred or occurred in a different way—has enjoyed something of a renaissance in recent research, though many historians remain skeptical that such an approach can rise to the level of serious scholarship.[5] This skepticism is understandable, for as with the history of life on Earth, we cannot rewind the tape of human history, induce a few choice perturbations while holding other variables constant, and observe how these perturbations affect the unfolding of the system. The worry is that counterfactual history is at best methodologically problematic and at worst a parlor game in which ideological penchants and flights of fancy substitute for sound science.

And yet counterfactual thinking is inescapable in historical scholarship. Historical accounts would be disjointed litanies of facts, not narratives, if they did not weave a tapestry of causal claims about beliefs, intentions, desires, actions, influences, zeitgeists, environmental conditions, economic factors, institutional contexts, and so on. And the grounding of causal claims (and for some, the metaphysics of causation itself) is widely thought to hinge on counterfactual analysis (see chapter 5 for further discussion). This is as true for history as it is for the physical sciences. When historians assert that Lee Harvey Oswald shot President John F. Kennedy, that Hitler was responsible for the Holocaust, or that Osama bin Laden funded and directed the 9/11 attacks, they mean that the assassination of JFK, the Holocaust, and the 9/11 attacks would not have happened, would have been far less likely to happen, or would have happened in a different manner were it not for the deliberate actions taken by Oswald, Hitler, and bin Laden in furtherance of their respective ideological goals. If historical narratives are to be meaningful and coherent, they have no choice but to indulge in causes, which in turn requires that they indulge (at least implicitly) in counterfactuals.

The value of counterfactual thinking in human history goes beyond substantiating specific causal claims. It is also necessary for establishing which historical outcomes and trends are robust against minor perturbations, and which are sensitive to small-scale events that could easily have been otherwise. Let us first consider outcomes. World War II historian Andrew Roberts dedicates the final chapter of a recent book, *The Storm of War*, to counterfactual analyses of the war in order to show that Hitler's defeat was guaranteed once Germany, fueled by Nazi racial ideology and buoyed by its lightning quick defeat of France and the British Expeditionary Force, invaded the Soviet Union.[6] In most historical accounts of the war, the Soviets would eventually have captured Berlin whether or not the D-Day landings had been successful and a second front was opened up against the Axis powers. Asking how alternative wartime decisions, operations, and allocations would have affected the shape of the war is a useful way of discerning the causal structure of Allied victory and teasing apart robust and contingent features of the war.

Second, let us consider trends. Historical trajectories that are driven by population-level or "deep structural" forces tend to be robust against micro-level perturbations. This is the explanatory approach taken by Jared Diamond in his justly famous book *Guns, Germs, and Steel*, in which he develops a deep structural explanation as to why peoples descending from agricultural populations in Eurasia ultimately displaced indigenous populations around the world.[7] Diamond argues that systematic differences in climate and geography resulted in technical advancements in plant and animal domestication in certain regions of the world and the absence of such innovations in others; these agricultural innovations, in turn, supported larger populations with highly differentiated systems of labor, advanced weapons manufacture, and unwitting pathogen transmission capabilities, all of which facilitated the "displacement" of indigenous populations. Whether it was the Spanish, Dutch, British, or Ottoman empires that did the displacing in any given case may be historically contingent, but the deep structural nature of Diamond's explanation means that the trend is highly replicable across reasonably close possible worlds. The same may be said for the rise of markets, currencies, military hierarchies, minimal states, and the like, which can be explained by "invisible hand"–like forces without appealing to the specific beliefs, desires, or actions of executive agents. As philosopher of biology Kim Sterelny has pointed out, trajectories driven by population-level mechanisms can be realized by numerous microstate configurations and pathways, and it is this massive multiple realizability that gives them their counterfactual stability.[8]

Another important function of counterfactual reasoning in history is to disabuse us of the fallacious notion that certain historical outcomes are inevitable. As historian Philip Tetlock and coauthors put it, counterfactual thinking about human history can liberate us

> from the cognitive tyranny of hindsight bias: to prevent the world that did happen from obstructing our view of the panorama of possible worlds that could have sprung up into being but for tiny twists of fate, to sharpen our appreciation of how uncertain almost everyone was about what would happen before they learned what did happen, and to sensitize us to the intricate complexity and probabilistic character of the causal processes that produced the world we happen to inhabit.[9]

In short, counterfactual thinking is inescapable if we are to recognize a contingently configured world. As we shall now see, this is as true for natural history as it is for human history.

2. Counterfactuals in Animal Evolution

In his acclaimed book *Wonderful Life: The Burgess Shale and the Nature of History*, Stephen Jay Gould proposes a series of macroevolutionary thought experiments designed to probe the contingent nature of life's history.[10] He imagines rewinding the "tape of life" to various critical junctures in the history of animal evolution and then considers how life's story would again unfurl. Gould argues that replaying the tape of animal life would result in a radically different set of macroevolutionary outcomes—a morphological menagerie bearing little resemblance to animal life as we know it. Not only would no humans, mammals, or vertebrates evolve, but neither would any creatures even approximating them. In fact, for Gould, contingency is not merely a characteristic of animal evolution; it is a pervasive feature of the evolutionary process. In one of his more hyperbolic moments, Gould exclaims that "almost every interesting event of life's history falls into the realm of contingency."[11] I will refer to this view of life, both here and throughout the book, as the "radical contingency thesis" (RCT).

Gould's most crucial replay concerned the dynamics and long-term aftermath of a critical episode in the history of animal life known as the "Cambrian explosion." This event marks the geologically abrupt emergence, about 545 million years ago, of the vast majority of animal body plans that comprise *Bilateria*: bilaterally symmetric animals with digestive tracks and a separate mouth and anus. The first detailed picture of this event came from the fossil *lagerstätten* of the Burgess Shale, located in Yoho National Park in the Canadian Rockies.

These fossil beds are located in one of the most sublime landscapes in North America (see figure 2.1). They are famous not for their idyllic setting, however, but for their remarkable preservation of the soft bodies of the first bilaterians, most of which had no skeletonized parts and whose physical forms had, as a result, long remained a mystery (only about 3% of Cambrian taxa had mineralized skeletons, and most of these were trilobites). The diagnostic hard parts of the major bilaterian clades—such as the exoskeletons of arthropods, the shells of mollusks and brachiopods, and the backbones of chordates—all had yet to form. The Burgess Shale provided the first clear window into the morphological and ecological diversity of the earliest bilaterians, as well as the tempo and mode of their evolution. Similar fossil assemblages have since been found in China and Greenland (most recently, the Qingjiang biota lagerstätte from south China[12]), showing definitively that the critters of the Burgess Shale were indeed a global fauna.

In addition to some familiar sights (like trilobites, brachiopods, and priapulid worms), many fantastical animal designs appear among the early Cambrian fauna and then vanish in the extinction crises that punctuate the same period. Gould was especially taken by the strangeness of *Opabinia*, a soft-bodied animal with five stalked eyes, a backward-facing mouth, and a hollow, flexible, vacuum cleaner–like proboscis with a grasping claw at the end (figure 2.2)—so much so, that he nearly named his book *Wonderful Life* "Homage to Opabinia" but was wisely talked out of this by his editor. Although *Opabinia* had some affinities

Figure 2.1
The Burgess Shale and surrounds. (*Top*) Waterfalls cascading down Michael Peak just after emerging onto the Burgess Highline from Yoho Pass. (*Middle*) View of Emerald Lake from Burgess Highline just past the switchbacks leading to the Burgess Shale. (*Bottom*) Guided group ascending to the Walcott Quarry outcrop, taken from the southern portion of the Burgess Highline. Photos by author.

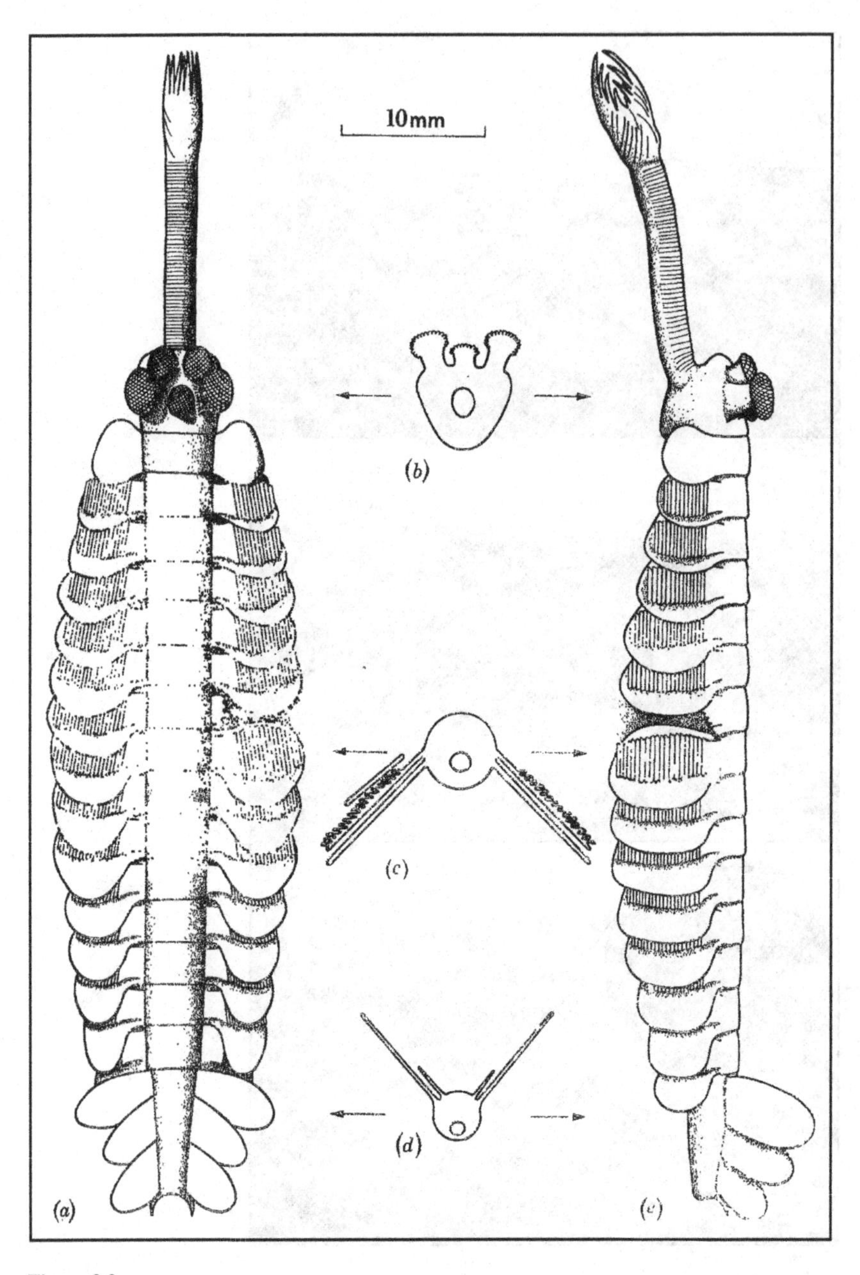

Figure 2.2
Dorsal and lateral views of *Opabinia*. From H. B. Whittington, "The Enigmatic Animal Opabinia Regalis, Middle Cambrian, Burgess Shale, British Columbia," *Philosophical Transactions of the Royal Society of London. B, Biological Sciences* 271, no. 910 (1975): 1–43.

to arthropods, it did not have a jointed exoskeleton, and there is nothing in existing arthropods that resembles its grasping hose-like appendage or its set of five stalked eyes. Legend has it that when Harry Whittington, one of the scientific heroes of *Wonderful Life*, unveiled the first reconstructed image of *Opabinia* to an audience of paleontologists, it was greeted with the laughter of disbelief.

Opabinia was not alone. The Burgess Shale pageant included many other science-fictiony forms, like *Wiwaxia, Hallucigenia*, and *Anomalocaris*, which like their contemporary *Opabinia* did not, at the time of Gould's writing, fit neatly into any known animal phyla. Taxonomically challenging critters of the Cambrian have emerged at a regular pace ever since. Recent additions to the pageant include two curious filter-feeding luolishaniid lobopodians, a paraphyletic group of wormlike creatures that fed on larvae in the water column; the first is a superarmored worm nicknamed "Collin's monster" after its eponymous discoverer,[13] and the second has been dubbed the "ovation worm" for its inferred upward limb-waving posture.[14]

These "mind-bending problematica," in the words of paleogeologist Andrew Knoll, shared the early Cambrian seas with many representatives of familiar body plans such as cnidarians, sponges, mollusks, priapulids, comb jellies, basal chordates, and traditional arthropods like trilobites.[15] This led Gould to surmise that an alternative set of body plans could very well have emerged as the evolutionary victors of the Cambrian—that if events had played out just a little bit differently, familiar animal forms might have been relegated to the realm of the fantastic while the weird wonders of the Burgess Shale inherited the Earth. This, for Gould, was the deep significance of these fossils. Whereas many have come to view the Cambrian explosion as the ultimate arena of animal experimentation, for Gould it represented an early crossroads pregnant with alternative macroevolutionary possibility. The Burgess Shale did not reveal an early phase of evolutionary experimentation in which bizarre and inferior designs were driven to extinction in the unrelenting crucible of natural selection; rather, they offered a window into what alternative—and equally functional—histories of life on Earth might have looked like.

There was one taxon in particular among this motley Cambrian menagerie that caught Gould's eye, one that was of particular significance for its apparent insignificance. This is *Pikaia*: a relatively understated creature in terms of its ecology and anatomical complexity, but one that is by most accounts a close relative to the ancestor of all modern vertebrates. *Pikaia* is a leaf-shaped animal that vaguely resembles jawless chordates such as the lancelet; it has blocks of segmented skeletal muscles similar to the myomeres of fish, with two tentacles on its eyeless, protohead. The macroevolutionary moral Gould draws from

Pikaia is this: had conditions at the end of the Cambrian been just a little bit different, then *Pikaia* and other nascent taxa might not have survived, and the morphological landscape of life on Earth would have assumed a markedly different shape.[16] If no *Pikaia* (and its contemporaneous ilk), then no vertebrates. If no vertebrates, then no tetrapods. If no tetrapods, then no mammals. If no mammals, then no humans. If no humans, then no self-consciousness. If this reading of animal evolution is correct, it offers perhaps the strongest basis yet for the SETI skepticism that swells among the ranks of evolutionary biologists (as discussed in chapter 1). If intelligent life is unlikely to be replicated were the tape of animal life to be replayed on Earth, then there is little reason to think it would arise in a predictable or law-like way throughout the cosmos.

Gould's counterfactual analysis of macroevolution was not limited to the origins and evolution of bilaterian phyla in the Cambrian. He also considered replays in relation to more fine-grained taxonomic events, such as the seemingly improbable survival of the lobe-finned fishes that gave rise to tetrapods (four-limbed vertebrates) as well as the extinction of the long-dominant nonavian dinosaurs, which cleared the way for the unlikely radiation of mammals and their eventual rise to macro-faunal dominance in the Cenozoic. Gould argued that were we to replay the tape of life in each of these cases, evolution would be channeled into a different pathway, resulting in outcomes that bear little resemblance to the history of life as we know it.

It is no accident that Gould entertained this particular set of evolutionary counterfactuals out of the vast number of evolutionary counterfactuals that he could have entertained. For both the evolution of tetrapods and the dinosaur–mammal succession were crucial junctures in the seemingly improbable chain of events that led to human beings. If any one of these links in the chain did not occur, so the argument goes, we would not be here, nor would any creatures even remotely like us. Had the understated *Pikaia* and its close chordate relatives been wiped out in the Cambrian, then vertebrates as we know them would never have arisen. Had a quirky lineage of lobe-finned fish in the Devonian drawn the short evolutionary straw, then tetrapods would have remained hypothetical forms of possible evolutionary worlds. Had a massive asteroid missed the Earth rather than careening obliquely into the Yucatan Peninsula some 65 million years ago, mammals would still be scurrying under the toes of sleeping dinosaurs, and no animals on Earth would be capable of reading these words or thinking these thoughts. By running these counterfactuals, Gould invites us to question the necessity of our living world and, by implication, the precariousness of our own existence.

Contingency is a ubiquitous feature of our personal worlds. My great grandmother, who fled pogroms in the Ukraine to settle in New York City as a teenager, landed her first paying job in the United States at the Triangle Waist Company factory in Greenwich Village. On March 25, 1911, a man she was casually dating (who would later become my great grandfather) invited her to lunch. She snuck out of work to meet him just after noon, and they had such a good time that she decided to skip out on the rest of the workday. The infamous Triangle Shirtwaist fire broke out in the mid-afternoon that day, killing 146 garment workers, most of them Jewish and Italian immigrants. Had my great grandmother not gone out for lunch, I would probably not be here today. This example is a particularly dramatic one, but similar stories of contingency, both existential and mundane, are part of everyone's life. A key question before us is whether natural history is amenable to similar sorts of contingency narratives, in which respects, and why.

3. The Radical Contingency Thesis

Let us now begin to unpack the various theoretical components of the RCT. When Gould looked at the distribution of existing animal body plans, he saw islands of form amid a vast and largely unoccupied theoretical "morphospace," with huge gaps between the islands. Mollusks, arthropods, vertebrates, echinoderms, annelids, brachiopods, nematodes, priapulids, and other bilaterian animal phyla are diverse groups (some more so than others), yet all this diversity is clustered within each respective "island" or body plan. We do not observe a smooth gradation of forms bridging the rather substantial morphological gaps between body plans. Gould's Cambrian thought experiment asks whether this clumpy distribution of form in an otherwise vast and uncharted morphospace is the result of a replicable optimizing process—one that drives evolution toward certain global attractors—or rather the product of quirky, causally formative events that took place early in the history of animal evolution.[17] How we answer this question will have profound implications for the astrobiological questions posed at the outset of this book.

Gould's contention is that small changes in early Cambrian conditions would have led to a very different initial occupation of morphospace and hence history of life. There are two theoretical pillars that undergird this claim.[18] The first relates to the stochastic (or pseudo-stochastic) nature of mass extinctions, with the Cambrian extinctions being the most causally formative of these culling episodes. The second concerns the developmental entrenchment or "freezing" of the body plans that survive these bouts of lineage sampling, which ensure that the gaps between morphological islands are never crossed. Neither subthesis is, on its own, sufficient to establish the RCT, but together they provide the bulwark for Gould's view of life.

3.1 Stochastic Extinction

The Phanerozoic Eon, which began 541 million years ago and continues to the present day, is punctuated by at least five major extinction pulses in which a large proportion of species and higher taxa were wiped out. In the aftermath of these extinction events, a previously dominant faunal assemblage is replaced by an entirely different biota, which radiates to fill the ecospace vacated by the decimated incumbents. Take, for example, the most extreme episode of extinction in the history of animal life: the end-Permian crisis, which took place about 250 million years ago, in which it is estimated that a staggering 95 percent of living species were extinguished.[19] This event, which coincided with the greatest volume of volcanism in the Phanerozoic,[20] eliminated some of the most successful groups of all time, such as the trilobites, while it brought to prominence other lineages such as bivalves, gastropods, fishes, and echinoids, which until the Permian extinction had been only marginal players in marine ecosystems dominated by brachiopods, crinoids, and bryozoans.[21] Many of the lineages that rose to dominance in the wake of the Permian catastrophe radiated from only a handful of surviving species. This pattern, which is replicated in other major extinctions, suggests that had things played out a little bit differently, these groups also would have perished, leaving the vacated niches to be occupied by other fortuitous and theretofore unassuming groups.

Reflecting on these patterns, paleobiologist Doug Erwin and his collaborators remark that mass extinctions "not only punctuate the history of life, they also forever alter its trajectory."[22] Part of the trajectory-altering power of mass extinctions comes from the fact that they are not merely temporary intensifications of background extinction rates.[23] Rather, they are macroevolutionary game changers—events that briefly but significantly alter the rules of survivorship that obtain in more halcyon times, upending successful incumbents and in some cases triggering a climatic overhaul that shapes the nature of postextinction recovery.

How precisely are the dynamics of mass extinction related to the RCT? Gould appears to reason as follows: the extinctions that filtered out the Cambrian *Problematica* either (1) were truly stochastic, in which case there was no rhyme or reason (or more precisely, no dominant explanatory cause or causes) for the observed patterns of taxa survivorship, and thus no reason to believe these patterns would be replicated if the tape of animal life were replayed under slightly different conditions; or (2) were selective and thus to some extent replicable, but in a sense that cannot be traced to any long-term functional superiority of the surviving groups.

Groups may have certain traits that make them more likely to survive the ravages of a given mass extinction, such as a wider geographic range, patchy population structure, smaller body size, burrowing capacity, generalist feeding ecology, higher fecundity, smaller generation time, higher intraspecific variation, or greater evolvability, to name a few candidates. But none of the traits that plausibly confer advantages during mass extinctions (or in selection among species to the extent this occurs) relate to the clusters of features that are associated with specific animal body plans.[24] Moreover, traits that confer differential survival in mass extinctions may not be properly characterized as adaptively superior traits or even as adaptations at all, because they are only causally connected to differential clade persistence in rare geological moments of mass perturbation, when the global biota is temporarily strained. The fraction of the Phanerozoic during which mass extinction conditions prevail is a miniscule proportion of the eon, spaced out by scores or even hundreds of millions of intervening years, during which time the ordinary "background rules" of clade survivorship govern. This, in turn, prevents adaptations to mass extinction conditions from accumulating.

Macroevolutionist David Jablonski has referred to extinction dynamics of this sort as "nonconstructive selectivity,"[25] a notion that is broadly similar to David Raup's concept of "wanton extinction."[26] I will refer to it here as "pseudo-stochastic" sampling because, although the sampling patterns are not truly random, they are both less obvious and more fleeting than those that might underwrite competitive replacements in normal times. Pseudo-stochastic extinctions are not truly stochastic like lightning strikes or David Raup's "field of bullets" scenario, in which all lineages have the same chance of being sampled (or persisting into some future geological period). The difference between pseudo-stochastic and merit-based models of extinction is not that one involves lineage sampling that is driven by fitness differences whereas the other is fitness neutral. The difference, rather, is that in one case the fitness conditions imposed are not stable for a sufficient amount of time to allow for a macroevolutionary response to selection—either at the level of individual organisms or at the level of lineages. The staggered and precipitous nature of these perturbations means that adaptations to mass extinctions will either fail to accrue or else come undone by countervailing selection and drift in the protracted intervening timespans in which ordinary selection regimes prevail.

Both mass-extinction scenarios—a truly random sampling of the biota and the episodic imposition of different rules governing clade survivorship—undermine the prospects for any long-term competitive replacement of functionally inferior lineages by superior ones, at least in terms of the cluster of traits that comprise animal body plans. If long-term competitive interaction is the most plausible driver of macroevolutionary replicability, then stochastic and pseudo-stochastic patterns of extinction cut deeply into the possibility of a law-like shape of life.

3.2 Faunal Turnover

There is a close relationship between macroevolutionary competition narratives and the progressivist readings of life that Gould had long sought to dethrone. Paleontologist Steven Stanley remarked that "paleontologists have almost universally accepted the idea that certain body plans have been rendered obsolete during modernization of the world's ecosystems."[27] Implicit in such talk of obsolescence is the notion that there are competitive interactions between taxa in which some have a distinct fitness advantage over others. These repeated rounds of interaction, so the logic goes, explains the waxing of one clade and the waning of another, until the functionally superior clade fully supplants the inferior one. Although plausible a priori, there is currently little evidence to indicate that major[28] and moderate[29] faunal turnovers—situations in which one faunal assemblage is succeeded by another—can be explained by the long-term competitive superiority of the succeeding faunas.[30]

What evidence is there that the faunal turnover that marked the end of the fantastic phase of animal evolution in the Cambrian was more like a lottery than a merit-based competition? Gould draws this ontological conclusion from an epistemic premise, namely that no biologist without the benefit of hindsight could have predicted the actual patterns of Cambrian survivorship on the basis of any plausibly relevant traits, such as anatomical complexity or ecological prominence. The fact that no evolutionary handicapper worth her salt would have predicted which designs would survive the end-Cambrian extinctions is then taken to indicate that early bilaterian extinctions were essentially haphazard—and thus counterfactually unstable.

For Gould, then, the Cambrian explosion does not simply recount "a unique and peculiar episode of possibilities gone wild"—it reveals a deeper truth about the contingent topography of complex life as we know it.[31] Because there was no necessity to the patterns of survivorship in the Cambrian, there is no necessity to the actual shape of animal life, whose parameters are bounded by the confines of surviving body plans. Before elaborating on this boundedness, let us consider other prominent faunal turnovers that might support a Gouldian view of life.

Just as no good evolutionary handicapper would have bet on the unassuming *Pikaia* and its kin to squeak through the Cambrian extinctions and go on to establish the vertebrate clade (but see section 4.2 of this chapter), so too would no late-Cretaceous handicapper have predicted that mammals would supplant the dominant *Dinosauria* as the Earth's prevailing megafauna courtesy of "one terrible day in the history of the Earth."[32] Mammalian diversity (species richness) and disparity (the breadth of functional anatomical designs) were largely suppressed throughout the Mesozoic era, an approximately 185-million-year period (from 250 million to 65 million years ago) during which time dinosaurs were by far the most salient terrestrial vertebrate fauna. The suppression of mammals is generally attributed to the success of the incumbent dinosaurs and other archosaurs (such as pterosaurs and crocodilians), which packed available niches and relegated mammals to nocturnal, insectivorous, rodent-like forms that generally weighed less than 4 kilos. After the sudden bolide-induced mass extinction that claimed the nonavian dinosaurs, pterosaurs, and apex marine reptiles (such as the mosasaurs and plesiosaurs) and nearly extinguished birds and crocodilians, therian mammals (which include placentals and marsupials) radiated rapidly and dramatically to fill the emptied niches.[33] Bats took to the sky and whales to the sea like pterosaurs and mosasaurs before them, and large predatory bear-like mammals (such as *Titanoides*) hunted in the same forests and grasslands that were stalked by tyrannosaurs not long before.[34] These magical worlds were separated by a geologically short span of time punctuated by a single apocalyptic event.

This now-canonical picture of the dinosaur–mammal succession serves as an effective antidote to the long-standing progressivist narrative that had wily mammals outsmarting and thus replacing their fiercer but dimwitted dinosaurian counterparts. It is rather like the classic fable of David and Goliath, but with Goliath dropping dead from a heart attack and David moving unopposed into his house. However, as is so often the case in science, the reality turns out to be more complicated. Multituberculate mammals—an extinct group of rodent-like mammals with molar-like teeth—appear to have radiated some 20 million years *prior* to the Cretaceous-Paleogene (or "K-Pg") boundary (formerly known as the "K-T boundary"). Multituberculates ate insects, and they appear to have diversified along with the proliferation of angiosperms (flowering plants) and their coevolving pollinators up to and even through the K-Pg boundary.[35] Nevertheless, all fossils that exhibit traits characteristic of modern mammalian orders appear after—not before—the K-Pg boundary.[36] There is no reason to think that multituberculates were outcompeting dinosaurs or played any role in their demise.

There are two additional wrinkles in the standard dinosaur–mammal succession picture. First, there are indications that therian mammals (not just multituberculates) may have begun to diversify before the end-Cretaceous extinction.[37] Second, there is now some data suggesting that nonavian dinosaurs had already begun their broad decline tens of millions of years before the K-Pg event. Data on dinosaur diversity and disparity during the last few million years of the Cretaceous comes almost entirely from the Hell Creek formation in the western interior of North America. Studies of Hell Creek have generally found many dinosaur clades (as well as other archosaurs, such as pterosaurs) doing well right up until the K-Pg boundary, beyond which no nonavian dinosaurs have ever been found.[38] However, a recent statistical analysis of speciation rates suggests that all three major dinosaur groups—sauropods, theropods, and ornithischians—were on the whole declining for more than 40 million years before the boundary, with net speciation rates being exceeded by extinction rates in the middle-early Cretaceous (this does not appear to be true for ceratopsians and hadrosaurs, however).[39]

Even if a broad dinosaurian decline were borne out, this still does not support a competitive replacement scenario in which ecologically superior mammals gradually dislodged their inferior dinosaur counterparts from their niches. Such a scenario predicts that we should see the allegedly superior group diversifying as its weaker competitor declines. This coordinated waxing and waning of fauna is pictorially depicted by Gould and Calloway's classic "double-wedge" pattern.[40] The double-wedge model contrasts with what paleontologist Michael Benton calls the "mass extinction" pattern,[41] in which the incumbent holds steady until the extinction boundary, with the successor remaining at low levels of diversity until the incumbent lineage goes extinct or suffers a severe depletion (figure 2.3).

Even if dinosaur diversity (net speciation) had long been in decline, at the time of the end-Cretaceous extinction event, they were still far and away the most ecologically diverse terrestrial vertebrates on Earth. Dinosaur abundance and functional diversity was thriving, and dinosaur ecosystems were bustling, right up to the boundary—until glass beads (tektites) and other impact ejecta are found clogging up the gills of fish struggling in the first hours after the impact.[42] It was only after the severe archosaurian depletion at the K-Pg boundary that mammals could radiate into their modern forms and massively increase in body size. Recent modeling work by Graham Slater supports the archosaurian suppression of mammalian diversity throughout the Mesozoic, followed by mammalian disparification at the boundary coinciding with the release of ecological constraints.[43] This picture does not fit the double-wedge model. Neither, however, does it perfectly fit the mass extinction pattern, given both the long-term diversity decline of *Dinsoauria* and the pre-K-Pg radiations of mammals.

Even if the impact event had never occurred, it is possible that the dinosaurs would have still declined to extinction over a longer time frame, due (say) to

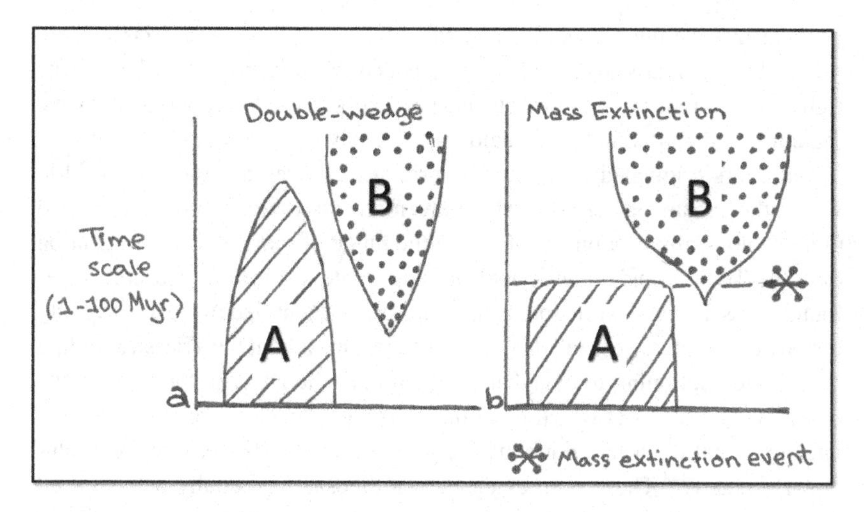

Figure 2.3
(a) Double-wedge diversity pattern, with one clade replaced by another through competitive interactions. (b) Mass extinction diversity pattern, in which one clade is extinguished or severely depleted in a major biotic perturbation and only then replaced by another clade. Redrawn from M. J. Benton, "The Late Triassic Tetrapod Extinction Events," in *The Beginning of the Age of Dinosaurs: Faunal Change across the Triassic-Jurassic Boundary*, ed. K. Padian, 303–320 (Cambridge University Press, 1986).

endothermic inefficiencies that made them poorly suited to a cooler world, while mammals would have continued to diversify during this drawn-out dinosaurian decline. It is also possible, perhaps even likely, that dinosaurs would have continued into the Paleogene to the present day—which would amount to only a small extension of their total reign—adapting to a new climatic regime and the restructuring of Mesozoic ecosystems thanks to the rise of angiosperms. In any case, other taxa, such as endothermically efficient birds, were also "waiting in the wings" and seem to have been equally amenable to a post-K-Pg radiation. Why then did mammals radiate when they did, whereas birds failed to re-evolve guilds of dinosaurian megafauna?

In the three decades since *Wonderful Life* was published, studies of faunal turnover have extended Gould's contingency counterfactual to the rise of *Dinosauria* itself. One might conclude from the remarkable evolutionary success—and astronomically bad luck—of the dinosaurs that the rise of this noble clade must have been due to its adaptive edge over the primitive reptile fauna that preceded it. In fact, the adaptive radiation of *Dinosauria* in the wake of the end-Triassic mass extinction, and its displacement of the formerly dominant crurotarsans (crocodile-like archosaurs), exhibits the same noncompetitive, contingent dynamics that we see with the rise of mammals 170 million years later. Like the Mesozoic mammals, dinosaurs during the Triassic were relatively marginal players that just happened to be in the right place at the right time with the right equipment and raw evolvability to support an adaptive radiation in the aftermath of a major biotic crisis.

Dinosaurs originated in the middle Triassic, entering a Mesozoic world in which the incumbent crurotarsans were more disparate, more diverse, and more ecologically prominent. Studying the range of morphospace occupation over the Triassic and Jurassic, paleontologist Steve Brusatte and colleagues found a classic mass extinction pattern: the crurotarsans had twice the diversity in body plans, lifestyles, and diets as dinosaurs throughout the Triassic, holding steady and continuing to overshadow them until the great perturbation at the end of the Triassic—at which point the morphospace curves reverse, with the once marginal dinosaurs doubling the crurotarsans in all measures of evolutionary success.[44] There is no evidence that dinosaurs gradually outcompeted the crurotarsans or otherwise competitively drove them to extinction. In a remark that is reminiscent of Gould's handicapper claim about the Cambrian, Brusatte noted, "If we were standing in the Late Triassic, 210 million years ago or so, and had to bet on which group would eventually dominate ecosystems, all reasonable gamblers would go with the crurotarsans." In a final twist, the crurotarsans themselves rose to prominence along with other archosaurs in the wake of the end-Permian extinction, where they replaced apex predators like the "stem" mammal gorgonopsians (therapsids), which perished along with 96 percent of species in the greatest biotic crisis of all time. Perhaps some understated fauna waiting in the wings will eventually replace therian mammals in the wake of the next mass extinction.

The dinosaur–mammal, crurotarsan–dinosaur, and therapsid–crurotarsan successions represent only a tiny handful of the faunal turnovers that have occurred in the history of animal life, though they are particularly relevant to the contingency of our own evolutionary history. Still, the overwhelming majority of animals are not vertebrates (let alone tetrapods), and the extent to which the above turnover dynamics generalize to other episodes in animal evolution is unclear. The aim here is not to provide an exhaustive review of this phenomenon so much as to appreciate the central role it played in the intellectual development of Gould's thesis.

3.3 Developmental Entrenchment

The stochastic or pseudo-stochastic sampling of lineages in mass extinction events, followed by the adaptive radiation of surviving clades into newly emptied ecospace, gets us only part of the way to the RCT. For it still leaves open the possibility that, in the intervening time between these perturbations, life will reliably gravitate back toward the optimal attractors in morphospace.

For instance, even if *Pikaia* and its proto-chordate ilk had perished in the Cambrian, perhaps "vertebrates" in some broad functional sense would have arisen again over deep time. And perhaps this "evolutionary overdetermination" goes for other bilaterian body plans. And perhaps within each of these body plan attractors there are still more fine-grained morphological attractors. For example, perhaps within the "mollusk" attractor there are "bivalve," "gastropod," and "cephalopod" attractors, such that once the mollusk plan emerges, these forms are, given enough time and a diversity of selection pressures, likely to emerge as well. Whether these sorts of attractors exist, how broadly defined they might be, and the crucial matter of their underlying causes will be investigated in the chapters to come. The point here is that the extinction/turnover dynamic, without more, does not rule out the possibility of global attractor repopulation.

Enter the second theoretical pillar on which the RCT stands: developmental entrenchment. It is the internal structure of development which ensures that bouts of non-merit-based sampling will have permanent effects on the future of animal life. To see how this might be so, let us return once again to the Cambrian. Gould's hypothesis is that once the early pages of the book of animal life were written in the Cambrian, they significantly constrained the form and content of subsequent chapters due to the entrenched nature of development.

"Development" refers to the patterns, processes, and mechanisms that characterize the maturation of a multicellular organism from zygote to death. One of the great mysteries of biology, now being slowly unraveled, is how a single cell—the embryo—can reliably give rise to an adult organism with trillions of cells, numerous cell types, and countless functionalities that comprise the tissues, organs, and organ systems of animals. The challenge is to understand how the descendants of a single cell are programmed to guide cellular differentiation and the formation of body parts depending on their location in space and time to the axes of the developing bilaterian body.

Gould's idea is that once the overarching developmental parameters of a lineage are laid down in the early history of a clade, they become highly impervious to selective modification due to their being causally bound-up with many interacting genes and functional pathways that lie "downstream" in the developmental cascade leading to the phenotype. Writing with fellow Harvard biologist Richard Lewontin in 1979, Gould surmised that "in complex organisms, early stages of ontogeny are remarkably refractory to evolutionary change, presumably because the differentiation of organ systems and their integration into a functioning body is such a delicate process, so easily derailed by early errors with accumulating effects."[45] In *The Structure of Evolutionary Theory*, Gould drew upon cutting-edge work in developmental biology to support this crucial claim about the conservation of body plans.[46]

3.4 The Causal Topography of Development

Gould's developmental hypotheses have been given a philosophical gloss by Jeffrey Schank and Bill Wimsatt, whose notion of "pleiotropic entrenchment" provides a conceptual model of the processes that might result in the conservation of "upstream" components in the developmental cascade.[47] These same developmental dynamics also bias evolutionary change toward the modification of factors that occur in the temporal tips, as it were, of the unfolding phenotype. Their rather plausible idea is that genes and regulatory networks that act early on in the developmental cascade tend to have collateral, knock-on effects vis-à-vis downstream structural and regulatory genes, and these "pleiotropic" effects make upstream components harder for selection to modify.

As a result of this causal topography, early developmental structures fail to meet the two necessary criteria for adaptation identified by Richard Lewontin in his classic paper in *Scientific American*: "quasi-independence" and "incrementality."[48] The quasi-independence criterion is violated: due to downstream pleiotropy, early ontogenetic structures cannot be modified independently of many other features of the organism, and the combination of blind variation and natural selection cannot coordinate the mutations necessary to modify all of these variables at once in a way that would result in a functionally viable outcome. The incrementality criterion is also violated: any alteration of upstream developmental components, again due to pleiotropic effects, are likely to result in morphological saltations (leaps), rather than incremental adjustments. And saltations are extremely unlikely to catch the gradient of a fitness peak in a "rugged adaptive landscape"; instead they are far more likely to land in a valley or plain. Any significant alterations of the upstream components of a gene regulatory network are likely to produce a "hopeless monster." Thus, while selection can *conserve* body plan features—and is probably a necessary force in their conservation—once these upstream structures are laid down and connected up to midstream and downstream components, they cannot be further *shaped* by selection even if such modifications would, *ceteris paribus*, be beneficial in any given case.[49]

Empirical research in the intervening years has more or less confirmed Gould's suspicions about the evolutionary developmental structure of animal body plans. Gene regulatory networks are among the most complex systems known in nature. They are informationally dynamic, controlling the expression and interaction of thousands of genes, proteins, and nongenetic components over the course of ontogeny to ensure the reliable unfolding of the phenotype. Precisely how phenotypic information is represented in "cis-regulatory" circuits, in which noncoding regions of the genome control the transcription of nearby structural genes by acting as binding sites for transcription factors, is not well understood and remains a source of ongoing investigation. Whatever the mechanisms involved, innovations in gene-regulatory networks undoubtedly played a crucial role in the evolution of complex multicellularity, including the emergence of bilaterian body plans.

Pioneering work in developmental biology has shown that gene-regulatory networks are hierarchical, just as Gould conjectured, with earlier or more central causal nodes having greater pleiotropic effects than the distal, more fine-grained terminal processes that guide cellular differentiation.[50] Upstream developmental components control the parameters of the body plan by specifying the spatial regulatory state of progenitor fields which, through localized cell signaling, guide the differentiation of specific organ systems and body parts. Upstream networks also guide the integration of phenotypic components at midstream positions in the developmental cascade. These upstream subcircuits—which biologists Eric Davidson and Doug Erwin have dubbed "kernels"—specify the general spatial configuration of the body parts of the developing organism.[51] Because kernels are "recursively wired"—that is, their cis-regulatory modules are linked together in feedback loops—interference with any single kernel gene can destroy the subcircuit's function altogether, resulting in catastrophic consequences for the phenotype such as blocking the development of major body parts.

In short, the regulatory output of an upstream circuit serves as a crucial component of the circuit below it and so on down the chain, making it difficult to modify upstream components without wreaking havoc on the phenotype. Because the phenotypic effects of mutations in developmentally upstream components like kernels are rarely linear or modular, they will often damage not only the structures that are directly implicated by the mutation but distant collateral traits as well.[52] This causal topography places a substantial and perhaps prohibitive evolutionary premium on the selective modification of kernel elements. This, in turn, explains the deep phylogenetic conservation of upstream components that specify limb coordinates, nervous system patterning, gastrulation, and many other key characteristics of bilaterian body plans. And it does so without adverting to the relative functional superiority of conserved body plans over hypothetical or extinct alternatives.

Take, for instance, the "through gut" that connects an anterior mouth to a posterior anus in all bilaterian animals. One might conclude from its ubiquity among animals that this design is optimal in some "global" sense—that it is a superior solution among all nomically possible animal designs and thus likely to be replicated on other living worlds where complex, multicellular life forms evolve. Yet the gut appears, on most counts, to have arisen only once in the history of life and to have been conserved in *Bilateria* ever since. Current debates over the phylogenetic position of ctenophores (comb jellies), discussed in chapter 9, leave open the possibility that the gut evolved twice in animals, which would substantially increase our credence in its replicability.

For the moment, however, let us presume that the single origin hypothesis is correct. Now add to this the additional premise that the causal topography of development is such that the gut will be conserved even if a functionally superior configuration is theoretically realizable, because any modification of the kernels that specify the axes of development will prove catastrophic for the phenotype. Given the conjunction of these two premises, there is no reason to suppose that this conserved design is optimal among all morphologically possible designs, or even that it is superior to most. Perhaps the gut is a frozen accident. This is why evolutionary iteration (of the right sort) is so important to the case against contingency.

In sum, the causal topography of gene-regulatory systems explains why bilaterian animal body plans have changed very little since their emergence in the base of the Cambrian, and why the diversification that has occurred has been restricted primarily to tinkering with downstream nodes of the morphogenetic-regulatory network.[53] "Body plans" refer, roughly, to the overarching morphological configuration of higher animal taxa, as quintessentially reflected by the "phyla" and "superphyla" Linnaean ranks (but potentially also referring to class and order-level organization). If body plans are specified by kernels, and if kernels are recalcitrant to modification once they become developmentally entrenched in a lineage, then it follows that the body plans of lineages that survive stochastic or pseudo-stochastic extinctions will effectively become locked in place.

Selection will thereafter be limited to tinkering with downstream or distal components of gene-regulatory networks, such as the "gene batteries" that control cellular differentiation, which can be altered incrementally (through changes in sequence, timing, intensity, etc.) without disturbing other critical elements of the organism. This allows for morphological divergence in lower taxa notwithstanding the conservation of phyletic parameters themselves. This is not to understate the amount of morphological change that can occur within phyla and even within midrange Linnaean taxa: mammals, you will recall, "disparified" from a small-bodied, rodent-like critter in the Cretaceous to the wide range of mammalian forms and lifeways we see today. But the fundamental organization of the body plan remains the center of gravity for vertebrate evolution ever since it congealed in the Cambrian, much as it has for all other animal groups.

3.5 Decimation and Diversification

We can now import this thinking about the causal structure of development back into Gould's original reading of the Cambrian. The fantastical parade of forms that emerged during the Cambrian explosion was only possible because upstream developmental components had yet to be hooked up to downstream batteries of genes that guide more fine-grained structural differentiation. Once these terminal portions of the circuitry were in place, the kernels of surviving animal body plans were locked in, stabilized by selection, and buffered by canalized developmental systems to maintain the integrity of the phenotype. The development entrenchment of body plans ensured that the effects of stochastic or pseudo-stochastic patterns of extinction would be felt for the remainder of life's history. Once the initial crop of body plans had been culled, large regions of evolutionary possibility were rendered permanently off limits. Postextinction recovery and diversification would forever after be confined to the body plans of surviving higher taxa. The accidental Cambrian survivors became, in effect, frozen accidents.

This picture supports the "decimation-diversification" theory that Gould sketched in *Wonderful Life.*[54] According to that hypothesis, animal life began with only loose developmental constraints, and thus it had the potential for an explosive bout of body plan–level diversification that Gould called "disparification." The explosive evolution of body plans could only occur once the genetic regulatory innovations necessary to support the development of complex phenotypic structures were in place. However, once the various body plan parameters were laid down, they became recalcitrant to modification due to their cascading causal connections to downstream components of the phenotype. Little to no disparity is therefore added in subsequent phases of the history of animal life, even while lineages continue to diversify within their plans, increasing in lower-taxonomic richness. Over time, mass extinctions whittled down the breadth of morphospace occupation that was established in the Cambrian, eliminating some islands of form and leaving increasingly large and unbridgeable gaps between the remaining islands (see figure 2.4). Virtually all post-Cambrian disparification and diversification, so the theory goes, would take place *within* surviving body plans.

In effect, Gould is proposing a "non-uniformitarian" theory of macroevolutionary change. The term "uniformitarianism," coined by the English polymath William Whewell, holds roughly that the fundamental mechanisms, laws, and

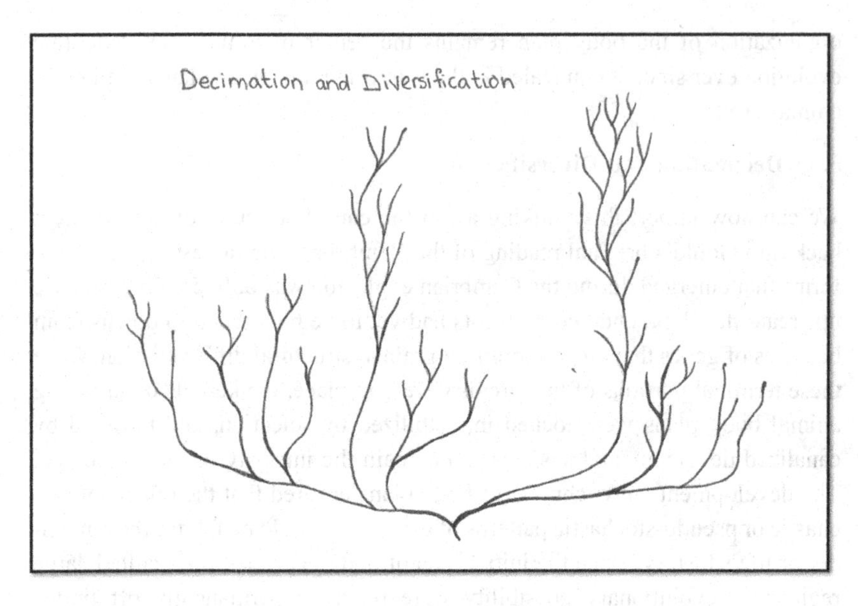

Figure 2.4
Decimation-diversification pattern showing large gaps in extant morphospace occupation due to early extinction events. Redrawn from S. J. Gould, *Wonderful Life: The Burgess Shale and the Nature of History* (Norton, 1989).

processes that presently govern our world are the same as those that governed our world in the past. Conceived methodologically, uniformitarianism holds that we should proceed as if the metaphysical assumption of uniformity were true. This assumption, which rings of Copernicanism, may have served science well at times in the past. As Doug Erwin has noted, "a uniformitarian approach may have been politically necessary early in the history of evolutionary thought as a counterweight to various non-Darwinian approaches to evolution."[55] But uniformitarianism should be treated as a working hypothesis, not a bedrock ontological or methodological commitment of any field of science. This is particularly so for biology, given the many empirical and theoretical reasons we have to doubt the uniformity of process and outcome in evolution (see chapter 1 for a discussion).

Studies of the fossil record, aided by quantitative analyses of morphological disparity developed by paleontologist Michael Foote and others, have more or less confirmed Gould's suspicions about disparity, lending credence to the notion that macroevolution is not merely "scaled-up" or aggregated microevolutionary processes.[56] At a very coarse grain of description, biology is uniformitarian: the same forces and tendencies, such as selection, drift, and mutation, are in operation at all biological times and in all biological places. However, this glosses over more contentful processes and patterns that exhibit a nonuniformitarian character, such as the evolution of morphological disparity. The present is not always a key to the past.

Current work on the geological and genetic fossil records of animals confirms that nearly all body plan–level characteristics—traits that are diagnostic of modern animal phyla—arose more than 500 million years ago, with subsequent morphological diversification occurring within, rather than among, those plans. Obstacles remain to the measurement of morphospace occupation, which make the decimation-diversification hypothesis difficult to test.[57] If the hypothesis is correct, however, it provides a formidable answer to the query with which we began the previous subsection: Why, in the intervening period between mass extinctions, does natural selection not return to globally optimal attractors in morphospace? Decimation-diversification is therefore both a signature of, and a plausible explanatory framework for, contingent dynamics at work in macroevolution.

Before moving on, it is worth underscoring a point about biological universality that is lurking in the background of this discussion. The constraints on body plan modifiability that are imposed by the causal geometry of development appear, on all counts, to be law-like. That is, they seem to be a general feature of the evolution of complex, multicellular life anywhere in the cosmos. Although Gould to my knowledge never made this point, if he is right, then it is due to universal features of biology that there are no specific laws of form, and the most we can infer about complex multicellular life elsewhere in the universe is that it will be as historically contingent as our own.

4. Two Empirical Critiques

How much of the evidence that Gould relied upon has withstood the test of time? We have already seen that Gould's core claims about the evolution of body plan disparity have been largely corroborated, as have his (and Lewontin's) perspicacious theories about the causal structure of development and the constraints it imposes on morphological evolution. Ditto for Gould's views on the dynamics of mass extinction and faunal turnovers. Other recent work in evolutionary biology, however, has begun to put pressure on various components of the RCT. We will take up what I think is the most formidable of these challenges—the critique from convergent evolution—after the concept of Gouldian contingency has been more systematically examined in the next chapter. Let us briefly consider here two current lines of criticism—one taxonomic and the other ecological—that take aim specifically at Gould's reading of the Cambrian fauna. Whether or not these criticisms are ultimately persuasive, they will help us to discern additional contours of Gould's thesis.

4.1 The Critique from Taxonomy

The first challenge to the RCT takes aim at Gould's taxonomic reading of the Burgess Shale. In *Wonderful Life*, Gould makes much of the fact that many of the Burgess Shale critters that had been "shoehorned" by their discoverer (paleontologist Charles Walcott) into familiar phyla were, upon closer examination by arthropod expert Harry Whittington and his team in the 1970s, determined not to fit neatly into any extant taxonomic groups. This classification conundrum led Gould to suppose that if *Opabinia* and its ilk had survived the Cambrian extinctions while chordates had perished—which seemed like an eminently plausible possibility—then the history of life would have taken a science fiction–like turn. Of course, to an observer emerging from any such alternative history of life, the sci-fi and the familiar would be reversed, and proto-vertebrates would be among the weird wonders whose reconstructed image would provoke laughter from an audience of five-eyed geologists.

In the light of modern cladistics, however, the Cambrian fauna are less recalcitrant to classification than Gould had believed. As paleontologist Graham Budd and his colleagues have convincingly shown, many of the seemingly bizarre Cambrian taxa that inspired the RCT can be recognized under modern phylogenetic classification systems as "stem" taxa—extinct basal lineages on the stems of crown groups.[58] "Crown" groups are the groups represented by extant phyla and all their ancestors, leading back to their common ancestor. The notion of stem and crown taxa were first introduced by paleontologist Richard Jefferies to little fanfare, only to be rejuvenated by Budd and his collaborators at the beginning of the new millennium.[59] Together, the stem and crown groups comprise a "pan" taxa, which includes the crown group and all extinct lineages that are more closely related to the crown group than to any other extant groups (see figure 2.5).

On this cladistic approach to phyletic classification, body plans are defined not in terms of broad morphological or developmental parameters but rather as "set[s] of features plesiomorphically [i.e., ancestrally] shared by extant taxa in a monophyletic clade."[60] There are two things to note about this definition. First, because it is neutral to taxonomic-level, "body plan" in this sense need not specify distinctive phyla-level traits; it could refer to ancestral similarities at any phylogenetic grain of description. Second, and as a result of this phylogenetic grain-neutrality, the cladistic conception of the body plan does little to convey the notion of developmental constraint that is so prominent in the traditional study of *Baupläne* (German (pl.), body plans) that influenced Gould's evolutionary thought. We will return to this point shortly.

Conceived cladistically, there is nothing about the origins of disparate body plans or their conservation that requires special explanation. The cladist can

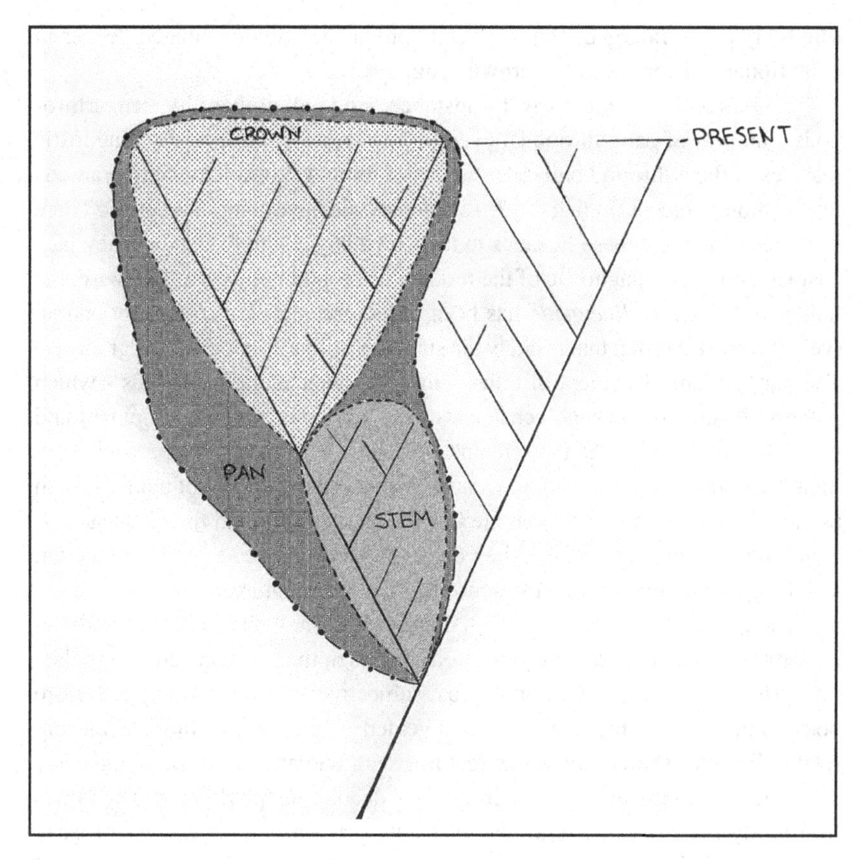

Figure 2.5
Illustration of the "stem taxa" concept in cladistic approaches to taxonomy, which can accommodate many of the Cambrian *Problematica* as members of the total (or pan) group that branched off before divergence of the crown group. Redrawn from R. P. S. Jefferies, "The Origin of Chordates—a Methodological Essay," in *The Origin of Major Invertebrate Groups*, ed. M. R. House, 443–477 (Academic Press, 1979).

acknowledge that there are few representatives of modern body plans *sensu stricto* among the Cambrian fauna, but there is nothing anomalous about this fact—indeed, it is an expectable result of the structure of clade diversification. Once morphological disparity and phylogenetic affinities are decoupled in our system of classification—a feat that cladistics achieves—the taxonomic challenges of the early Cambrian can be "explained away" as an artifact of clade geometry and intuition-based approaches to classification that rely on subjectively chosen phenotypic traits. In essence, the stem/crown group formalization allows us to establish phylogenetic affinities of the Cambrian fauna before true body plans emerged. What we call "phyla" actually originated deeper in evolutionary history than the crown group.

Opabinia and *Anomalocaris*, for instance, are in all probability stem arthropods, and part of panarthropoda, even if they lack the traditionally diagnostic features of the arthropod body plan such as a jointed exoskeleton and biramous limbs (though recent findings suggest that they did have compound eyes). There is no need to assign these lineages to their own phyla simply because they lack characters that are diagnostic of the modern arthropod body plan. Likewise, the famously bizarre *Hallucigenia* has been reinterpreted as a stem onychophoran (velvet worm), a group that is likely the sister taxa of [arthropods + tardigrades].[61] The pan-phylum placement of other stem taxa, such as the wiwaxiids—which possess features of both polychaete worms (chitinous scales and spines) and mollusks (foot and radula)—remains unclear.[62] The cladistic approach may illuminate some even larger phylogenetic puzzles in the origins of animals, such as the nebulous relation between the late Ediacaran fauna and the metazoans. It could suggest, for example, that the Ediacaran fauna constitutes a stem bilaterian lacking gastrulation and other diagnostic metazoan features.[63]

Although Cambrian taxonomic challenges remain, the point is that with the concepts of stem and crown group at our disposal, there is no need to populate the early history of life with numerous, subjectively delineated phyla. Before body plans in their modern forms congealed in the later Cambrian, closely related lineages shared numerous features with one another even though they do not fit tidily into any modern groups. By decoupling phylogenetic affinities from body plans, cladistics allows us to construct a treelike picture of early animal diversification, crown and stems and all, before modern developmental organizations arose. No further evolutionary processes are needed, so the cladistic argument goes, to explain the putative Cambrian anomalies. And thus, one major pillar on which the RCT rests has been explained away by modern phylogenetic analysis. Taxonomic mystery solved. RCT refuted?

Not quite. *Wonderful Life* can still be a "homage to *Opabinia*" even if this creature and other oddballs of the Cambrian turn out to be stem taxa of familiar clades. Whereas Gould was fascinated by the *Problematica* because of their unique features and combinations, cladistic analysis ignores these unique features and instead uses shared derived characters to situate the Cambrian taxa unproblematically in relation to modern phyla. Yet in doing so, as philosopher Keynyn Brysse has pointed out, the cladistic reconstruction of Cambrian phylogeny simply bypasses the big theoretical questions that preoccupied Gould, such as in relation to patterns of morphological disparity and extinction and their implications for the contingent nature of macroevolution.[64] In essence, cladistics conceptually guts the *Bauplan*. This is a mere nominal victory, however, because it leaves intact the key research questions that the Bauplan construct is designed to probe. The cladistic reinterpretation of the early Cambrian fauna solves a problem, but not the one that Gould had underscored.

Simply stated, cladistics is pursuing a different set of research questions than Gould was pursuing. Gould uses taxonomy as a proxy for morphospace occupation in exploring possible evolutionary worlds—an exercise that is orthogonal to the cladistics project. Gould's claims are not about taxonomy per se but about the traits that traditional taxonomy is intended to capture. Trilobites, anomalocarids, opabinids, and the like may all be diagnosable arthropods on the cladistic scheme, but there is no reason to think that the modern arthropod body plan (which includes a jointed exoskeleton and biramous limbs) was an inevitable or robustly replicable result of early arthropod evolution. The sheer dominance and diversity of arthropods in the Cambrian seas makes the survival of this clade into subsequent geological periods highly likely, but this does not mean that the distinctive characters of the arthropod body plan are themselves stable across possible Cambrian worlds.

The point is not simply that cladistics is concerned with establishing patterns of genealogy rather than the mechanisms or processes underlying those patterns. The point is that cladistics misses important patterns themselves. Recognizing this can help reconcile discord between molecular and fossil data about when major animal groups diverged. Crown groups, for example, appear to have diverged genetically as early as the end of the Ediacaran (about 545 million years ago), but this genetic divergence did not result in morphological innovations that reflect our modern range of animal phyla until later in the Cambrian. The same pattern is true, for example, of therian mammal diversification between the Cretaceous and Paleogene. This further underscores the importance of distinguishing taxonomic divergence from the morphological and ecological innovations that are relevant to Gouldian questions. In short, the cladistic reinterpretation of the Burgess Shale leaves the RCT unscathed. The fact that the Cambrian *Problematica* can be accommodated as stem taxa (rather than freestanding phyla) does no damage whatsoever to Gould's decimation-diversification hypothesis, to his claims about the developmental mechanisms that underwrite nonuniformitarian patterns of morphological evolution, or to his readings of mass extinction and faunal turnover.

This brings us to a point that has been glossed over in critical discussions of the RCT. Making this point requires telegraphing material that will be elaborated on in chapters to come, so I ask that the reader forgive the brevity of the treatment here. Let us assume, for the sake of argument, that the early Cambrian fauna did not reflect the broad range of forms that Gould believed them to. This would not, in fact, support the antithetical view of life, namely the notion that macroevolutionary outcomes are robustly replicable. The reason for this is that the latter requires a "global" merit-based competition among early animal lineages—and such a competition could not have taken place if only a small, contingent subset of possible forms arose. It is not enough for selection to optimize body plans that did arise; if it is to achieve anything approaching global optimality, selection must cull functionally inferior plans from a wide assortment of alternative designs, many of which need to be vetted and eliminated by selection as doomed experiments in living. Thus, the lack of an early experimentation phase—especially when combined with the absence of later disparification—does little to refute the RCT.

4.2 The Critique from Paleoecology

A second recent challenge is aimed at Gould's ecological reading of the Burgess Shale fauna. Recall that the main metaphysical conclusion of Gould's Cambrian thought experiment is that the shape of animal life is radically contingent. By this, he means that the actual distribution of body plans occupies only one of many equally functional configurations in a vast morphospace of evolutionary *possibilia*. Remember also that Gould drew this inference on the basis of an epistemic assertion: that no evolutionary-ecological handicapper worth her salt would have successfully predicted which lineages would survive the Cambrian extinctions. Although Gould was impressed by many of the Cambrian critters, he was particularly enamored of *Pikaia*, a close relative of the last common ancestor of all chordates, mainly for reasons of anthropocentric relevance: had *Pikaia* and other proto-chordates been eliminated in the luck of the draw, we—and perhaps nothing like us—would be here today.

At the time of Gould's writing, *Pikaia* was portrayed as a relatively understated lineage that could easily have gotten lost in the blooming buzzing confusion of the Cambrian seas. Our knowledge of early Cambrian fauna at that time was limited to the Burgess Shale. Since that time, however, similar fossil lagerstätten have been found in the People's Republic of China (and also in Greenland), including the now famous Chengjiang assemblage of the Maotianshan Shales. Like the Burgess Shale, the Chengjiang assemblage contains exquisitely preserved, non-mineralized (soft) tissues of Cambrian animals, including all the Burgess Shale groups in addition to many previously unknown lineages. The rich Chengjiang record confirms that the Burgess Shale biota was indeed a global fauna, and it strengthens the case for the geologically abrupt origin of animals—a fact that is not critical to the RCT, but which is often incorrectly thought to be. It also paints a fuller picture of the structure of Cambrian ecosystems, including the ecological role played by early chordates. Eight new candidate chordates have been discovered, including possible 530-million-year-old craniates—chordates with a notochord and distinct head—such as *Myllokunmingia* and *Haikouichthys*.[65] Similar basal chordates have recently been found in newly uncovered deposits in the Canadian Rockies, such as in the Marble Canyon site of Kootenay National Park, which includes among its spectacular assemblage a primitive jawless fish with discernable tail and camera-type eye (*Metaspriggina*).[66]

What are the implications of these findings for Gould's thesis? The first thing to note is that Gould's analysis of *Pikaia* and its implications for the contingency of vertebrate history do not rely on the claim that *Pikaia* was a *direct* ancestor of modern vertebrates—such that, had *Pikaia* gone extinct in the Cambrian, vertebrates would never have come to be. Gould was well aware of the incompleteness of the fossil record, especially of the earliest soft-bodied animals, and he no doubt presumed that *Pikaia* was not the lone proto-vertebrate lineage in the Cambrian. Nor does Gould's account rely on the claim that *Pikaia* and its ilk were ecologically *irrelevant*; the taxon's repeated preservation among Cambrian assemblages indicates that it was commonplace enough to be recorded in the fossil record, if comparatively less frequently than some other lineages. Instead, Gould's rhetorical use of *Pikaia* hinges only on the assertion that vertebrates were an *evolutionary undercard* in the Cambrian.

As Doug Erwin noted in his recent reevaluation of *Wonderful Life*,[67] new findings from the Canadian and Chinese assemblages show that basal chordates were in fact a more diverse clade, with a cosmopolitan distribution and a greater ecological presence, than the original Burgess Shale data led Gould to believe. Because prevalence is a significant determinant of survivorship in mass extinctions, the new Cambrian chordate data calls into question Gould's epistemic premise that no good evolutionary handicapper would have bet that chordates would go on to become a permanent fixture of the Phanerozoic Eon.

Even so, I do not see this adjustment to the subjective probability of early chordate elimination doing any real damage to the extinction component of the RCT. Erwin is right that it would not be unreasonable to bet on chordate survival in the Cambrian. Yet while geographical range and ecological significance are statistically relevant to the chances of survival, they are not decisive determinants of survival in the end-Cambrian or any other mass extinction. For as we have seen, many of the most diverse and dominant clades on Earth either perished or were permanently cut down in mass extinctions, paving the way for the rise of less prominent (if not ecologically negligible) clades.

The Cambrian extinctions culled lineages that were more prominent than basal chordates, including many of the critters that Gould conjectured could have been early nodes of alternative histories of life of Earth. The fact that *Pikaia* turns out to be part of a more diverse clade may make it more likely that chordates survive into the Ordovician—but it does little to support a merit-based account of Cambrian survivorship. Gould's point about contingency in faunal turnovers remains, even if chordates were not quite the "Cinderella taxa" that Gould imagined them to be. Once surviving Cambrian phyla underwent substantial diversification in the Paleozoic, it became unlikely that any would vanish entirely in a subsequent extinction event. Not so, however, for the weird wonders of the Cambrian, which lived during the most perilous time for body plans in the history of life.

3 A Philosophical Theory of Evolutionary Contingency

Stephen Jay Gould was one of the great popular science essayists of his day. His voluminous, wide-ranging, and baroque body of scholarship is known for many things, but understatement is not one of them. His views were often presented in bold terms, sometimes to the ire of his peers who cringed at the language of "revolution" or chaffed at what they saw as characterizations of evolutionary science that were either uncharitable or vulnerable to exploitation by creationist political forces. On occasion, Gould was compelled to walk back some of his more provocative assertions while still clinging to scientific vindication (see, for example, the multidecade controversy over "punctuated equilibria"). Other bastions of Gouldian thought were defended categorically and unapologetically to the very end.

The emphasis on historical contingency in evolution is one such unyielding commitment. Gould never abandoned his self-professed crusade to free biology from the yolk of physics envy by encouraging paleontologists to unabashedly seek narrativistic explanations of macroevolution. More than this, he insisted that contingency dominated at the level of scale and detail that is of the greatest interest to macroevolutionary biologists, namely at the level of form.

Yet after more than two decades, the radical contingency thesis (RCT) remains underspecified. Ambiguities in its formulation have caused ongoing confusion about the core commitments of the theory and consequently the evidence that might bear on its adjudication. Not only did Gould not attempt to define contingency in any systematic way, but he also used the term inconsistently, at times conflating the metaphysics of contingency with the epistemic consequences of contingent causal dynamics, such as in relation to prediction and explanation.

Further ambiguities abound. For instance, it is not clear which types of evolutionary outcomes the RCT is intended to capture and at what level of description. Nor is it obvious how dissimilar outcomes must be across replays of life's tape for them to support the RCT or how this outcome dissimilarity should be measured. A further wrinkle is that contingency can only be assessed relative to initial conditions; hence, whether an outcome is contingent or robust will depend on what we take the replay starting conditions to be. Are we contemplating replaying the tape from the origins of the first cell, the first eukaryotes, the advent of complex multicellularity, the emergence of tetrapods, the demise of the nonavian dinosaurs, and so on? An outcome may be stable across the evolution of mammals, even while it is unstable across the evolution of tetrapods, chordates, bilaterians, metazoans, or eukaryotes. Gould's Cambrian thought experiment, discussed in chapter 2, is structured in this manner, as it assumes the existence of complex multicellular organization among its initial conditions and queries whether the evolution of specific body plans and their associated regularities hold across replays of the complex multicellular tape of life.

Finally, and even more daunting from a methodologic perspective, the contingency question is one of relative frequency, in that it may apply to some evolutionary outcomes but not to others. Adjudicating the RCT requires, therefore, that we make some global judgment about the relative significance of contingency in the history of life, and as a result there is no single crucial experiment that can be performed that would refute Gould's thesis. A frustrating implication of this is that a handful of putative counterexamples can at best refute subclaims of the RCT, not the RCT as a whole. For instance, even if turns out that the mind is evolutionarily robust, as will be argued in part II, this does not preclude other major aspects of life, such as body plans, from being radically contingent. Despite these difficulties, it is nonetheless meaningful to ask whether contingency is a dominant theme in macroevolution and to investigate the specific outcomes to which it does and does not apply.

How does acknowledging outcome ambiguities, contingency gradations, spheres of application, and initial condition-relativity affect the ways in which we understand and evaluate the RCT? By sketching out an answer to this question, we can begin to develop a better sense of how the RCT hangs together with Gould's other theoretical commitments and whether it stands up to our current scientific understanding of life. Although one goal of this chapter is to tease apart the nuances of Gould's view, the primary aim here is not exegetical. It is one thing to consider what Gould might have had in mind when he described evolutionary outcomes and processes as "contingent"; it is quite another to consider which reading of contingency best conforms to the existing body of contemporary biological theory. The task of this chapter is also critical, therefore, in that it underscores a number of conceptual problems with the RCT both as Gould framed it and as it has been elaborated on in more recent philosophical scholarship. As we shall see in the next three chapters, these conceptual problems have seeded misinterpretations of Gould's thesis, which in turn have impeded evolutionary investigations that are designed to put that thesis to the test.

1. Conceptions of Evolutionary Contingency

Contingency is a conceptually fraught subject, bristling with different meanings in different literatures and contexts. For instance, in traditional philosophical discussions of laws of nature, a common view is that to satisfy the criteria for lawhood, a universal generalization must be a "contingent necessitation," by which it is meant that laws must not be logically necessary truths. In evolutionary biology and other historical sciences, the term "contingency" takes on a very different meaning—something closer to a "formative happenstance," or an accident that shapes or influences some relevant outcome or set of outcomes. There are many nonequivalent definitions of contingency on offer in the biological literature, and many interpretations of what Gould meant by the term. Existing accounts capture important aspects of evolutionary contingency, but, as we shall see, these accounts are ultimately incomplete.

1.1. Two Senses of Gouldian Contingency

Because Gould did not attempt a definition of evolutionary contingency, students of Gould are left to piece together a conception out of fragmented passages in Gould's prolific corpus of scholarship. Our reconstruction project begins with philosopher John Beatty, who argues that Gould equivocates between two compatible but importantly different conceptions of contingency.[1] Although Beatty is right that these different senses of contingency can plausibly be attributed to Gould, we shall see why, taken both individually and collectively, they do not adequately capture the concept of contingency as it figures in the Gouldian view of life.

The first sense of contingency that Beatty attributes to Gould is what he calls "contingency as causal dependence." In brief, this implies that a series of prior events (E_1, E_2, … , E_i) in a chain (or dynamic web) of causation are each necessary with respect to the production of an outcome O, such that if any of these events had not occurred or had occurred in a different way, O would not have occurred or would have occurred in a different way. Where such a causal structure obtains, we can say that O is contingent on E. Beatty is right to say that evolutionary contingency involves the causal dependence of evolutionary outcomes on prior events. But causal dependence *simplicter* is vastly overinclusive, for two reasons.

First, it fails to rule out nomically expectable outcomes. If all events (E_1, E_2, … , E_i) along a causal chain or in dynamic causal web are highly likely to repeat, say, due to constraints of the laws of physics, then O will be virtually certain to repeat, given a replay of the system. Star formation may be contingent on inhomogeneous concentrations of matter brought about by the shock waves of supernovae or, more distally, by quantum fluctuations in the early expanding universe, but stars cannot plausibly be considered radically contingent phenomena in the Gouldian sense because they are law-like outcomes that will reliably occur across countless replays of the tape of the cosmos. So, on Beatty's first definition, *O* may be contingent on *E* even if *E* is nomically necessary and *O*'s occurrence is, due to its lawful connections to *E*, highly replicable. This notion of contingency does not get at the causal dynamics the RCT is after.

A second problem with conceiving of evolutionary contingency as mere causal dependence, in the way that Beatty does, is that it does not enable us to locate the source of the disagreement between the RCT and its detractor theories. It cannot be that contingency in the Gouldian sense entails that *any* change in initial conditions will tend to produce *any* change in outcome. Without saying something more definitive about the kind of causal dependence that is actually in dispute, we are liable to wind up talking at cross purposes or addressing straw man versions of opposing views—which, as we shall see in the next chapter, has in fact occurred.

The second sense of contingency that Beatty attributes to Gould is what Beatty calls "contingency as unpredictability." This entails that identical initial conditions do not suffice to produce the same outcome. This definition seems to accord with Gould's various "rewind the tape" thought experiments, whereby we go back in time to key junctures in animal evolution and let life march forward once again, only to find that it does so to a very different macroevolutionary tune. As we saw in chapter 2, Gould does gloss the RCT in terms of unpredictability, an epistemic state that he attributes to the "ecological handicapper" in his macroevolutionary thought experiments. Although there is a textual basis in Gould's work for interpreting contingency in terms of unpredictability, there is a major exegetical problem with this interpretation, as well as an even more serious conceptual problem with such an account.

The exegetical problem is that this notion of contingency would seem, on its face, to commit Gould to the metaphysical thesis of indeterminism because by definition "contingency-as-unpredictability" requires that the *same* initial conditions produce *disparate* outcomes—a physical impossibility if determinism obtains for biological systems. Beatty expressly disavows the inference from contingency-as-unpredictability to indeterminism,[2] and rightfully so given that Gould explicitly divorced randomness from contingency.[3] The difficulty for Beatty's reading, however, is that if determinism is true, then Gould's rewinding-the-tape thought experiments are trivial exercises, for they will always play out in precisely the same way. In a "deterministic" universe, any state of the world is necessitated or determined by a previous state together with the laws of nature. French mathematician Pierre-Simon Laplace eloquently describes such a universe:

> We ought to regard the present state of the universe as the effect of its antecedent state and as the cause of the state that is to follow. An intelligence knowing all the forces acting in nature at a given instant, as well as the momentary positions of all things in the universe, would be able to comprehend in one single formula the motions of the largest bodies as well as the lightest atoms in the world, provided that its intellect were sufficiently powerful to subject all data to analysis; to it nothing would be uncertain, the future as well as the past would be present to its eyes.[4]

Like Laplace, Gould entwined the metaphysical thesis of determinism with the epistemic state of predictability. As the philosopher Yemima Ben-Menahem points out, the literature on historical contingency tends to conflate chance (indeterminism) and contingency, even though the two concepts are clearly severable.[5] Indeed, not only are the concepts of contingency and predictability severable, they describe *entirely different categories of thing*: metaphysical states of the universe, on the one hand, and knowledge states about metaphysical states of the universe, on the other. Although determinism and predictability bear important relations to one another, it is also easy to see how the two come apart and why their conflation is problematic.

For instance, chaotic dynamical systems are deterministic, yet they are in principle unpredictable; quantum mechanical systems are irreducibly indeterministic, yet they support the greatest predictive precision ever achieved by a human science. Even comparatively simple deterministic systems will support prediction only to the extent that the laws of nature can be known, present states ascertained, and future states computed by the cognizer in question. The "*n*-body problem" in physics shows that even for Newtonian systems involving only three bodies moving solely under the effects of mutual gravitation, attempts to derive future states of the system can be intractable. Thus, whether the universe or some relevant subset of it is deterministic is a metaphysical question that is wholly distinct from the question of whether future states can be predicted by any given cognizer. It follows that whether macroevolution is radically contingent is a metaphysical question that is decidedly not determined by the knowledge state of any observer. Were it so determined, then a system could at the same time be radically contingent for one observer and not for another, depending on their respective computational capacities, knowledge of initial conditions, and understandings of nature's laws.

Imagine, for example, a late-Cretaceous observer—whom we will refer to as the Cretaceous Daemon—who has a profoundly deep understanding of Mesozoic marine and terrestrial ecosystems, as well as an awareness of the position and velocity of all kill-grade celestial objects in the solar system. Imagine also that our idealized observer, upon witnessing a gravitational disturbance in the Kuiper Belt or the Oort Cloud, calculates the projected impact angle and energy of the bolide that will deterministically careen thousands of years hence into the shallow, sulfur-rich sea beds of the Yucatán Peninsula. From there our Daemon derives the likely geological, climatological, and ecological aftermath of this event, predicting the global wildfires borne of raining impact ejecta, the blanketing of the planet in sulfuric acid aerosols that would cause global temperatures to plummet and trigger acidification of the oceans, the plunging of the Earth into global darkness and cooling that shut down photosynthesis and result in catastrophic losses to phytoplankton and plant species, which in turn would trigger a collapse of Cretaceous ecosystems and usher in a new climatic regime on Earth. What's more, our Cretaceous Daemon can calculate the extinction probability distribution for lineages that are likely to be severely depleted or extinguished in this biotic crisis based on factors such as their geographic distribution, trophic position, biomass, speciocity, feeding ecology, body size, species-typical behaviors, coevolutionary interactions, and so on, allowing the Daemon to foretell the radiation of surviving mammals into the vacated archosaurian niches.

Even if our idealized Cretaceous observer were capable of such preternatural epistemic feats, the dinosaur-mammal succession *would still be radically contingent*. For it is not the amenability of this outcome to prediction but rather its causal structure that makes it contingent. This means that the RCT is at bottom a metaphysical thesis, not an epistemic one. It holds that certain macroevolutionary outcomes are sensitive to low probability events that are unlikely to be replicated across the vast majority of alternative evolutionary histories. In Gould's own words, "alter any early event, ever so slightly and without apparent importance at the time, and evolution cascades into a radically different channel."[6] Here again Gould can be found tying the metaphysics of contingency to epistemic states ("apparent importance"). But whether the metaphysics of contingency translates into *ex ante* unpredictability is simply irrelevant for purposes of characterizing these causal dynamics.

In *The Structure of Evolutionary Theory*, Gould abetted the confusion on these points by contrasting the "contingent phenomenology" of natural history with Laplacean-style determinism.[7] Yet, as we have seen, if determinism is true, then rewinding the tape of life would be a trivial exercise because the trajectory of life would unfold in precisely the same manner. If determinism obtains, then the same set of initial conditions *must* result in the same outcome. How then can we make sense of Beatty's claim that what Gould meant by "contingency" is that different outcomes follow on from the same initial conditions? It seems we have two options.

The first option is to maintain that Gould was, despite protestations to the contrary, committed to the metaphysical indeterminism of evolutionary biological systems. Biological indeterminism is not, substantively speaking, an implausible position, as irreducibly chancy events may very well "percolate up" from the quantum level to affect evolutionary trajectories. This percolation may occur, for example, through proton tunneling that affects which mutations arise or in what order they do so. Or it may occur through quantum alterations of microscopic initial conditions on which chaotic geophysical systems (such as weather, climate, tectonics, etc.) are sensitively dependent, with chaotic dynamics magnifying these events to the point that they influence large-scale selective environments. Nevertheless, because Gould was explicit about not equating contingency with indeterminism, it is best to opt for another interpretation.

According to this different interpretation, Gould is either explicitly excluding certain evolutionary boundary conditions from the "same initial conditions" or else referring to an epistemically equivalent (but metaphysically nonequivalent) set of boundary conditions, wherein the undetected nonequivalent conditions are responsible for the disparity in outcomes. Let us call this latter view the "hidden variables" reading of contingency-as-unpredictability. It bears noting that Gould at no time explicitly excludes any classes of initial conditions—and neither does Beatty in his interpretation of Gould—and it is unclear what principled rationale could justify such an exclusion. But the hidden variables account is the more plausible reading of the two more plausible readings of Beatty's second formulation of Gouldian contingency.

Nevertheless, there is a fundamental problem with this interpretation as well. The problem is that the issues of determinism-versus-indeterminism and predictability-versus-unpredictability are red herrings because they tell us nothing about the accidental or law-like nature of macroevolutionary outcomes—which is what the RCT is arguably all about. Accidental outcomes remain accidents, and law-like outcomes retain their nomic necessity, regardless of whether they are part of a deterministically or indeterministically configured universe, and regardless of whether they are amenable or recalcitrant to prediction. The RCT is best understood, I submit, as a "modal" thesis that describes the sensitive causal dependency of evolutionary outcomes on small changes in initial conditions. If this is correct, then the RCT hypothesizes a causal structure of life that can only be revealed through analyses of possible evolutionary worlds.

1.2 Contingency, Stochasticity, and Path Dependence

Before canvassing this positive view, let us briefly consider two other philosophical accounts of Gouldian contingency and their limitations as doing so will further motivate the view that I want to defend. In chapter 2, we saw that stochastic or pseudo-stochastic extinction plays an important role in the RCT. The sampling of Cambrian body plans for reasons unrelated to long-term functional merit is one of the two major pillars on which the RCT rests, the other being the developmental entrenchment of the body plans that fortuitously survived these perturbations. Philosophical accounts of Gouldian contingency have been influenced by each of these pillars, though as we shall now see they come up short of a full-fledged theory.

For instance, the philosopher Derek Turner has argued that we should conceive of Gouldian contingency as a claim about macroevolutionary stochasticity. Turner maintains that "for Gould, evolutionary contingency is the random or unbiased sorting of entire lineages. It just is the macroevolutionary analogue of random drift."[8] In a similar vein, Beatty is puzzled by Gould's decision not to include stochastic processes in his concept of contingency, given that "Gould acknowledged [these phenomena] as sources of historical contingency."[9] Turner's recommendation that we think of Gouldian contingency in explicitly macroevolutionary terms is on the right track. But focusing exclusively on stochasticity is problematic for several reasons.

First, as we saw in the previous chapter, many lineage culling episodes may be selective and thus not genuinely stochastic, even if the selective regimes imposed are too fleeting and sporadic to permit the construction of macroevolutionary adaptations. The key characteristic of these sorting episodes is not that they are stochastic, but that the sorting is unrelated to the relative *long-term* functional merits of the sorted lineages. Another way of putting the point is that these sorting events are stochastic *with respect to body plans*, which merely come along for the ride with other population-level or organism-level traits that are fortuitously connected to clade survival during mass extinctions.

Second, as we also saw in the last chapter, the RCT presupposes constraints on the ability of evolution to "correct" for mass extinction perturbations in the enormous intervening timespans between major extinction events. It thus requires a developmental constraints component to ensure that stochastic or pseudo-stochastic episodes of sorting have permanent effects on the shape of life. And constrained evolutionary trajectories, whether due to selection or to the internal biases of development, are inconsistent with genuinely stochastic patterns like random walks because the latter entail equiprobable sampling probabilities at every branching point.[10] In contrast, the very gist of constraints is that they bound the space of evolutionary possibility, making some outcomes more likely than others. Thus, stochasticity alone cannot explain the bounded variation observed in between mass extinction perturbations, which is a crucial element of Gould's thesis.

A third problem with equating contingency and stochastic lineage sorting is that doing so fails to distinguish the sources of radical contingency from the contingent dynamics themselves. Drift, mutation, and stochastic fluctuations of ecological environments are all potential sources or causes of evolutionary contingency (more on these causes later); however, the *causes* of evolutionary contingency should not be incorporated into our *definition* of contingency, lest they be prevented from serving in explanations of contingent dynamics. If contingency is a multiply realizable phenomenon, then no single cause or causes should be incorporated into its definition, lest this obscure its diverse causal base. So, although stochastic or pseudo-stochastic sampling at the lineage level may be a source of contingency and figure in explanations of the same, these processes should not be equated or conflated with contingent phenomena.

A similar problem arises for philosophical accounts that equate contingency with path dependence. Philosopher Eric Desjardins (a former student of Beatty's) conceives of contingency in terms of "historicity," which he cashes out in terms of causal dependence on the past.[11] Evolutionist Eörs Szathmary describes the notion of path dependence this way: "Path dependence entails that the probability of going back to some previous state decreases with time, or that switching to a state that could have easily been reached, had the population taken a different turn previously, is becoming increasingly improbable as time goes on."[12] Desjardins's contribution is to formalize the notion of path dependence, or the causal structure that, he argues, makes history matter.

The explanatory primacy of history figures prominently in Gould's work, from his decimation-diversification hypothesis about the early origins and subsequent canalization of body plan disparity (discussed in chapter 2), to his advocacy of concepts like "exaptation" and "spandrels" that figure in his broader critique of adaptationism. Path dependence explains why species that adapt to the same external environments never evolve the exact same solutions, and why traces of the unique, contingent histories of converging lineages are never fully erased (see chapter 4).[13] Here is a simple analogy: In the early building phases, a house can take on many different configurations. Halfway through construction, however, the possible functional outcome space has narrowed considerably; because of accumulated and convoluted structural interdependencies, certain designs are now off-limits—unless one scraps the entire project and starts over, which unlike foresighted human engineers natural selection cannot do. For example, in the early stages of their evolution, plants came to rely on growth, rather than on neural-muscular systems, as their primary means of movement. This irreversible evolutionary "choice" imposed significant constraints on how plants could respond to predation, foraging, and mating tasks that have since arisen in the course of their evolution.

I argued in the last chapter, and in publications going back to 2007, that a crucial component of the RCT is the developmental entrenchment of stochastic (or pseudo-stochastic) bouts of lineage sorting—particularly at the body plan level of morphological organization. The notion of path dependence can reasonably encompass internal constraints that are imposed by the causal topography of development (as discussed in chapter 2). However, without the stochastic or pseudo-stochastic culling component, the path-dependency view glosses over why the RCT presents as the antithesis of progressivist views of macroevolution. If globally optimal animal designs arose in the Cambrian and steadily out-competed their functionally inferior counterparts, then it would be of little theoretical consequence if these forms exhibited path dependence within the parameters of their own evolution. When path dependency is combined with non-merit-based culling at the lineage level, however, it begins to capture the Gouldian view of life.

Although macroevolutionary stochasticity and path dependence in combination go a long way toward *explaining* contingent dynamics, they are not equivalent to them. It is only when contingency is framed not in terms of mechanisms or processes but as a modal claim about the stability of evolutionary outcomes across initial conditions that we can then go on to formulate hypotheses about the causes and relative frequency of contingency. Stochasticity and path dependence are both plausible causes of modal instability, but they are not equivalent to it.

Defining contingency in terms of stochasticity or path dependence or both misses the mark for another crucial reason: evolution is neither wholly stochastic nor wholly path dependent, and the extent to which it is either depends on the grain of resolution at which we examine its processes and products. This issue proves critical to understanding the RCT, for reasons that will be unpacked over the next three chapters. For now, it is enough to note that evolutionary systems can exhibit path *in*dependence (and hence modal stability) at finer grains of phylogenetic resolution, while exhibiting path dependence (and hence modal instability) at coarser grains. Accounts that reduce the RCT to stochasticity or path dependence or to other features of the evolutionary process overlook this textured feature of Gould's view of life. I will argue later that such a mixed contingent/convergent system is not only consistent with but actually a direct entailment of the RCT.

1.3 Contingency as Sensitive Dependence

I will now sketch an alternative conception of contingency that I think better suits the role that the concept plays in the RCT, and that can make sense of its variable usages both in Gould's work and in the broader literature on the topic. The conception of macroevolutionary contingency that I will defend is drawn from an account I have developed in a series of papers.[14] Once this account has been laid on the table, I will go on to consider how it interacts with other areas of philosophical research, such as biological laws and explanation. I will refer to this account, both here and throughout the remainder of the book, as *radical contingency*. The working definition is as follows: Outcome O is radically contingent if and only if a marginal change in some initial condition $I_1, \ldots, I_n$ would tend to result in Outcome O^* in any physically possible world, where O^* is radically disparate from O. There is much to unpack here and many ambiguities to tackle, but let us begin with an exegetical defense of this formulation.

The origins of radical contingency can be found in some of Gould's earliest writings on the subject, where he describes the quintessential case of contingency as one in which "*small and apparently insignificant changes* … lead to cascades of accumulating difference," yielding entirely different evolutionary outcomes.[15] Gould hints again at this interpretation of contingency when he draws an analogy to the dynamics of human history, surmising that if we were to rewind the tape of the American Civil War, "with just a few *small and judicious changes* (plus their cascade of consequences), a different outcome, including the *opposite resolution*, might have occurred with equal relentlessness."[16] Consistent with Beatty's pluralism thesis, however, Gould can also be found associating contingency with "an *unpredictable* sequence of antecedent states, where any *major change* in any step of the sequence would have altered the final result."[17] Instead of quibbling over interpretations of Gould's voluminous work, it is more productive to focus on the virtues of conceiving of contingency in the way outlined here, as compared to the alternative accounts discussed in the previous two sections.

The first thing to note about this account of radical contingency is that it describes a counterfactual causal structure. This accomplishes four useful things. First, it separates the metaphysical dimensions of contingency from the epistemic consequences of contingent causal dynamics, thus eschewing problems that confront epistemic characterizations of the thesis (such as Beatty's second definition). Second, by focusing on evolutionary modality, or the stability of evolutionary outcomes across nomically possible biological worlds, radical contingency has the luxury of remaining agnostic to metaphysical debates over determinism versus indeterminism in biology and beyond. The reason it can remain agnostic to those metaphysical questions is that on the account proposed, the perturbations in initial conditions can be either actual (e.g., via percolating quantum events) or idealized (e.g., had such and such initial developmental conditions been different, a disparate outcome would have ensued). Third, radical contingency allows contingent phenomena to be explained by various mechanisms and processes, such as stochasticity and path dependency, without being identified with or reduced to those mechanisms and processes. Any attempt to define contingency in terms of mechanisms and processes will forego these explanatory virtues.

Fourth, radical contingency does a good job of explaining, and goes some way toward unifying, the two versions of contingency that Beatty has identified, in addition to the accounts of Turner and Desjardins, all of which capture important dimensions of Gouldian contingency but are, in and of themselves, incomplete. Radical contingency accounts for Beatty's notion of "contingency as causal dependence" because it describes the causal dependence of particular outcomes on changes in particular initial conditions. However, radical contingency picks out a *particular sort* of causal dependency—namely, a sensitive dependence on small changes in initial conditions, which allows it to home in on the sort of counterfactual instabilities that lie at the heart of the RCT. This sensitive dependency implies that relatively small changes in boundary conditions, including geophysical, biochemical, developmental, phylogenetic, and ecological conditions, will substantially affect long-term evolutionary outcomes. Radical contingency also helps to make sense of Beatty's notion of "contingency as unpredictability," as well as the common entanglement of the distinct but causally related metaphysical and epistemic aspects of contingent systems. The behavior of radically contingent systems will be hard to predict because their trajectories are easily derailed by small perturbations along the way. Radically contingent systems magnify small differences in initial conditions, resulting in disparate outcomes from initial conditions that are *for all practical purposes identical*. Radical contingency can thus account for the hidden variables reading of contingency as unpredictability.

If narrativistic explanations are to be intelligible, they must provide causal explanations (see chapter 2), and making good on causal claims may require appealing, if tacitly, to laws that explain the interactions identified. But any laws lurking in the background of narrativistic explanations will not significantly constrain the long-term behavior of the system, and thus will not provide the predictability that is associated with nomological styles of explanation. As Gould put it, "the contingency of history guarantees that any body of theory will underdetermine important details, and even general flows, in the realized pageant of life's phylogeny on Earth."[18]

Much ambiguity remains. It is unclear, for example, how small the changes in initial conditions and how large the outcome disturbances must be in order for a system to count as radically contingent. Some cases are clear-cut; the ordering of mutations in a lineage surely counts as a small perturbation, and if this ordering substantially affects the stability of a given evolutionary outcome, then that outcome is radically contingent. Other cases of radical contingency involve large perturbations that are themselves contingent on small perturbations. The K-Pg impact event, for example, clearly constitutes a large perturbation of late Mesozoic ecosystems, but the impact can be traced to negligible changes and interactions in the positions of objects orbiting the sun, which ensured that a 10- to 15-kilometer bolide would collide with the Earth at a particular place, time, and angle—details that would determine the shape of the resultant biotic crisis and its aftermath. If events that occur with some regularity over the course of any history of life, such as magmatic activity, bolide impacts, or continental collisions, have the potential to substantially and permanently alter large-scale evolutionary outcomes depending on when and how they manifest, and if the occurrence or nonoccurrence of these events can be traced to small changes in initial conditions, then the system can be properly characterized as radically contingent.

1.4 Characterizing Radical Contingency

In short, the RCT is best understood as a universal biological claim about the sensitivity of large-scale evolutionary outcomes to initial conditions. The claim is universal because it cites causal mechanisms and processes that are applicable to the history of life on Earth and to all nomically possible histories. These include stochastic or pseudo-stochastic bouts of extinction and a causal topography of development that makes upstream components of development resistant to modification, resulting in the entrenchment of non-merit-based episodes of lineage-level sorting. There is no reason to think that the causal geometry of development and the constraints it imposes on macroevolutionary search space would be different on other life worlds. Evolution everywhere will build incrementally on existing gene-regulatory networks, resulting in the entrenchment of phylogenetically and ontogenetically earlier components.

Also likely to be universal is the non-meritorious culling power of mass extinctions. Life will tend to evolve on geophysically dynamic planets,[19] and the downside of living on a geophysically dynamic planet is that it comes with the occasional mass perturbation due to volcanism, tectonic movements, asteroid impact, climatic change, and the like. And once we add strategic evolutionary interactions and ecological dependencies among living lineages into the mix, the prospect of mass perturbations increases substantially. Note that the *content* of background fitness conditions on any given life world does not matter for the point being made here—what matters is that this content will be rapidly and dramatically altered for brief punctuations, resulting in biotic crises and faunal turnovers which, when combined with developmental entrenchment, give rise to radical contingency. Thus, although the RCT issues a profound challenge to the project of a contentful universal biology, it also offers some properly cosmic lessons of its own.

Let us elaborate further on the notion of radical contingency by returning to the modal dimensions of the concept. The set of initial conditions upon which an evolutionary outcome is sensitively dependent must contain "non-replicable" members. That is to say, it is necessary that this set contains events that are essentially accidental. For instance, the origin of life may be contingent (in the simple causal sense) on the precise positioning of a planet within the "goldilocks zone" of its host star or of the solar system within the so-called galactic habitable zone. Yet causal dependence on goldilocks positions does not make the origins of life radically contingent. For although worlds within the goldilocks zone may be somewhat rare, current astronomical research suggests that an appreciable fraction of rocky planets are likely to fall within the habitable range. This is our first clue that radical contingency is closely connected to cosmic frequencies, a connection we will return to shortly.

Likewise, it is important to be clear about the nature of the outcome at issue. For instance, the formation of a *token* star, such as our own Sun, may be radically contingent: had a supernova not occurred or occurred in a different way, then the ensuing collapse of the interstellar molecular cloud that led to the formation of our solar system would either not have occurred or would have taken a different form, and a different type of star in a different place and time would have formed. But the formation of stars and planets qua *entity types* is highly robust, much like the positioning of worlds within the stellar or galactic habitable zones.

1.5 A Problem with a Solution

This brings us to a problem with the notion of radical contingency as it has thus far been framed. The problem is that as a thesis about evolutionary patterns throughout the Phanerozoic, radical contingency is quite obviously false. Much of evolution is clearly not sensitively dependent on initial conditions. Indeed, it is largely through the efforts of Gould and his collaborators that the nonrandom clustering of organismic form within discrete body plans was foregrounded as an important explanandum in macroevolutionary theory. It is precisely this explanandum—bounded variation within body plans—that is underscored in Gould's decimation-diversification hypothesis (see chapter 2). The shape of life is not incomprehensibly chaotic as one might expect if evolution was thoroughly radically contingent. How do we square this with the RCT?

A closer look at Gould's view of life reveals that it does not entail evolutionary chaos at all. On the contrary, the developmental pillar of the RCT holds that internal constraints due to the causal topography of development impose significant order and predictability on the actual occupation of morphospace once the initial crop of body plans had congealed. Take any two animal clades, such as mammalian vertebrates and bivalve mollusks: no matter what their ecological conditions and selective regimes, these lineages will not, according to the RCT, escape the architectural confines of the vertebrate and mollusk body plans, respectively. Nor are these lineages capable of transcending the developmentally "midstream" anatomical parameters of their respective mammalian and bivalve organizations. In other words, they will never venture down "radically different" evolutionary paths, where this refers to alternative occupations of morphospace at phyletic and perhaps class-based levels of organization. These are the crucial contrast classes against which outcome similarity and dissimilarity should be measured.

On this view, the reason why lineages do not venture beyond their initial body plan parameters is not because the existing clumps in morphospace are perched atop globally optimal fitness peaks, with natural selection having culled suboptimal forms that wandered into what paleobiologist Simon Conway Morris memorably calls "the howling wilderness of the non-adaptive."[20] Rather, it is because internal constraints prevent lineages from traversing these regions of morphospace even when doing so would (*ceteris paribus*) be evolutionarily beneficial.

In short, the "sensitive dependence" component of the RCT is not intended to describe the evolutionary behavior of existing lineages *within the confines of their own body plans*. On the most plausible reading of Gould, much of evolution is constrained and predictable—but these constraints are essentially "internal" and embedded in a deeper causal framework of sensitive dependence. This claim will be fleshed out and defended in the chapters to come.

2. Radical Contingency and the Laws of Life

Contingency is frequently contrasted with necessity, and necessity is often associated with laws of nature. Given this, one might think that adjudicating the RCT will boil down to whether there are biological laws describing the relevant sorts of evolutionary outcomes—that if such laws can be found, then

Gould's thesis will have been refuted. Indeed, Gould attributed evolutionary contingency to the lack of invariant biological laws.[21] Nevertheless, the RCT is not in tension with the existence of evolutionary laws *tout court*. And as we shall see, depending on how one conceptualizes laws of nature, the RCT may even be consistent with the existence of specific laws of form and function. The upshot is that questions about contingency ought to be decoupled from questions about laws.

2.1 The Nomological Vacuum of Biology

The notion that there is a "nomological vacuum" in biology—a stark absence of laws of life—was first systematically argued by John Beatty.[22] Beatty's core contention is that unlike thoroughly nomological disciplines like physics and chemistry, biology has no distinctive generalizations that are both exceptionless and universal. The reason for this, Beatty contends, can be found in the contingent nature of the evolutionary process. Even biological generalizations of the widest scope—such as the "laws" of Mendelian inheritance or the ubiquitous citric acid cycle (the series of chemical reactions involved in carbohydrate metabolism in aerobic respiration)—appear to be merely accidental. Although these traits are pervasive on Earth due to their unbroken descent from a common ancestor that possessed them, they are not subject to lawful generalizations in the strict sense for several reasons.

First, they are riddled with exceptions, whereas laws of nature are supposed to describe necessary relations that are invariant. For instance, not all organisms, nor even all animals, have Mendelian-style diploid inheritance, and those that do have numerous exceptions. Meiotic drive, for example, breaks Mendel's fair (50/50) segregation rule. The same is true of aerobic respiration: not all organisms on Earth use aerobic respiration; among those organisms in which the citric acid cycle is conserved, there is significant variability in the enzymes that are deployed in the cycle. Why are there invariably exceptions to even the most ubiquitous biological regularities? The reason for this, Beatty contends, is that all biological regularities are the product of an evolutionary process, and the same process that produces these regularities (via directional selection) and sustains them over evolutionary time (via stabilizing selection) will inevitably break them down by generating counterexamples. Beatty refers to these tendencies as the "rule-making" and "rule-breaking" capacities of evolution.

A second reason for the nomological vacuum is provided by philosopher Alexander Rosenberg, who notes that biological fitness and function are multiply realizable. Fitness, or expected reproductive success, "supervenes on" (roughly, is determined by) a vast heterogeneous base of intrinsic and relational properties of the organism.[23] Fitness, in other words, is massively multiply realizable—and a consequence of this massive multiple realizability is that there is no handful of traits that we can identify as being the fit ones. For instance, sometimes being more intelligent increases survival, but in other developmental and ecological contexts selection favors reduced intelligence (as with the evolution of herbivory in a formerly carnivorous lineage); and ditto for morphological complexity (for example, in the context of the evolution of parasitism). In addition, specific biological functions may themselves be multiply realizable, and selection is unable to discriminate among different structures that realize the same function. Perhaps the citric acid cycle is only one of many possible chemical pathways that can extract energy from carbohydrates. If there are many structural realizations of a given function, then this bodes poorly for the prospect of developing a workable set of contentful biological laws.

A third reason for the nomological vacuum, also discussed by Rosenberg, concerns the ubiquity of evolutionary arms races that can destabilize evolutionary outcomes. Adaptations are constructed in local selective environments that are subject to fluctuation; these fluctuations arise not only from changes in the abiotic environment but also from biotic interactions between coevolving lineages. Even the most successful adaptations are destined for obsolescence in the unrelenting arms race of natural selection, wherein one lineage's solution becomes another's design problem and vice versa.[24] It is only a matter of time (barring constraints and trade-offs) before lions evolve trichromatic color vision that makes the zebra's black and white stripes pop out as clear as day on the African Savannah—and if that day did come, then zebras would likely respond by evolving less conspicuous patterns of pigmentation. And if that happened, then so much for the "law-like" generalizations "zebras have black and white stripes" and "lions have dichromatic vision."

The macroevolutionary implications of interbiotic arms races are illustrated by Leigh Van Valen's "Red Queen" hypothesis. According to the Red Queen hypothesis, lineages must keep "running in place" (much like the theory's eponymous Lewis Carol character) merely to remain where they are—that is, to maintain their present fitness levels. "Running" here is a metaphor for adaptively evolving, and this evolvability is necessary if lineages are to respond effectively to strategic moves and countermoves of coevolving lineages (e.g., predator and prey, host and parasite, competitors, etc.). Van Valen offered the Red Queen hypothesis as an explanation of his "Law of Extinction," or the stunning observation that the stochastic probability of a lineage going extinct does not vary with its taxonomic age, even in stable abiotic environments.[25] In other words, Red Queen dynamics might explain why it is that lineages do not get better at not going extinct over time as one might expect they would if evolution were a gradual optimizing process and lineages evolved to better match their niches. In sum, due to the multiple realizability of fitness and function and the ubiquity of evolutionary arms races, biological generalizations tend to lack the necessity, stability, and projectibility that is characteristic of natural laws.

Some laws of nature are formulated with explicit or implicit *ceteris paribus* qualifications, or provisos, which specify disturbing conditions under which the law does not apply, thereby preventing its falsification when application conditions are not right. For instance, we might say that *ceteris paribus*, planets have elliptical orbits, demand increases the price of goods, meiosis exhibits fair segregation, and so on. But the "*ceteris paribus*" strategy does not work for biology because we cannot even begin to parse the application conditions that would need to be packed into the proviso in order to make contentful biological generalizations true, exceptionless, and, most importantly, nonaccidental.

One reason for the failure of the *ceteris paribus* strategy in biology is the evolutionary metaphysics of radical contingency. If a minimally different set of initial conditions would have resulted in substantially different patterns of inheritance, then any forms of inheritance that are currently ubiquitous in our biological world will be accidental regularities. On the traditional view, laws do not describe true accidental generalizations but rather outcomes or relations that are *necessarily true* given the nomic structure of the universe. Philosopher of science Bas van Fraassen illustrates the distinction between laws and accidents by comparing the generalization "there are no solid spheres of gold greater than one mile in diameter" with the generalization "there are no solid spheres of enriched uranium greater than one mile in diameter."[26] The first is an accidental generalization, as nothing prohibits large gold spheres from existing (e.g., intelligent spacefaring aliens could compile one if they were so inclined); the second is a lawful necessity due to the critical instability of uranium bodies that exceed a certain mass. Beatty and Rosenberg contend, in essence, that all biological generalizations are like van Fraassen's first generalization: they are only accidentally true. Thus, biological generalizations formulated in terms of universals are either false, analytically rather than empirically true (such as idealized models of population genetics), or else describe historically contingent outcomes of the evolutionary process and thus lack the necessity of physical laws.

2.2 The Case for Biological Laws

Although these are all good reasons for thinking that the structure of evolutionary theory precludes the existence of law-like regularities in biology, they do not make a slam dunk case for the nomological vacuum—even if we assume strict, traditional criteria for lawhood modeled on physics. For starters, there are strong cases to be made that the principles of natural selection[27] and drift[28] are both laws in the strict sense. There is ongoing debate about whether the principle of natural selection, when formulated as a general law, has empirical content, and whether drift is a disturbing condition for natural selection.[29] But the generic conditions for natural selection and drift (which refers to deviations from expected frequencies in the sampling of finite populations) will exist wherever life is found because these are the very conditions that make the origin and evolution of life possible in the first place.

Moreover, as we saw in chapter 2, the RCT is predicated on universal biological properties, namely the dynamics of mass extinction and the causal geometry of development. It is also consistent with a wide range of candidate biological laws that are equally nonspecific in that they do not predict any details of form and function. For instance, the RCT is consistent with the Zero Force Evolutionary Law (ZFEL) as described by Robert Brandon and Daniel McShea, which refers to the tendency of biological systems to depart from a preexisting trajectory or to increase in diversity over time in the absence of biological forces or constraints.[30] It is also consistent with laws governing the trophic structure of ecosystems, such as those which explain why apex predators will always be relatively rare in a community.

Thus, it is not the existence of macroevolutionary laws per se that would pose a problem for the RCT. What would pose a problem is the existence of generalizations describing the evolution of specific form and function that are universally projectible and hold up across deep rewinds of the tape of life. At least two robust morphological trends in the history of life on Earth have been convincingly identified: a trend toward increasing maximum body size,[31] and a trend toward increasing maximum hierarchy (or the "nestedness" of parts within parts, which serves as a rough marker of complexity).[32] It is unclear whether these trends are driven by selection or the result of "diffusion" from a minimal complexity boundary leaving nowhere to go but "up," so to speak. Regardless of their causes, generalizations like "over the course of evolution, life will tend to increase in body size and hierarchy" are strong candidates for biological laws; as we shall see, however, they are insufficiently specific to fall within the crosshairs of the RCT. A greater threat to Gould's thesis comes from laws intimated by patterns of convergent evolution, which we will investigate throughout this book. The point for now is that not all strict biological laws are equally in tension with the RCT.

In any case, not all philosophers of science operate with the strict sense of lawhood that Beatty, Rosenberg, and others have presupposed. There are many philosophical accounts of laws of nature, and these differing conceptions of lawhood have different implications for the logical relation between the existence of biological laws on the one hand, and the viability of the RCT on the other. Many theorists maintain that stricter conceptions of laws modeled on those in physics are a poor fit for the production of scientific knowledge in biology, psychology, and the social sciences. Some have argued, for instance, that generalizations do not need to be exceptionless or even have empirical content to be proper scientific laws. Others have advocated laxer accounts of lawhood that allow radically contingent antecedent conditions to be incorporated into law-like statements. The latter sorts of biological laws do not pose a problem for the RCT even if they describe highly specific evolutionary outcomes.

For instance, philosopher of biology Elliott Sober has argued that generalizations formulated in terms of counterfactual conditionals with historically contingent antecedents can satisfy the desiderata for lawhood because, so formulated, they are exceptionless and universal.[33] Sober offers the following law-like schematic (*L*): If *I* (a given set of contingent initial conditions) obtains at one time, then the generalization [if *P* then *Q*] will hold thereafter. Framed in this manner, law-like generalizations state that wherever such and such application conditions are met (e.g., where diploid genetics obtain), then some rule will hold (e.g., fair meiosis, the Hardy Weinberg equilibrium, etc.). The fact that the application conditions are themselves radically contingent does not show that the broader generalization, *L*, is accidental. Sober is right to say that the relations described by such counterfactual conditionals are not in themselves radically contingent, but if they contain antecedent conditions that are radically contingent (such as the evolution of diploid genetics), then this will undermine the modal stability (and hence cosmic projectibility) of the outcomes they describe. The existence of laws on these more permissive accounts of lawhood is thus not in conflict with the RCT.

Sober's analysis shows that many nonaccidental biological generalizations obtain only if certain accidental conditions obtain; yet the accidental nature of these antecedents infects the whole generalization, such that laws framed in this manner are, in effect, *accidental nonaccidental generalizations*. For instance, there is a sense in which it is no accident that all life on Earth shares the same DNA code, for this is both the result of faithful replication from a common ancestor and the expectable result of communal innovation pools dominated by lateral gene transfer in the early phases of life (see chapter 1). But the specific code that all known life shares may be a radically contingent, frozen accident that is not replicated on any other epistemically accessible life world. The fact that humans share a specific genetic code with plants, protists, and prokaryotes is an accidental nonaccidental generalization. In contrast, the fact that all known life has a digital (nonanalog) molecular basis of inheritance is, in all probability, a nonaccidental generalization full stop. Framing both of these as nonaccidental generalizations would be misleading. Given the diversity of thinking about laws, linking the soundness of the RCT to a verdict on the nomological vacuum of biology is liable to confuse more than it is to illuminate.

2.3 Modal Frequencies

Let us move the discussion away from laws of nature and their associated notions of inviolability and inevitability to focus instead on *modal frequencies*—the distribution of particular outcomes across possible evolutionary histories. Low-probability outcomes become probable given a sufficiently large population of chance setups. "Chance setups" are iterations of a process that produces token outcomes, such as coin flips (heads or tails), lottery draws (win or lose), and habitable worlds (life or no life). On standard interpretations of chance, frequencies apply to outcomes, whereas explanations of those frequencies lie in the causal structure of the chance setups in conjunction with the laws of nature. From the fact that it has occurred, we know that the history of life as it has unfolded on Earth is, at the very least, *consistent* with the laws of nature. But due to observer selection effects (see chapter 1), we do not know how replicable particular aspects of that history are likely to be.

If the universe is effectively infinite, as standard cosmologies assume—this means that there are an effectively infinite number of chance setups, and thus an effectively infinite number of life worlds, and an effectively infinite histories of life that play out in precisely the same way as our own. Whether there are histories of life somewhere in an infinite universe that mirror the history of life on Earth is not the crux of the issue. The issue is whether a single history of life constitutes a sufficiently large population of chance setups to make certain low-probability outcomes probable. In other words, radically contingent traits should be vanishingly rare in the cosmic horizon—not merely *within* life worlds, but also *across* them.

Gould appears to be making a frequency claim of this sort when, for example, he discusses the contingency of human evolution:

> If life started with all its models present, and constructed a later history from just a few survivors, then we face a disturbing possibility. Suppose that only a few will prevail, but all have an equal chance. The history of any surviving set is sensible, but each leads to a world thoroughly different from any other. If the human mind is a product of only one such set, then we may not be randomly evolved in the sense of coin flipping, but our origin is the product of massive historical contingency, and we would probably never arise again even if life's tape could be replayed a thousand times.[34]

Gould is arguing that due to the metaphysics of radical contingency, many evolutionary outcomes would occur only very infrequently across alternative evolutionary histories, whether hypothetical or actual.

Focusing on claims about modal frequency allows us to home in on what is actually at stake in the contingency debate. The existence of law-like evolutionary outcomes with very low modal frequencies—that is, outcomes that obtain necessarily or with high probability but do so only under conditions that are astronomically uncommon in the universe—do not undermine the RCT or support its antithesis (see below) because such features cannot be "mainstreamed" or projected onto a broad spectrum of epistemically accessible life worlds. A central task of the next three chapters is to determine whether, and if so under what circumstances, convergent evolution can be taken as evidence for the higher modal frequencies of evolutionary outcomes. Before we can tackle this question, however, we first need to have a clearer picture of the RCT's antithesis and its own peculiar set of theoretical commitments.

3. Framing the Antithesis: Robust Replicability

The foregoing analysis of radical contingency helps to frame its contrast class. Gouldian contingency is often framed in opposition to repeatability (a common misinterpretation of the RCT that will be discussed in chapter 4). Yet recall that if determinism is true, then the tape of life would be perfectly repeatable and hence a trivial thought experiment. It is therefore not repeatability per se but rather *robust repeatability* that properly characterizes the antipodal view of life, where "robustness" relates to the stability of an outcome over a wide range of initial conditions. In other words, the antithesis to the RCT is the view that rerunning the tape of life from disparate initial conditions and under diverse perturbing influences would nevertheless produce highly similar macroevolutionary outcomes (figure 3.1). Let us call this view the "robust replicability thesis" (RRT).

3.1 The Robust Replicability Thesis

Beatty rightly suggests that any notion of evolutionary contingency should rule out a "robust equilibrium" explanation of macroevolution, in which disparate starting points lead reliably to a small set of attractors. Note, however, that the radical contingency debate is not about the evolutionary robustness of *specific taxa* at particular spatiotemporal locations in the history of life. The RRT does not claim, for instance, that replaying life's tape from different starting points would produce vertebrates per se, with all the traits that vertebrates possess, at a particular spatiotemporal location. The claim, rather, is that what would be reproduced is not vertebrates (sans quotation marks, denoting a spatiotemporally restricted clade on Earth), but "vertebrates" (with quotation

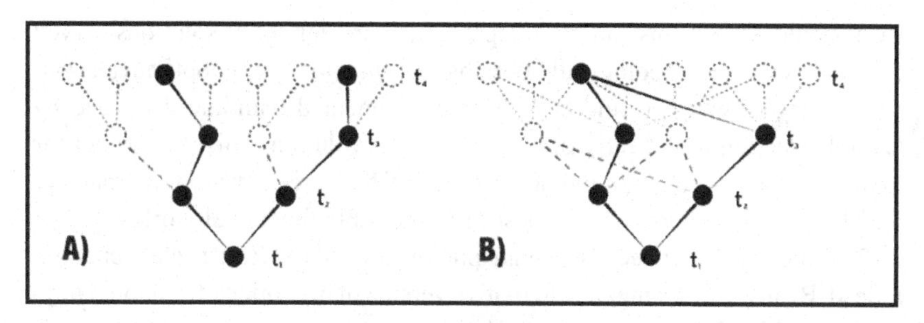

Figure 3.1
(a) A radically contingent system in which paths not taken at earlier times (dotted lines) cause evolutionary outcomes to become inaccessible at later times. (b) A convergent system in which many outcomes remain accessible from even distant evolutionary trajectories. From R. Powell and C. Mariscal, "Convergent Evolution as Natural Experiment: The Tape of Life Reconsidered," *Journal of the Royal Society Interface Focus* 5, no. 6 (2015): 1–13.

marks, denoting a morphofunctional kind). The idea is that, over the long haul, the evolutionary crank will tend to churn out highly similar animal forms. Thus, according to the RRT, even if some features of lineages are radically contingent, enough of their basic elements will be reproduced so as to result in forms that are fundamentally familiar.

We will examine this view of life and its evidential bases in the next chapter. The lesson for the time being is that the RCT-RRT debate does not turn on genealogy or identity, but rather on whether there are nontrivial bundles of specific properties that, like their celestial counterparts, exhibit a truly robust range of counterfactual stability. These two views of life sit at opposing ends of what is surely a continuum of views on the balance of contingency and robustness in the history of life. The aim here is to set up the dialectical space necessary to stake out a position that falls in between the extremes.

3.2 Robust Adaptationism

The only known mechanism that could repeatedly drive lineages toward complex functional attractors in morphospace and keep them there is natural selection. For this to happen, it is not enough that selection is an important evolutionary mechanism. Rather, it must overwhelm other evolutionary factors that tend to divert away from the paths leading to these attractors. The RRT is therefore committed to "strong adaptationism": the view that in the long run selection will tend to overpower perturbing evolutionary forces, constraints, and tendencies that would otherwise undermine robustness.

On this view, successful animal body plans are not the fortunate winners of a macroevolutionary lottery locked in by developmental constraints, as the

RCT would have it. Instead, they represent a superior set of solutions carved out of the vast set of ecologically possible but functionally suboptimal alternatives, many of which would have been experimented with and discarded by natural selection in the early phases of animal evolution. This view is not far removed from one that Gould spent much of his career excoriating, namely the "Panglossian" notion of the best-of-all-possible functional worlds.[35]

The connection between strong adaptationism and the RRT is clear enough. Indeed, Beatty even frames an alternative version of his notion of "contingency-as-predictability" in terms of a denial of strong adaptationism. But even if strong adaptationism were true, and selection were in fact sufficient to guarantee a particular outcome in a given case, this does not get us to the RRT. For in addition to showing that selection reigns supreme and that macroevolutionary success is determined by a merit-based competition, the RRT must also convince us that *there are only a limited number of ways to succeed in this competition.* Without the latter qualification, there is no reason to think that selection would proceed to the same set of outcomes from disparate initial conditions, and thus strong adaptationism alone gives us no reason to think that macroevolutionary pattern would be robustly replicable. The mere predominance of selection does not imply that there is a limited number of solutions to common design problems, even if selection ruthlessly adjudicates on the merits between solutions that happen to arise. In addition, therefore, the RRT must show that the set of optimal solutions is manageably small—a premise that is not contained in the thesis of strong adaptationism.

Strong adaptationism has been subjected to its fair share of criticism, yet the evidentiary burden is even weightier for the RRT. For it turns not only on whether selection is at the helm of macroevolution but also on whether there is a limited number of destinations to which it can proceed. If there are such limitations, they will not be set by natural selection alone. For as we saw in chapter 1, the principle of natural selection, when formulated as a general law, has nothing specific to say about the sorts of evolutionary outcomes we can expect unless we plug in extensive—and for all we know, accidental—developmental details about the lineages in question. What reason, then, do we have to believe that there are contentful morphological regularities that hold up across deep rewinds of the tape of life? This brings us to the striking phenomenon of convergent evolution.

4 The Critique from Convergent Evolution

Darwin's mentor and the founder of modern geology, Charles Lyell, was skeptical of the notion that species changed over time and that new species arose from existing ones through natural, rather than miraculous, means. Yet Lyell had begun to recognize, as much as anybody of his day, that many of the creatures represented in the fossil record have no analogs among living species. The signature of extinction grew ever stronger in Lyell's time, to the point that its theoretical importance became impossible to deny. In an effort to reconcile clear patterns of extinction in the fossil record with the doctrine of divine creation, Lyell postulated a theory of "cyclical return."[1] On this view, species are immutable for the duration of their existence but bound for extinction as their habitats inevitably deteriorate until they can no longer support their particular lifeways. Lyell hypothesized that the origin of taxa would track global cycles in climate, with the reiteration of species corresponding to ecologically propitious stages in the cycle. Because he believed that climatic change was cyclical, and because the origin of species was thought to depend on the presence of environmental conditions that corresponded to specific ways of life, Lyell believed that extinct forms were destined for reiteration.

Writing in a colder, drier period of Earth's history (the Cenozoic Era), Lyell argued that if the conditions of existence were to return to the temperature and moisture levels that typified the Mesozoic, we might then expect "those genera of animals return, of which the memorials are preserved in the ancient rocks of our continents. The huge iguanodon might reappear in the woods, and the ichthyosaur in the sea, while the pterodactyle might flit again through umbrageous groves of tree-ferns."[2] Lyell was well aware that the Earth's climate did not cycle precisely, so he conceded that iguanodon-the-resurrected would be sufficiently different from iguanodon-the-original to recognize it as a distinct species—though he thought the two taxa would be similar enough to fall under the same genus (based on overall similarity, of course—cladistic principles would not emerge for more than a century). In contemplating a naturalistic ground for cyclical return, Darwin wrote to Lyell in 1859 (the same year that the *Origin of Species* was published) that although it is possible in theory that natural selection could produce identical forms over deep evolutionary time, this is so astronomically improbable that it can safely be ignored by evolutionary theory.[3] For Darwin, the stamp of phylogenetic history is indelible, even when lineages are subject to the same selection pressures.

Lyell's theory of cyclical return has a certain aesthetic appeal to the nostalgic among us who long to commune with living worlds that have been lost forever to the strata of geological time. At the time of Lyell's writing, however, there was no evidence for cyclical return. To the contrary, there was every indication that extinction is forever. Nevertheless, Lyell's view, so viciously mocked by his contemporaries and so thoroughly eclipsed by the Darwinian revolution, may have some nuggets of truth after all. The aim of this book is to determine how substantial these nuggets may be and to gauge the pressure they put on the Gouldian view of life.

1. Is Extinction Forever?

At the very outset of this investigation, we flagged the legitimate worry that the contingency dispute, though philosophically intriguing, might be empirically intractable. Some of this intractability stems, as we saw in chapter 3, from problems that are conceptual in nature. But even if these conceptual problems could be ironed out, adjudicating the radical contingency thesis (RCT) would still face sizable methodological challenges. The most glaring among these is the sample size problem noted in the introduction: we are working with a single history of life.

Because we cannot literally rewind the tape of life to see how outcomes hold up against perturbations in initial conditions at different junctures in life's history, it is only by consulting extraterrestrial data sets that we could decisively distinguish accidental from law-like features of the living world. We would begin by observing numerous alien histories of life, each beginning from a different set of initial conditions with respect to geophysical and climatic variables, the ordering of mutations, developmental configurations, ecological relations, and so forth. With these data in hand, we would then search for nonaccidental regularities that allow us to infer cosmic frequency distributions and the macroevolutionary processes that underpin them. To the extent that such regularities existed and could be identified, they would allow us to make predictions about how life will tend to unfold on the grandest of scales.

Unfortunately, the "$N = 1$" situation prevents us from investigating the stability of evolutionary outcomes in this way. Even just one example of an alien tree of life would probably tell us more about the prospects of cosmic biology—and about the status of biological laws—than any amount of theorizing on the basis of a single sample. What could we possibly learn about the nature of other living worlds by studying life as we know it on Earth?

1.1 Finding Empirical Traction

Gould's macroevolutionary thought experiments, explored in the previous two chapters, are designed to circumvent the $N=1$ problem. But such exercises of the imagination are empirically inconclusive. Perhaps in the vast majority of close possible Cambrian worlds, the vertebrate clade survives and thrives in the wake of the end-Cambrian extinctions; perhaps it survives but does not thrive, its elimination simply delayed rather than indefinitely staid; or perhaps it is eliminated and nothing like it ever arises again. We simply do not know the frequency distribution of these outcomes with any reasonable levels of confidence. The same is true for the persistence and replicability of other animal body plans. Gould's thought experiments are effective when it comes to posing critical research questions, but less so when it comes to answering them.

Other investigations of evolutionary contingency tackle the question of replication more directly. These experiments include controlled manipulations of evolving microbial populations,[4] longitudinal studies of selection in the wild,[5] and evolutionary simulations.[6] As ingenious as these studies are, they do not generalize to patterns in eukaryotic evolution that only manifest over immense timescales, and so they do not speak to key claims of the RCT.

Take, for instance, Richard Lenski's pioneering Long-Term Experimental Evolution (LTEE) project. Lenski and his collaborators have investigated evolutionary contingency by cloning twelve populations of *E. coli* bacteria and observing how they respond to identical selective environments over tens of thousands of generations (the experiment is currently at 60,000 generations and counting). As part of the study, researchers froze a time slice of each population every 500 generations, so that replays from various stages of evolution could be carried out. If selection were the dominant force acting on these separate populations, they would be expected to evolve in parallel; if chance or history were dominant, then the populations would be expected to diverge even under common selective regimes. This setup accords with Beatty's "contingency-as-unpredictability" formulation of Gouldian contingency, as discussed in chapter 3, wherein the same starting conditions result in different outcomes or, alternatively, the same selection pressures are insufficient to guarantee the same outcomes. It also accords with "radical contingency" as positively formulated in the same chapter, wherein small and in some cases imperceptible perturbations (such as stochastic events like mutation and drift) cause substantial differences in outcomes.

The findings thus far have been mixed, with some bacterial populations evolving in parallel along a number of dimensions (e.g., metabolic capacity, cell size, growth rates, gene expression, etc.), and others diverging from one another due to a combination of chance and history that is difficult to parse.[7] For example, researchers found that after some 30,000 generations of constant exposure to a challenging metabolic substrate (e.g., citrate), only a tiny fraction of the populations evolved the ability to metabolize the substrate, even though doing so was significantly fitness enhancing and clearly within the evolvability space of all the lineages exposed (as a few managed to achieve it). The LTEE researchers inferred from these nonreplications that numerous mutations must be in place at the same time for certain salutary traits to emerge, whereas each of these necessary mutations is not selectively beneficial on its own nor is some subset of them. This "macromutation problem" could prevent incremental selection from reliably driving lineages toward advantageous outcomes. It may also explain the significant delay that is often observed in the fossil record between the presence of a particular ecological regime and successful evolutionary responses to it.[8]

Does the LTEE project actually test the RCT, even in a highly localized bacterial setting? As just noted, the Lenski studies do seem to probe Beatty's conception of contingency-as-unpredictability, wherein the same initial conditions result in disparate evolutionary outcomes. This is achieved by isolating cloned populations and subjecting them to identical ecological regimes. The differences in outcome could then be explained by the vagaries of chance or the constraints of history. As we saw in chapter 3, however, there are good reasons not to conceive of Gouldian contingency in this way, and there are several reasons why the LTEE study does not in fact test the RCT, properly construed.

First, it is important to note that these experiments do not support the robust replicability thesis (RRT). To do so, they would need to show not parallel evolution *geometrically* conceived—wherein two populations evolve along some character dimension in the same direction from the same starting point, producing a geometrically parallel set of trajectories—but rather, proper *convergence*, wherein two lineages arrive at a structurally similar solution from *highly disparate starting points* (particularly as it relates to their initial developmental/phylogenetic conditions). Because the LTEE observes the evolution of bacterial lineages from identical or highly similar starting points, it does not provide evidence for the RRT.

Neither, however, is the study clearly in tension with Gould's thesis. For as we saw in the previous two chapters, the RCT is a non-uniformitarian theory that does not treat all time frames in the history of life, at all phylogenetic grains of resolution, as chaotic or unpredictable. In chapter 2 we saw that the developmental component of the RCT holds that the causal topography of gene-regulatory networks constrains the space of evolutionary solutions that a lineage can deploy in solving its ecological design problems. If any such Gouldian constraints are in operation, they will not be detected by subjecting identical populations to the same selection pressures, as the LTEE has done. Studies of convergence that are designed to test the RCT are vulnerable to a similar critique, as we shall soon see.

Another reason that the LTEE studies fail to test the RCT's core claims is that their findings are not generalizable to macro-morphological evolution—though, to be fair, they never claimed to be. The marvelous advantage of experimental evolutionary work on prokaryotes is that tens of thousands of bacterial generations can be explored within a single human scientific career (indeed, within a single graduate student's fellowship)—a feat that is impossible for studies of animals or even unicellular eukaryotes, which have vastly longer generation times and cannot be cloned as readily as bacteria. Exploiting this advantage, however, limits our ability to generalize from these experiments to the kinds of patterns that Gould was trying to explain.

This is not to say that the evolution of microbial metabolic innovation is a trivial feat; but at bottom, the contingency debate concerns the broadest brush strokes on the canvas of animal form, not the capacity to digest a challenging substrate. Just as importantly, it is plausible that historical constraints will be far more pronounced in macro-morphological evolution than they will be in the evolution of bacterial metabolism, given the more intricate interlocking of gene-regulatory networks and phenotypic components that are involved in the production of body plans. If the RCT is to be put to the test, it will have to be through the examination of large-scale patterns in the history of life on Earth.

1.2 The "*N* = Many" Scenario

Chapter 1 argued that purely statistical approaches to cosmic biology, such as applications of the Copernican principle, are theoretically unmoored, infected with observer selection biases, and too coarse-grained to draw any meaningful conclusions about life on other worlds. Chapters 2 and 3 showed that the nomic structure of biological science does not, on its face, support any specific biological laws of form and function; in addition, the universality of radically contingent dynamics would seem to preclude the possibility of making any specific projections from the shape of life as we know it to life as it might exist on other worlds. Very generic universal predictions are easy to make: life will metabolize, it will reproduce, it will have a digital genetic code, it will form adaptations under the guidance of natural selection, it will exhibit certain trophic ecosystem structures, and so on. But any more contentful predictions about macroscopic form, in the absence of an extraterrestrial data set, will be on far shakier epistemic ground.

Not everyone agrees with this gloomy characterization. Some theorists reject the seemingly uncontroversial claim that $N=1$. Their assertion is not that we have actually discovered instances of extraterrestrial life from which we can begin to glean the laws of life. Nor do they assert that we have found independent trees of life that arose on Earth. Rather, the "$N=$many" claim is premised on the idea that the history of life on Earth contains within it countless replays of the tape of life that hint at the existence of evolutionary regularities that, taken together, cast doubt on the RCT. Thus, a distinctively macroevolutionary argument against the RCT is now taking shape.

Evolutionary biologists and philosophers of science have begun to pay increasing attention to the theoretical importance of a phenomenon known as "convergent evolution"—the independent origination of similar biological forms and functions. Convergence has been interpreted by many theorists as tantamount to natural experimental replication in the history of life, and thus as a promising source of evidence for investigating the contingency question. If this interpretation is correct, and our single history of life contains numerous replays, the results of which can be analyzed and generalizations drawn therefrom, then the sample size is actually far greater than 1. Perhaps the contingency dispute is empirically tractable after all.[9]

Unlike controlled evolutionary experiments on prokaryotes in the laboratory or observations of natural selection in the field, studies of convergent evolution can draw upon a voluminous database of natural history to make inferences about the robustness of evolutionary processes operating over vast timescales and across immense phylogenetic gaps. We cannot literally replay the tape of life on Earth from different starting points, but we can infer the existence of natural replays by consulting "phylogenetic reconstructions" that reflect our best current hypotheses about the evolutionary relationships among animal groups. Sophisticated methods of evolutionary tree reconstruction (including parsimony, likelihood, and Bayesian statistical frameworks) are painting an increasingly precise picture of how evolutionary outcomes are distributed in space and time. The question that will preoccupy us in this and the next two chapters is this: Can evolutionary repetitions revealed through phylogenetic analyses be marshaled into an argument against the RCT?

1.3 The Case for Convergence

Several biologists have compiled expansive evidence bases of convergent phenomena with the aim of debunking, or casting doubt on, Gould's thesis. The seminal treatment of convergence as evidence against the RCT is Simon

Conway Morris's *Life's Solution: Inevitable Humans in a Lonely Universe.*[10] A more recent monograph defending the same theoretical reading of convergence came from George McGhee, whose *Convergent Evolution: Limited Forms Most Beautiful* makes a comprehensive case for the ubiquity of convergent evolution.[11] Both of these authors have documented an impressive body of convergence to be mined by future researchers, and their books are brimming with many fascinating and underappreciated examples of convergence. Convergence has been established at all levels of the biological hierarchy, from molecules and morphology to functions and lifeways. McGhee, for instance, reviews convergence on animal swimming, flying, walking, burrowing, and other locomotion-related morphologies, sensory modalities, masticatory apparatuses (e.g., teeth, beaks, and claws), poison-injection systems, digestive capabilities, defensive structures, reproductive strategies (e.g., live birth), and so on. He does the same for plants, where he documents convergence in tree and leaf morphologies, water-transport modes, root systems, seeds, and seed-dispersal mechanisms. How should we interpret this body of convergence?

"All observation must be for or against some view if it is to be of any service," Darwin remarked in an 1861 letter to economist Henry Fawcett. Convergent events are compiled by the above authors with the aim of testing a theory, namely the RCT, which the convergence data is then interpreted as refuting. This "critique from convergence" (CFC), as we shall call it, strikes back at the RCT and the supposed "nomological vacuum" of biology by arguing that patterns of evolutionary repetition demonstrate the modal robustness of highly specific evolutionary outcomes across vast timescales and phylogenetic histories. The signal of convergence is taken to suggest that over the macroevolutionary long haul, convergence will overcome the constraints of history and the derailing tendencies of stochastic processes to ensure the reliable (re) production of evolutionary outcomes. And *contra* the nomological vacuum characterization, the CFC argues that we can infer from patterns of evolutionary iteration that there are, in fact, specific (nongeneric) biological laws of general, perhaps even universal, projectibility lurking beneath the surface of what would otherwise appear to be a nomically unconstrained domain. Although these specific laws of life have yet to be articulated with any rigor, patterns of convergence intimate that they exist and are waiting to be described.

Some cases of convergence are not only striking, but even "eerie," as Conway Morris describes them. Two naturalistic reasons for this eeriness come to mind. One is that on the face of things, there is nothing in the known laws of biology that predicts the occurrence of such repetitions or explains why they should occur. In this respect, convergence presents as a scientific "anomaly," or an observation that is in tension with our existing body of theory, requiring either the modification of existing theories or the development of new ones.

Selection to a common ecological regime is undoubtedly *part* of the explanation of convergence; but why should this result in *similar functional forms* rather than *similarly functional but morphologically divergent* forms? The principle of natural selection, when framed as a general law, does not enlighten us. There must be other causes and laws operating in the background that account for convergent outcomes. A key question for the next two chapters is what if anything can we say about these other causes and how they bear on the contingency debate. A second reason why some cases of convergence might present as eerie is that evolutionary replications are often close but rarely exact, resulting in an "uncanny valley" experience that we interpret as "creepy."[12] In the case of eerie convergence, it is the signature of history that is responsible for the partial failure of resemblance, though the outcomes are close enough to fall within the uncanny valley.

The CFC is not merely on a negative mission to defeat Gould's thesis, however. It is also marshaled in support of a positive theory, namely the claim that patterns of convergent evolution indicate that rerunning the tape of life, even from disparate starting points, would produce a set of macroevolutionary outcomes that is substantially similar to the shape of life as we know it. This, of course, is the "robust replicability thesis" (RRT) discussed in chapter 3. The RRT is most prominently defended by Conway Morris, both in his pioneering work on convergence and in his public debates with Gould.[13] Similar appeals to patterns of convergence in support of the broad replicability of evolutionary history are made by biologists George McGhee,[14] Geerat Vermeij,[15] and Jonathan Losos,[16] as well as paleontologist Larry Martin[17] and philosopher Daniel Dennett,[18] to name a few.

In his book *The Crucible of Creation*, Conway Morris argues that patterns of convergence show that contingent processes are irrelevant so far as the history of life is concerned. He contends that major evolutionary outcomes will, despite the meanderings of their actual sequence, inevitably manifest in the unfolding of macroevolutionary time. He concludes with a bold cosmic prediction: "Although any history is necessarily unique, the resultant complex end-form is not simply the contingent upshot of local and effectively random processes. On any other suitable planet there will I suggest be animals very much like mammals, and mammals much like apes. Not identical, but surprisingly similar."[19] Although the phrase "surprisingly similar" is not quantified by Conway Morris, we can take it to mean something like a cosmic uncanny valley. Philosopher Daniel Dennett likewise holds that "convergence … is the fatal weakness in [the] case for contingency."[20] In discussing the Cambrian experiment in animal evolution, Dennett contends that "whichever lineage happens to survive will gravitate toward the Good Moves in Design Space.…

Replay the tape a thousand times," Dennett claims, "and the Good Tricks will be found again and again."[21]

On these views, successful animal groups are not the fortunate winners of an early extinction lottery, as Gould's theory would have it. Instead, surviving lineages represent a globally optimal set of solutions among the set of theoretically possible alternatives that were experimented with and discarded by natural selection, particularly and most spectacularly in the early phases of animal evolution when the great branches of animal evolution congealed.

Some evolutionists, like George McGhee, go even further to defend a quasi-essentialist reading of convergence, analogizing between the periodic table of atomic elements and the limited set of organismic forms that can be inferred from patterns of convergent evolution. McGhee suspects that "the modern scientific discipline of evolutionary biology is in a similar position as the scientific discipline of chemistry before the discovery of the periodic table of elements."[22] Given certain physical conditions, we can expect the law-like emergence of particular atomic elements with predictable sets of physical properties. Likewise, given certain ecological conditions, we can expect the emergence of particular biological forms with predictable sets of morpho-functional properties. On this view, historical contingencies recede to the explanatory background and convergence dominates over deep evolutionary timescales.

Darwin for his part believed that high degrees of convergence are so unlikely as to warrant ignoring this possibility for the purposes of taxonomy. But if convergence is as powerful and ubiquitous as the above authors suggest, then it could impede our ability to construct phylogenetic histories, for it would call into question approaches to phylogenetic reconstruction (such as parsimony) that operate on the assumption that the single origin of a complex trait is more probable than multiple origins. Indeed, it could lead one to be suspicious of the cladistics enterprise itself. This may sound like a radical inference to draw from convergence data, but it would explain why in Conway Morris's leading study of convergence there is not a single cladogram to be found—and why McGhee's work, though it makes extensive use of cladograms, defends an "atomic elements" analogy that could call into question the motivations for cladism. Nevertheless, the signatures of history that result in the uncanny valley make phylogenetic reconstruction possible, and this in turn permits reliable inferences of iterated origins.

Even theorists who are largely sympathetic to Gould's view of life and to the role of narrative explanation in the sciences more broadly, such as John Beatty,[23] interpret convergent evolution as undercutting the RCT. Cosmologists, too, have jumped on the convergence bandwagon, which they view as lending distinctively biological support to the justification of SETI-like programs.[24] In short, the CFC argues that, given enough time and the presence of certain environmental conditions, similar biological forms will emerge over and over again. If so, then perhaps in a meaningful sense Lyell was right: extinction is not forever.

1.4 Convergence as Evidence

If destiny and predictability are balms to human psychology, then contingency chaffs at the soul. Psychological palatability is not, however, a virtue of theories that aim to describe the causal structure of the world. Scientific theories stand or fall on their empirical adequacy. Although much research has been devoted to documenting the phenomenon of convergence, the logic of convergence as evidence remains critically underexplored.

On its face, the evidential relation between convergence and the RCT/RRT debate seems straightforward and compelling: convergent evolution is tantamount to natural experimental replication in the history of life. And to the extent that macroevolutionary replication is ubiquitous, this would seem to cast doubt on the RCT and corroborate the RRT. Even if our explanations of convergence in general are not fully fleshed out, and even if the adaptive motivations for particular cases of convergence are unclear, patterns of convergence can nonetheless support the evolutionary robustness of certain outcomes, which is ultimately what the RCT/RRT debate is all about.

For instance, the fact that the "saber-toothed lion" ecomorph evolved at least four times in mammals, three times within the placentals, and once between placentals and marsupials, gives us prima facie reason to think that this complex suite of cranial modifications is adaptive and evolutionarily replicable, even if we are not entirely certain about what its function was (e.g., predation, sexual selection, etc.). The "sabertooth syndrome" includes a generally catlike appearance with the elongation and lateral compression of the upper canines (which are bladelike in contrast to the conical canines of modern cats), robust forelimbs for grasping prey, and dozens of iterated cranial features (such as a massively reduced coronoid process of the mandible) that allow for increased gape and bite force.[25] The balance of evidence suggests that the sabertooth morphology is a highly specialized feeding adaptation. The point, however, is that we can be confident in an inference of *replicability* without being confident in any particular adaptive hypothesis.

Indeed, there are a host of challenges that confront specific functional attributions, even for cases in which we are confident that a trait is an adaptation. These challenges arise from the fact that adaptationist claims are claims about selection histories, and selection histories can be difficult to piece together due to our limited epistemic access to the past. As Richard Lewontin notes, we may never know whether Stegosaur plates *originally* evolved to deflect the teeth of their theropod predators, to regulate heat, or to signal to mates and other conspecifics, because the information necessary for making such inferences may be irretrievably lost to the crumbling geological record.[26] We should not underestimate the diverse epistemic toolkit that paleobiologists have at their disposal to test and adjudicate such theories,[27] but the point is we can be confident that complex structures like stegosaur plates are adaptations while being uncertain as to their specific function.

Nevertheless, convergent evolution is generally thought to offer some of the strongest evidence we have for adaptation, and indeed for specific functional attributions. The fact that ichthyosaurs, dolphins, plesiosaurs, pinnipeds, sea turtles, and sea snakes all evolved flippers or paddles in an aquatic environment overwhelmingly suggests that these features were shaped by selection to meet a similar functional demand: locomotion of a macrobe in an aqueous medium. A more detailed swimming convergence is the iterated evolution of the high-powered "thunniform" swimming style in four distant groups of vertebrates, including (in order of evolutionary appearance): ichthyosaurs (~250 mya), tuna (~55 mya), lamnid sharks (~50 mya), and dolphins (~45 mya) (see figure 4.1).[28] The same goes for other strongly functionally constrained structures, such as wings and eyes, which like paddles are unambiguous cases of adaptive convergence.

How reliable the inference from convergence to adaptation is will depend on the complexity of the underlying trait and how clearly the trait is matched to a common ecological regime.[29] With respect to simple (low-dimensional) traits, some degree of convergence can be expected *as a simple matter of chance*—that is, under the influence of stochastic processes alone.[30] The more structurally complex a trait, however, the less likely it is to arise *repeatedly* due to chance, particularly if it solves an ecological design problem. Adaptation is thus implicated in cases of complex functional match that have been arrived at independently, and many cases of convergence documented in the literature fit this bill.

Recall, however, that adaptationism alone does not get us all the way to the RRT. Selection could be a dominant force in evolution and yet the multiple realizability of fitness and function could undermine evolutionarily replicability. Thus, it is critical that convergence support three additional, more philosophically onerous assertions that underpin the RRT. First, convergence must indicate that certain design problems are pervasive in any history of life. If the conditions that precipitate convergent episodes are restricted to extremely rare circumstances, they will have very low replicability across life worlds (more on this in chapter 6). Second, convergence must indicate that the set of

Figure 4.1
The great white shark (*Carcharodon carcharias*, top sketch, a fish) and the ichthyosaur (*Stenopterygius quadricissus*, bottom photo, a Mesozoic marine reptile) have converged on numerous features of the body plan, including a teardrop shape, a heavy dorsal fin and high-aspect-ratio caudal fin for swimming at speed, and a crossed-fibered architecture of the skin composed of the same chemical fibers, as well as a specialized caudal peduncle and ligament force-transmission system. Redrawn from T. Lingham-Soliar, "Convergence in Thunniform Anatomy in Lamnid Sharks and Jurassic Ichthyosaurs," *Integrative and Comparative Biology* 56, no. 6 (2016): 1323–1336. Photo of specimen in Senckenberg Museum, Germany. Courtesy of Wikimedia Commons.

adaptive solutions to these pervasive ecological design problems is highly circumscribed. If there is an unmanageably large number of equally good solutions to pervasive ecological problems, then there is no reason to think that any particular outcomes will (re)occur at predictable frequencies. Third, and most crucially, convergence must indicate that these few optimal solutions to pervasive design problems are accessible to selection notwithstanding, and irrespective of, the internal constraints of phylogeny.

In other words, convergence must not only show that lineages can navigate to these solutions from very distant developmental starting points, but also that the solutions themselves are not contingent on body plan parameters that could easily have been otherwise. If Good Tricks (in Dennett's terminology) are not accessible from distant evolutionary trajectories, or if they are only

"Good" in the context of accidental developmental parameters, then there is no reason to think that they would be subject to repetition across deep replays of the tape of life on Earth or independent replays of the tape on other worlds. As we will see in the next two chapters, drawing each of these inferences from existing data on convergent evolution can be highly problematic.

2. Matters of (Mis)interpretation

Before delving into these evidential matters further, however, let us foreground several interpretive confusions that have prevented the CFC from making contact with key claims of the RCT. Some of these exegetical misreadings of Gould are understandable, given the lack of detail and rhetorical variability in Gould's expressions of the thesis (see chapter 3). Still, it is a useful exercise to consider what claims the RCT might be making and determine which of these claims is subject to refutation by convergence data. At the end of the day, however, the problems we are wrestling with are not exegetical but epistemic and ontological: they concern our best understanding of the nomological structure of the biological world regardless of what Gould or anybody else thought on the matter.

2.1 Misconception 1: Contingency ⇒ Nonrepetition

A general problem with the CFC is that it has often mischaracterized the RCT, dismantling "straw man" versions of Gould's thesis. By engaging with easy-to-refute characterizations, proponents of the CFC have tended to gloss over important nuances of Gould's view of life and how it might accommodate convergent phenomena. For example, McGhee in his leading review of convergent evolution aims to refute several claims that he attributes to Gould, but which Gould almost certainly did not hold.[31] The first is "the view that the evolutionary process is nonrepeating."[32] This misinterpretation of the RCT, which we shall refer to as "M1," is by no means unique to McGhee. We can infer that other convergence proponents, such as Conway Morris and Dennett, also endorse M1 because they take the entire unqualified body of convergence data, including any and all evolutionary repetitions, to militate against the Gouldian view of life. Biologist Zachary Blount, a researcher on the LTEE studies (see section 1.1 in this chapter), likewise holds that contingency should preclude evolutionary repeatability, citing Gould's *Wonderful Life* and *The Structure of Evolutionary Theory*.[33] This (mis)reading might explain why the LTEE studies, which are designed to test the RCT, are structured in the way that they are (see section 1.1). Much of the philosophical attention to

Gould's thesis has likewise focused on its implications for the nonrepeatability—and, as we shall see, unpredictability—of evolutionary outcomes.

The first thing to say about M1 is that it is demonstrably false. If the RCT did entail M1, then, as McGhee argues, it too would be demonstrably false. Yet numerous impressive examples of convergence have long been documented. If M1 accurately described Gould's thesis, then McGhee's systematic review of convergent phenomenon would not be needed to refute it. Rather, a single or a few well-documented cases of convergence would suffice. There is simply no way that Gould, well aware of the phenomenon of convergence, would embrace such an obviously falsified thesis. We can make M1 more plausible by rephrasing it to express a *relative frequency* claim rather than a *categorical denial* regarding evolutionary repetition. This revised (mis)interpetation, which we can refer to as "M1.1," might hold that "the evolutionary process is *rarely* repeating." Refuting this claim is no longer a trivial exercise and justifies the systematic review that Conway Morris, McGhee, and others have diligently carried out. However, M1.1 still misses the mark because at bottom the contingency debate is not a relative frequency dispute about *evolutionary repetition per se*—rather, it is a dispute over the *nature and causes of those repetitions*. Or so I will argue over this and the next two chapters.

There is in fact ample room within the Gouldian view of life for a great deal of evolutionary iteration and predictability. What is glossed over by M1 and M1.1 is the role of internal developmental constraints in the RCT and how they figure in the possibility space of life as we know it. By giving theoretical primacy to developmental constraints, the RCT in fact predicts certain kinds of repetition, namely repetitions that result from entrenched developmental systems that make certain adaptive outcomes more likely due to their accessibility to selection.

One might reply that I am drawing entailments of Gould's theory that he personally did not entertain, even if these entailments are theoretically justified. This is not so. In *The Structure of Evolutionary Theory*, Gould's last and most comprehensive monograph, Gould maintains that "Homologous developmental pathways can also be employed … as active facilitators of homoplastic adaptations that might otherwise be very difficult, if not impossible, to construct in such strikingly similar form from such different starting points across such immense phyletic gaps."[34] Thus, although many authors cite *Structure* for the claim that Gouldian contingency is inconsistent with evolutionary repeatability, this is an erroneous reading of that text and of Gould's overarching view of macroevolution. It is through the causal explanatory role of developmental constraints that Gould's view of life can accommodate certain kinds of evolutionary repetition. In the rest of this book, we will explore just how far this accommodation can go before the weight of convergence data bends Gould's theory to the breaking point.

2.2 Two Readings of Developmental Constraint

To understand how the RCT might accommodate evolutionary repetition, it is important to distinguish between negative and positive readings of developmental constraint. As we saw in chapter 2, the RCT maintains that once animal body plans congealed in the early stages of animal evolution, the causal topography of development restricted the subsequent exploration of morphospace, leaving ever larger and unbridgeable gaps between body plans as some were culled in stochastic (or pseudo-stochastic) extinction events. In this account, developmental constraint is read in the "negative" sense as confining or restricting the adaptive search of morphospace, preventing selection from journeying beyond the confines of the body plan. This negative framing is consistent with the canonical definition, which holds that "developmental constraint" refers to "biases on the production of variant phenotypes, or limitations on phenotypic variability, caused by the structure, character, composition or dynamics of the developmental system."[35]

We can distinguish these "internal" developmental constraints on form from the "external" constraints imposed by the optimizing agency of natural selection working to solve ecological design problems within the confines of physical and chemical laws.[36] The basic Gouldian idea is that internal developmental constraint, not the external optimizing force of natural selection, is the fundamental cause of "clumpy" morphospace occupation. *Within* these clumps (or islands) of form, natural selection can be a difference-making cause of evolutionary change; yet the ultimate bounds of that restricted space are determined not by selection but by the frozen accidents of developmental parameters. On this picture, developmental constraints are more causally important than selection in determining morphospace occupation, because they are a greater determinant of where in a vast morphospace a particular lineage lies. Whereas selection explains, for example, why a lineage occupies one of two close regions of morphospace within a body plan, developmental constraints explain why selection is confined to discriminating between those close regions out of the vast theoretical space of possible forms. This appears to be the logic behind Gould and Lewontin's conjecture that "developmental constraints … may hold the most powerful rein of all over possible evolutionary pathways."[37]

So conceived, the role of developmental constraints is glossed entirely in the negative—it is intended to explain where selection cannot go, and why. Yet as Gould's remarks in *Structure* suggest, not all biases of development should be thought of in terms of hindering selectively superior outcomes or rendering them off-limits. As developmental evolutionary biologist Günter Wagner has shown, by restricting the space of possibility to a handful of locally optimal solutions, internal constraints make certain solutions more accessible to selection, allowing them to be arrived at over and over again whenever suitable environments arise.[38] Although it is true that selection is partly causally responsible for these iterations, the reason why these particular solutions are locally optimal and thus repeatedly accessible to selection is due to the internal biases of developmental systems. Indeed, given that genetic search space is hyper-astronomical,[39] with more gene combinations than subatomic particles in the visible universe, how is it that any adaptive variation is ever present for selection? Developmental biases must be an integral part of this constructive story.[40]

This "positive" reading of internal constraint as channels facilitating adaptive change can be found in Gould's work on snails in the late 1980s, where he defines constraint as the "channeling [of] phenotypic change in a direction set by past history or formal structure rather than by current adaptation."[41] The idea is that internal parameters of development impose strong biases on the probabilities of different evolutionary outcomes, making some propitious outcomes highly likely (the positive reading) and others unlikely or even astronomically improbable (the negative reading). Later, in *Structure*,[42] Gould illustrates the positive reading of constraint-as-internal-facilitation with the repeated evolution of maxillipeds in crustacean arthropods. Maxillipeds are feeding appendages that evolved multiple times from anterior walking legs via the repeated selective deployment of homologous developmental pathways (in particular, *Hox* genes). Developmental constraints make evolutionary iterations such as these possible, even if the crustacean body plan parameters that set local optimality are themselves radically contingent.

In his discussion of path dependency in evolution, Eric Desjardins (see chapter 3) considers the example of stick insects (phasmids), which evolved from a flying ancestor, subsequently lost their wings, and in a few cases regained them.[43] It may only take a single mutation in a complex gene network to render these insects wingless, whereas producing flight from scratch is likely to require numerous coordinated mutations affecting major limb structures as well as nervous and muscular functionalities. If a large proportion of the necessary genetic-developmental machinery for wings is conserved in wingless stick insects, then there is a reasonable likelihood that wings—and even a particular type of wing—could repeat in this clade. Given certain conserved developmental parameters in phasmids that shape the probability distribution of evolutionary outcomes, wing iteration may be highly accessible

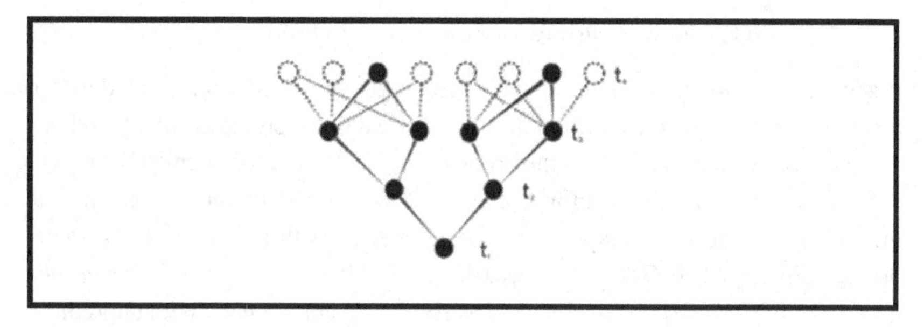

Figure 4.2
A mixed contingent-convergent system that exhibits path independence and predictability at finer grains of taxonomic resolution, but path dependence and unpredictability at coarser grains. For comparison and discussion, see figure 3.1. From R. Powell and C. Mariscal, "Convergent Evolution as Natural Experiment: The Tape of Life Reconsidered," *Journal of the Royal Society Interface Focus* 5, no. 6 (2015): 1–13.

to selection and thus more likely to occur, and to occur in particular ways, than functional alternatives.

The stick insect example hints at what it would mean for certain phenotypic potentials to be genetically conserved in a clade, an important aspect of iteration that we will return to in the next chapter. The key point here is that radically contingent developmental parameters can set the channels for evolutionary iteration. Such a mixed contingent-convergent scenario, inspired by a positive reading of developmental constraint, can be illustrated in schematic form by a system that exhibits convergence at finer grains of phylogenetic resolution but path dependence at coarser grains (see figure 4.2).[44]

We can now distill the crux of the first misconception. The contingency dispute turns not on the existence of evolutionary repetitions per se, but on the *causes* of evolutionary repetition and whether they support the deep evolutionary robustness of the outcomes observed. What contingency theorists like Gould reject is the proposition that the driving forces behind convergence transcend the contingently entrenched developmental plans of particular lineages. The main question, therefore, is whether instances of evolutionary iteration reflect this transcendence—and the problem is that whether they do, as we shall see in chapter 5, is underdetermined by convergence data as it has thus far been collected and analyzed. A more general problem is that it is not entirely clear what such "transcendence" would look like. The very notion of external constraints on form may be incoherent unless it is indexed to the particular developmental systems of evolving lineages. We will return to these matters shortly. For the moment, let us press on to the second misconception that has prevented the CFC from properly engaging with the RCT.

2.3 Misconception 2: Contingency ⇒ Unpredictability

A second common mischaracterization of Gould's thesis, which is related to but distinct from the first, is the notion that the RCT entails unpredictability. McGhee, for instance, attributes to Gould the proposition that "evolution is entirely historically contingent and thus unpredictable."[45] We saw in chapter 3 that it is a mistake to equate the metaphysics of contingency with epistemic facts about predictability, even if Gould was occasionally guilty of doing so. However, we might charitably interpret McGhee not as conflating contingency with unpredictability or attributing this conflation to Gould, but instead as purporting to test an apparent entailment of Gould's theory: if macroevolution is indeed sensitively dependent on small changes in initial conditions, then evolution should proceed in an unpredictable manner.

Yet studies of convergence clearly show that selection does often drive lineages to similar evolutionary outcomes in ways that plausibly admit of prediction. For instance, tetrapods making the "fin-to-limb-to-fin" transition, such as ichthyosaurs and whales, evolve in predictable ways. Precisely why it is that some lineages end up making this transition but others do not, and why some (such as ichthyosaurs and whales) make the full transition but others (such as plesiosaurs and pinnipeds) make only partial transitions, may be harder to predict and may hinge on the contingent quirks of their respective histories. But *when* these full and partial transitions are made, the CFC argues, they will tend to come with a predictable suite of features. Thus, an entailment of the RCT—that evolutionary trajectories will not admit of prediction—is demonstrably false, and hence the RCT is refuted.

This characterization of the RCT, call it "M2," also misses the mark. Like M1, M2 is framed in implausible categorical terms that make it vulnerable to refutation by a single counterexample. This problem can be cured, as with M1, by modifying it to address a relative significance claim. But this more plausible reading, call it "M2.1," falters as well for a simple reason: the lack of predictability is not an entailment of the RCT. As we have seen, there is plenty of room in the Gouldian picture for certain kinds of repeatability. Indeed, the RCT is consistent with, and arguably confirmed by, what we might call "bounded repeatability"—or *repetitions within the bounds of, and caused by, internal developmental constraints*. Simply put, the RCT does not imply chaos and unpredictability at all phylogenetic and temporal scales in macroevolution, as M2 and M2.1 would suggest.

Nor does Gould intend that his contingency claims apply to all levels of trait description. In a passage from *Wonderful Life* worth quoting at length, Gould states in no uncertain terms that he is not arguing that all of evolution is historically contingent and unpredictable:

Am I really arguing that nothing about life's history could be predicted, or might follow directly from general laws of nature? Of course not; the question that we face is one of scale, or level of focus. Life exhibits a structure obedient to physical principles. We do not live amidst a chaos of historical circumstance.... Much about the basic form of multicellular organisms must be constrained by rules of construction and good design.... Invariant laws of nature impact the general forms and functions of organisms; they set the channels in which organic design must evolve. But the channels are so broad relative to the details that fascinate us! ... When we set our focus upon the level of detail that regulates most common questions about the history of life, contingency dominates and the predictability of general form recedes to an irrelevant background ... almost every interesting event of life's history falls into the realm of contingency.[46]

It seems that Gould would be happy to grant that, for example, the fusiform (tapered) body shape is an evolutionarily robust feature of fast-moving, large-bodied aquatic life wherever it evolves. What he would deny is that the specific parameters of body plans and the locally optimal iterations that hinge on them are robust features of complex multicellular evolution. Convergence on fusiform morphology does little to detract from this conclusion.

There is a weak point in the above excerpt that is worth drawing attention to, however, and that is Gould's relative interest claim. Why should universal biomechanical constraints on the evolution of form not be as interesting to biologists as the quirky, more detailed outcomes of evolution? Gould offers no argument to support this assertion. Indeed, Gould's push for an autonomous, law-like paleontology shows that he was committed to the idea that a central goal of science, and indeed, of paleobiology, is to uncover spatiotemporally invariant laws.[47] Whatever one makes of Gould's advocacy of narrativistic explanation, it is hard to defend the claim that universal constraints on the evolution of form are objectively uninteresting, even if they are not among the details that some find most fascinating.

In effect, Gould is making a rhetorical move here similar to one that he (writing with Richard Lewontin) famously excoriated in connection with adaptationist explanations:

In natural history, all possible things happen sometimes; you generally do not support your favoured phenomenon by declaring rivals impossible in theory. Rather, you acknowledge the rival, but circumscribe its domain of action so narrowly that it cannot have any importance in the affairs of nature. Then, you often congratulate yourself for being such an undogmatic and ecumenical chap.[48]

Gould seems to be engaging in the very mode of argument for which he chastises adaptationists. He acknowledges there are some robustly replicable outcomes in evolution, but he relegates these to theoretically uninteresting phenomena in the history of life. It is best to avoid quibbling over which biological phenomena are the most "interesting," and instead pose the question this way: Given our current understanding of convergence, which features of the shape of life are likely to be radically contingent, and which features are likely to be robustly replicable?

3. Macroevolutionary Overdetermination

How might proponents of the RRT make sense of the ostensibly contingent patterns of extinction and faunal turnover on which much of the Gouldian view rests? One way they might attempt to do so is by arguing for the "overdetermination" of major evolutionary outcomes. For example, Dennett claims that even if *Pikaia* and its proto-chordate ilk had been felled with other *Problematica* in the Cambrian, some other lineage in the future history of life would have hit upon the good "vertebrate" trick and thereby refilled the "vertebrate" attractor in morphospace. With vertebrates surviving and thriving throughout the Phanerozoic Eon, however, the vertebrate niche has remained packed and thus unrefillable, in accordance with incumbent advantage and ecological exclusion theory. Were the vertebrate niche to empty due to some unprecedented biotic crisis, then it would, over the long haul, be refilled by something close enough to warrant the "vertebrate-like" label.

The macroevolutionary overdetermination thesis is not incoherent, but it is underevidenced. There are simply no indications of convergence on the bundle of traits that comprise phyla or subphyla such as "vertebrates." And relying on an incumbent advantage theory (discussed in chapters 1 and 2) to explain this lack of repetition is also problematic. If incumbent advantage did not prevent the staggering multitude of convergences documented by the CRC, why then would it preclude iterations of major body plans? The RRT has no obvious rejoinder to this query. In contrast, the developmental pillar of the RCT (see chapter 2) offers a non–ad hoc explanation of this asymmetry in iteration: the causal topography of development precludes certain iterations but permits (and even facilitates) others.

It is now time to tie the threads of the argument together. Certain exegetical problems have prevented the CFC from making contact with the key theoretical framing assumptions that underpin Gould's thesis. As we have seen, the RCT is more capable of handling repetition and predictability than its detractors (and even its sympathizers) have acknowledged. First, critics have failed to recognize that the crux of the contingency dispute turns not on repeatability or predictability per se in evolution, but on the *causes* of observed iterations and whether they indicate the evolutionary robustness of the regularities observed. Second, the RCT is aimed at specific levels of morphological description, and thus pointing to convergence on more generic evolutionary phenomena does not cut to the heart of Gould's argument. Finally, RRT proponents have failed to provide an adequate explanation of the lack of repetition at the level of animal body plans. The next two chapters will elaborate on each of these points by taking a closer look at the evidentiary significance of convergence.

5 Convergent Evolution as Natural Experiment

George Gaylord Simpson, one of the most influential evolutionists of the twentieth century and an architect of the modern biological synthesis, begins his skeptical paper on the search for extraterrestrial intelligence (SETI) by proclaiming that "we can learn more about life from terrestrial forms than we can from hypothetical extraterrestrial forms."[1] Simpson's argument would be elaborated on extensively several decades later by macroevolutionist Peter Ward and planetary scientist Donald Brownlee in their influential book *Rare Earth: Why Complex Life Is Uncommon in the Universe.*[2] Lamenting that we are unlikely to discover an extraterrestrial instance of life, Simpson concludes that SETI monies would be better spent on terrestrial concerns. Whatever one thinks of this triaging recommendation, Simpson's methodological point stands: in the absence of extraterrestrial data, studies of earthly evolution are our best bet for assessing the prospect of a cosmic biology.

We saw in the last chapter that the radical contingency thesis (RCT) is addressed to big-ticket philosophical questions about large-scale evolution that controlled experiments simply cannot answer. This leaves open the possibility that "natural experiments" in macroevolution might play this crucial evidentiary role. In what sense might convergence constitute "experimental" evidence that bears on the contingency debate? How should the validity of natural experiments in convergent evolution be assessed? In this chapter, we will explore the evidential logic of convergence in greater detail.

1. The Experimental Logic of Convergence

1.1 The Nature of Natural Experiments

We normally think of scientific experiments as involving the controlled manipulation of independent variables to assess their causal influence on dependent variables. But to conduct an experiment, broadly construed, is simply to put oneself in an epistemic position from which to make observations that affect our confidence in hypotheses about the causal, nomological, or historical structure of the world. Given such a broad framing, it is not easy to distinguish between observational studies that make use of conditions already found in nature and experiments that involve the controlled manipulation of nature, be it in terms of their methods, epistemic goals, or ability to confirm (or refute) theories.

As philosopher of science Samir Okasha points out, many classic experiments in physics, such as the crucial tests of general relativity, are essentially observational and do not involve human intervention in nature.[3] The solar eclipse of 1919 made it possible to test Einstein's prediction that starlight would bend twice as much under the gravitational influence of the sun than is predicted by Newtonian physics because the light of stars traveling close to the sun could be measured at that time. In such cases, nature presents conditions that make it possible to test rival hypotheses, and it seems unproblematic to think of tests that take advantage of these natural conditions as "experiments." At the same time, studies in historical sciences, such as paleontology, are rarely purely observational—they often involve controlled, systematic searches for traces of the past that draw on a wealth of background theory, employing refined methods of data collection and setting out with the aim of testing hypotheses. There is thus no clear conceptual difference between observational and interventionist studies.

Nor is there an obvious epistemic difference between the two. One might be inclined to distinguish manipulation-based experiments from observational studies on the grounds that the former generate more reliable inferences about the causal structure of the world. Yet this is far from evident. In fact, in evolutionary biology, unlike in physics, controlled laboratory studies can create highly artificial conditions that are only weakly projectible to the natural living world. This is in part because the natural living world is complexly causally configured in ways that are not reflected by idealized laboratory conditions, and these differences in causal complexity can result in substantially different results under controlled and natural conditions, respectively.[4]

Nevertheless, "natural experiments" are in certain ways more like proper manipulationist experiments than they are like observational studies. This is because natural experiments are fortuitously structured in ways that resemble interventionist experiments that might be designed by rational agents were they capable of manipulating large-scale natural conditions. In natural experiments, researchers select and analyze samples that differ naturally in one or more independent variables but are similar with respect to other variables, with the aim of developing, corroborating, or refuting hypotheses. Although natural experiments in evolutionary history lack the control of classic experiments, and thus risk internal validity, they have the advantage of allowing biologists to gather data across numerous taxa and habitats reflecting vast timespans of evolution.[5] This breadth and depth of study, impossible in the laboratory or field, permits inferences about the robustness of macroevolutionary patterns.

For instance, isolated islands are often described as "natural laboratories of evolution" that can reveal law-like evolutionary patterns and processes. "Dwarfing," for example, can occur when large mainland vertebrates (especially mammals and birds) invade or become trapped on isolated islands. Dwarfing has been observed for a wide range of animals, from dinosaurs to humans. Fossils of dwarf sauropods and ankylosaurs have been found on an ancient offshore island in what is now modern-day Romania[6]—may they have been hunted by similarly dwarfed carnosaurs? Likewise, dwarf populations of *Homo erectus* have been found on the island of Flores in Indonesia, where they apparently hunted dwarfed elephants, known as stegodons.

Studies of ecosystem recovery after the removal of human influence, such as in the demilitarized zone between North and South Korea, also have the structure of a natural experiment. Most recently, the Japanese tsunami of 2011, triggered by a magnitude 9.0 earthquake, carried a staggering diversity of sea creatures on tsunami debris from the coast of Japan to the West Coast of the United States in the greatest maritime migration ever recorded. Researchers have described this event, too, as a natural experiment that probes the stability of local ecosystems to large-scale invasion. These events and others like them amount to replays of the tape of life from which biologists can glean law-like generalizations about ecology and evolution.

1.2 Evaluating Natural Experiments in Convergent Evolution

Natural experiments in convergent evolution can probe deeper into the causal structure of the living world. But like laboratory experiments, they can be contaminated, poorly structured, and misleading. A crucial feature of any experiment, whether natural or artificial, is that it is isolated in ways that control for the relevant confounding variables. The variables that must be controlled in any given experiment are determined by the research questions being posed and the hypotheses being investigated. It is critical, therefore, when it comes to natural experiments in convergent evolution, that we are crystal clear about the claims that the experiments are designed to test, as this will determine how to assess their validity.

Our goal here is to establish the validity criteria for natural experiments in convergence for purposes of testing the RCT. Evaluating the setup of natural experiments in convergence for this specific evidentiary purpose requires, at a minimum, that we assess the "independence" of observed replications by controlling for the influence of conserved developmental constraints. We saw in the previous section that certain types of evolutionary iteration are consistent with, and perhaps even corroborative of, the RCT. It is critical, therefore, that we distinguish the underlying causes of the repetitions observed. Failure to do so has resulted in flawed experimental setups, even in paradigmatic studies of convergence.

One such paradigmatic natural experiment in convergence is the series of pioneering studies conducted by Jonathan Losos and his collaborators documenting the iterated evolution of *Anolis* lizard "ecomorphs," or distinct forms adapted to specialized microhabitats on isolated islands in the Caribbean.[7] Arriving on floating debris, these beautiful lizards established new populations on numerous islands in the Greater Antilles. These populations then evolved in an iterated, predictable way on each "laboratory island," producing at least six distinct ecomorphs, each associated with a suite of anatomical and behavioral properties that are matched to a specialized ecological zone.[8]

For example, the "crown-giant" ecomorph exhibits a cluster of traits adapted to life in the highest tree canopies, whereas "trunk" ecomorphs are anatomically and behaviorally suited to life on tree trunks; "grass-bush" and "twig" ecomorphs, meanwhile, are adaptively matched to selective regimes with their own distinct modes of predation, predator-avoidance, terrain, climate, and so forth (see figure 5.1). Each ecomorph has a quantifiably distinct body plan, including limb and tail length, head and toe-pad size, and coloring, and each ecomorph is more closely related to the other ecomorphs on their own island than they are to the similarly adapted anoles on other islands. For instance, tokens of the twig ecomorph found on Cuba, Puerto Rico, Hispaniola, and Jamaica, despite their affinities, are more distantly related to one another than they are to the very different ecomorphs on their own respective islands.

These patterns of convergence appear to suggest that there is a limited set of "attractors" in *Anolis* morphospace, and that selection reliably overcomes any contrary forces or tendencies in driving the repeated differentiation of lizard populations toward these respective ecomorph attractors. This, in turn, may be taken to indicate not only that selection is a dominant force in macroevolution but also that selection *works in highly repeatable and circumscribed ways*—two key elements of the robust replicability thesis (RRT). *Which* particular ecomorph a subpopulation is driven toward may be highly contingent on numerous factors, such as the ecological zone in which it first becomes established, the ordering of its genetic mutations, the effects of drift and migration, and so on. But the fact that isolated island populations will

Figure 5.1
Recurrent evolution of Anole ecomorphs on isolated Caribbean islands. Photograph panel courtesy of the Luke Mahler laboratory at the University of Toronto. From left to right, the top row depicts twig specialists *A. garridoi* (Cuba) and *A. occultus* (Puerto Rico); the second row depicts trunk and ground specialists *A. cybotes* (Hispaniola; photo by B. Falk) and *A. lineatopus* (Jamaica); the third row depicts grass specialists *A. alumina* (Hispaniola; photo by M. Landestoy) and *A. alutaceus* (Cuba). Photographs not otherwise marked are by Luke Mahler.

differentiate into these respective *Anolis* equilibria appears to be robustly replicable across natural island experiments.

The Losos studies have been widely taken to be "valid" natural experiments in the sense that they employ the necessary controls of confounding variables in order to adjudicate the hypothesis being tested. Clearly, these studies have ruled out the hypothesis that ecomorphs are more closely related to one another than they are to other ecomorphs on their own island. However, this is not the hypothesis being tested; rather, it is the RCT that is under scrutiny. Both biologists and philosophers of science have assumed that the geographic isolation of islands ensures the independence of the replications observed.[9] The problem, however, is that *geographical* isolation is not the only kind of isolation that is necessary for establishing the validity of natural experiments in convergent evolution. *Developmental* isolation is equally important. The *Anolis* ecomorph experiments demonstrate unequivocally that selection can optimize form within a given set of developmental parameters. This is an important finding, to be sure, but *as tests of the RCT*, the *Anolis* experiments are invalid. For despite their geographical separation, the observed evolutionary systems are not isolated in crucial developmental respects. Although the target systems end up in similar ecomorphological attractors, they begin from highly similar initial developmental conditions. As such, they fail to rule out the possibility that the observed attractors are caused by shared developmental constraints that facilitate the repetitions.

At this point, one might respond with a shrug of exasperation: "Of course selection is working in these cases with a common developmental plan! What else could possibly account for these outcomes?" Yet this is precisely the thrust of Gould's thesis, and it explains why observations such as these fail to speak to the RCT. Conceptually speaking, there may be no way of delineating ecomorphological attractors, such as the ones observed in the *Anolis* studies, absent a stipulated set of developmental parameters within which selection can optimize form.[10] Yet if these parameters are radically contingent, then the ecomorphological iterations that hinge on them will be radically contingent as well. That is to say, the replicability of ecomorphs will extend no farther back than the evolution of the specific developmental parameters on which they causally depend and with respect to which they are defined. If a pattern of iteration is due to selection acting on conserved developmental parameters, then the question of contingency falls back on the evolutionary robustness of the parameters themselves, and this can only be established by observing iterations *of* (rather than *within*) the developmental plans at issue.

A general problem with appeals to natural experiments in convergent evolution is that the sense of "independence" that is operative in dominant definitions of convergence does not support some of the evidential uses to which the phenomenon has been put in the contingency debate. To see this, it is not necessary to navigate the dizzying array of concepts that cluster around "convergence."[11] A few key distinctions will suffice. We can begin by distinguishing "homology" from "homoplasy." On the standard "taxic" account, *homology* refers to a character resemblance or similarity (S) between two lineages (L1 and L2) that is present in their last common ancestor (LCA), with the inference being that S arose once in LCA and was continuously transmitted in L1 and L2 since they diverged from LCA. *Homoplasy*, on the other hand, describes (using the same placeholders) a case of S between L1 and L2 that was not present in LCA, and thus is inferred to have arisen at least twice.[12] Whereas some terms, such as "analogy," describe similarity that results from adaptation to a common selective regime, the term "homoplasy" is causally neutral and makes no claims about adaptive motivations. Whether a character resemblance is a homoplasy is determined solely in virtue of its *phylogenetic pattern*. For example, the rodent-like common ancestor of marsupial and placental saber-toothed lions (see chapter 4) did not have a sabertooth morphology, let alone one that was transmitted continuously for the hundred-odd-million years since the divergence of marsupial and placental mammals. Thus, the sabertooth suite is homoplasious, not homologous. End of story.

The taxic account, though dominant, is not the only account of homology/homoplasy on offer in the literature, nor does it exhaust the relevant terminology. Some authors treat homoplasy as synonymous with convergence, while others treat convergence as a causally distinct type of homoplasy. A minority view holds that for a homoplasy to count as a convergence, the developmental machinery causally responsible for the relevant character resemblance must not have been continuously inherited from a common ancestor.[13] On this view, "parallelisms" (see section 3) would count as homologies. Other authors sympathetic to the idea of distinguishing between iterations based on their underlying causes have chosen to retain standard definitions of homology and homoplasy, instead distinguishing between parallel and convergent homoplasies, where "parallelism" refers to homoplasy that is underwritten by conserved developmental generators.[14] The latter approach is the one that will be adopted here.

Our focus for the moment, though, is on the limitations of the taxic approach, because it is the account most often relied upon by proponents of the "critique from convergence" (CFC) discussed in chapter 4. On the taxic account, convergence/homoplasy can ensue from both highly similar and highly disparate initial developmental conditions. So long as S—delineated phenotypically—was not present in the LCA of L1 and L2, then S is deemed to be homoplasious. This pattern-based approach to homoplasy is not problematic in and of itself; but when convergence is called upon to serve as evidence in the contingency debate, the failure to parse the category of homoplasy can undermine the validity of convergence experiments.

As noted in this and the previous chapter, the contingency/convergence dispute turns on whether evolutionary repetitions have the kinds of causes that bespeak their evolutionary robustness. Because the taxic approach does not distinguish among homoplasies on the basis of their underlying causes, convergence data gathered in this way cannot adjudicate the RCT. The problem is not that convergence data fail to provide a "crucial" test that could decisively confirm or refute the RCT; that is much too demanding a requirement, given the kinds of questions we are investigating. The problem, rather, is that because the CFC has operated exclusively with the taxic conception of homoplasy, it fails to provide any clear evidence against the RCT at all. Without a more targeted analysis of the causes underlying cases of convergence, the CFC is unable to distinguish natural experimental setups that fail to speak to the RCT from those that genuinely undercut it. In essence, studies of convergent evolution carried out with the intent of testing the RCT have been working with tainted experimental setups.

1.3 The Lumping Problem

The failure to parse convergence data in this way results in a heterogeneous class of convergent events with varying implications for evolutionary robustness. Because it does not control for the internal developmental determinants of iterated evolution, the CFC is unable to differentiate biological regularities that reflect deep truths about the living universe from those that are essentially accidental. Let us refer to this as the "lumping problem."

For instance, George McGhee's state of the art review of convergent evolution (discussed in chapter 4) is a significant improvement over previous efforts in part because it includes extensive phylogenetic information.[15] However, because McGhee is operating with a taxic notion of convergence, he does not use this information to control for the developmental facilitators of iterated evolution. To take but one example, consider McGhee's inclusion of more than twenty iterations of "cantharophilous" flowers, which have evolved specialized morphologies to facilitate pollination by beetles. This specialized "pollination syndrome" includes large dish-shaped flowers that are typically dull in color and heavily scented, have easily accessible pollen, and protect their ovaries from beetle mouthparts. Bees, flies, lepidopterans, beetles, and birds have acted as unwitting couriers of male angiosperm gametes since the Cretaceous. Thus, it is not surprising that similar observations have been made for other angiosperm–pollinator pairings.[16] Predictions such as "If there are beetles, then there will be cantharophilous flowers" may be law-like on weakened accounts of biological laws that presuppose radically contingent antecedents (as discussed in chapter 3). But if the clusters of traits that compose "beetle" and "angiosperm" ground plans are highly contingent accidents of Earthly evolution, then this adds little to our understanding of the deep structure of evolution.

The same goes for Simon Conway Morris's discussion of ant mimicry, which is so ubiquitous—evolving more than seventy times in insects and spiders (see figure 5.2)—that it even has its own name in the literature: "myrmecomorphy."[17]

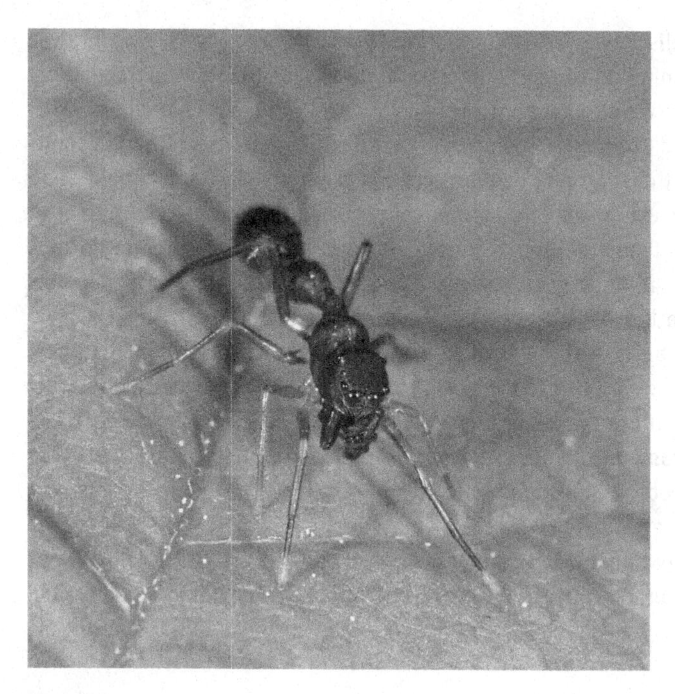

Figure 5.2
Synemosyna formica, jumping spider ant-mimic. Note the two large, forward-facing camera eyes. Photo courtesy of Tom Murray.

If the "ant" and broader "arthropod" phenotype is radically contingent, then ant mimicry will not be stable across deep replays of the tape of life. In other words, ant mimicry is only replicable across shallow replays of the tape that begin with the developmental parameters in question—which in this case, includes the arthropod parameters of the mimic and the ant-specific parameters of the subject of mimicry. Regularities such as these may indicate the deep evolutionary robustness of very generic phenomena like "coevolution" and "mimicry," which can be found across the whole of complex multicellular life. But as noted earlier, and as will be discussed in more detail in the next chapter, this is not the level of detail at which the contingency debate takes place.

The take-home message is that the sheer number of iterations, without further analysis, tells us very little about the evolutionary robustness of the regularity described. Dozens of coevolutionary convergences on flower and ant morphologies offer far weaker evidence against the RCT than do convergences that are fewer in number but that can be shown to arise from broad physicochemical constraints on life. For the latter do not hinge on the peculiar—and singularly evolved—developmental parameters of particular lineages.

For instance, gills have evolved countless times in arthropods, vertebrates, mollusks, and annelids for the universal biochemical function of mediating gas exchange between an organism and its surrounding fluid medium. The same may be said of image-forming eyes, a topic that will be explored in some detail in the next chapter and again in part II. It is only once the "quality" of a convergence is established that the "quantity" can become very telling.

The upshot is that the high frequency of an evolutionary outcome in this history of life, without further analysis of its underlying causes, cannot be taken to imply a high frequency across alternative histories of life. The fact that certain structures or functions recur in evolution is not evidence that they are cosmic in scope or even robust across alternative histories of life on Earth. Lumping is harmful to the case against contingency because it causes the CFC to overplay its hand, while obscuring the strongest evidence against the RCT. In compiling a body of undifferentiated data on convergence, the whole is less than the sum of its parts. In short, the greatest asset of the CFC—its impressive inventory of convergent events in the history of life—is also its greatest flaw. The task before us is to begin the work necessary to sharpen this promising class of evidence.

2. Parsing the Reference Class

As we saw in the last chapter, the Gouldian disagreement with convergence proponents turns not on *whether* evolutionary iteration is theoretically important, but *why* it is important, and what it signifies about the causal—and in particular, modal—structure of the biological world. Understanding the theoretical import of convergence requires that iterations be distinguished in accordance with their underlying causes. What does it mean to be a cause of an iteration, and since iterations may be said to have multiple causes, what makes one cause or type of cause comparably more important than another? These are challenging questions, but they must be answered if convergence is to serve as evidence, one way or the other, in the contingency debate.

2.1 Gouldian Repetitions versus True Convergence

The challenge is to distinguish iterations that are indicative of deeper modal robustness from those that are evidence for shallower degrees of replicability. How can this be achieved in concept and operationalized in practice? This is the million-dollar question, and I do not pretend to offer a definitive answer to it here. My less ambitious aim is to sketch a preliminary picture of what such an account might look like, though this sketch will no doubt be incomplete.

If, when all is said and done, one remains unconvinced that this distinction can be drawn, then one must accordingly resign to the fact that the contingency debate is permanently underdetermined by convergence data—and that we must look elsewhere, perhaps beyond Earth, to make any headway on it. I hope, however, to show that this distinction is not incoherent, and that it can be drawn in an epistemically accessible way.

The first thing we need to do, if we are to make good on this distinction, is break down "iteration" or "homoplasy" into further, more finely tuned categories. Let us call it "Gouldian repetition" when evolutionary iteration results from selection acting on conserved developmental components or their sequelae. Let us refer to all other iterations as cases of "true convergence." Where we are epistemically neutral as between these two types of homoplasy, we shall refer to biological "iterations," "repetitions," or "homoplasies" to signify this neutrality. Although there are clear-cut cases of each type of homoplasy, there is also likely to be a gray zone in between.

The existence of gray areas should not deter us from drawing the distinction between Gouldian repetition and true convergence any more than it prevents us from drawing useful common language distinctions like "night and day." Gray areas should be expected in the living world, where fuzzy ontologies are the norm rather than the exception. The notion of "evolutionary individual," for instance, is important and meaningful for biological theory, even if there is a fuzzy borderland of pseudo-individuals (such as holobionts[18]) between clear-cut cases of individuals (such as organisms) and nonindividuals (such as populations).

Before proceeding, let us reiterate why this distinction is so important to the evidential value of homoplasy. Iterations that are properly characterized as Gouldian repetitions do not support the CFC because they do not lend credence to the three additional philosophical claims, over and above strong adaptationism, that underpin the RRT (see chapter 4). First, Gouldian repetitions fail to show that certain design problems are pervasive in any history of life, because the design problems that prompt this type of iteration may be shaped by the internal developmental parameters of the lineage in question (in a form of organism-niche codetermination). Second, Gouldian repetitions fail to show that the set of solutions to the design problems they solve is highly circumscribed in a way that would support their law-like projectibility, because in such cases accidental internal parameters are constraining (and facilitating) the solution space. Third, Gouldian repetitions fail to show that certain solutions are accessible to selection notwithstanding the internal constraints of phylogeny, because internal constraints are integral to explaining the morphospace that is repeatedly explored by Gouldian repetitions. This last point warrants further elaboration.

Researchers may in some cases take developmental parameters as a given in attempting to isolate the effects of selection. This is a legitimate mode of evolutionary inquiry, but it is orthogonal to Gould's project.[19] The optimization of form within a given set of contingent developmental constraints may be a useful way of assessing adaptationist theses,[20] but it does not speak to the RCT/RRT debate. Even if we were to achieve a "periodic table of forms" based on a thorough understanding of vertebrate development and how it interacts with natural selection—as first envisioned by the Soviet biologist Nikolai Vavilov and echoed in George McGhee's work (see chapter 4)—this would not show that the parameters of vertebrate development are themselves modally robust. And if they are not, then the periodic table analogy, with all its nomological undertones, is misleading.

Jonathan Losos, director of the *Anolis* lizard studies discussed earlier, acknowledges that iterations work on a common developmental platform, but he denies that this indicates a significant role for internal constraints in explaining *Anolis* evolution.[21] Losos maintains that because all the key features of *Anolis* ecomorphs are continuous, quantitative characters, they are subject to standard selection explanations without requiring any appeal to a lack of variation on which selection can act. Losos's remarks are correct, but they are talking at cross purposes with the RCT. If one takes the overarching developmental parameters of *Anolis* as fixed, then selection does figure most prominently into explanations of these iterated outcomes. Further, if one interprets "developmental constraint" in the exclusively negative sense, then a finding of repeated adaptive optimization within these stipulated developmental parameters undercuts an internal constraints-based explanation. But as discussed in the previous chapter, this overlooks the positive role of internal constraints in facilitating iterative adaptive evolution within the highly contingent parameters of existing body plans. Selection can explain why a particular range of morphospace has come to be populated within a restricted set of internal parameters, but this says nothing about the modal robustness of the internal parameters themselves.

How can we determine whether a given iteration is a case of Gouldian repetition or a case of true convergence? This launches us face first into a thicket of thorny conceptual and methodological problems. The first hurdle is to carve out the elements of underlying development that are causally relevant to a given iteration. This is not easy to do. The trouble is that because all known life on Earth is related to one degree or another (see chapter 1), there will always be crucial developmental mechanisms underlying cases of iteration. As we proceed down taxonomic ranks from eukaryotes to metazoans, from metazoans to bilaterians, from bilaterians to deuterostomes, from deuterostomes to vertebrates, from vertebrates to tetrapods, from tetrapods to mammals, and so on, more and more homologous genetic-developmental machinery will be shared. There is a sense, therefore, in which all iterations must be caused by shared development because all traits in any two lineages will be produced by a partially homologous developmental apparatus.

Does this make all repetitions necessarily Gouldian and thereby gut the value of the term? Does any amount of shared development undermine the independence of the evolutionary replications observed? If so, then natural experiments in convergence that aim to test the RCT would be incurably confounded. These questions become even more pressing in light of the growing body of work on "deep homology," or upstream regulatory networks that are conserved by animal groups that diverged more than 600 million years ago. Deep homologs are implicated in the development of some traits that are the subject of impressive repetitions across very distant groups, such as the evolution of eyes and various body plan modifications. Does deep homology undermine the independence of iterations in which they are involved?

Distinguishing Gouldian repetitions from true convergence will require that we identify shared developmental characters that bear the right sort of causal relation to the iterated outcome. In other words, if we are to distinguish repetitions that are consistent with the RCT from those that undermine it, we must parse cases of convergence in accordance with their underlying developmental causes—and to do this, we must parse the developmental causes themselves.

One workable approach would be to infer a degree of independence that is proportional to phylogenetic distance. Let's call this the "taxonomy heuristic." On this approach, the greater the evolutionary distance between converging lineages, the greater the independence of the iterations observed. Evolutionary distance is used here as a proxy for developmental independence, much as the number of higher-level taxa has been used as an indirect measure of morphological disparity. The evolutionary distance of convergence admits of degrees: for example, converging classes of mammals bridge a much smaller phylogenetic gap than do converging phyla of animals. Similarly, the independence of natural evolutionary replications would admit of degrees, with independence proportional to the phylogenetic distance of convergence.

The taxonomy heuristic explains why a small number of convergences across kingdoms, phyla, and classes (or clades that roughly map on to these Linnaean categories) tend to indicate deeper modal stability than numerous convergences confined to families, genera, and species. The trouble with this approach, however, is that phylogenetic distance is not always a good indicator of developmental difference when it comes to evolutionary iteration. As biologists Jeff Arendt and David Reznick point out, evolutionary iteration between closely related groups—even populations within species—can be produced by different developmental mechanisms, while evolutionary iterations between distantly related groups can sometimes be produced by the same developmental mechanisms (the "same" developmental mechanisms here means homologous developmental mechanisms, or those that have been transmitted continuously from a common ancestor).[22]

For instance, evolutionary iterations of pigmentation in populations of mice within the same species have been shown to be produced by different developmental pathways. Convergence between higher taxa, meanwhile, can involve the repeated activation of deep homologs. *Pax6*, for instance, is involved in the development of eyes in numerous eye-bearing phyla (such as vertebrates, mollusks, and arthropods), and "homeobox" genes are implicated in the anatomical evolution of groups as distant as animals, plants, and fungi. Thus, we must develop ways of assessing independence that do not rely entirely on the blunt instrument of phylogenetic distance.

3. The Problem of Parallelism

One way of glossing the distinction between Gouldian repetitions and true convergence is by falling back on a conception of "parallelism," which does a reasonable (though far from perfect) job of tracking these phenomena. In his comprehensive analysis of convergent evolution, Conway Morris states in a footnote that he will "avoid that old chestnut of whether it is convergent evolution as against parallel evolution,"[23] reasoning that the difference is merely one of degree rather than of kind. Conway Morris is in good company here, as evolutionists the likes of G. G. Simpson have held that parallelism and convergence "intergrade continuously and are often indistinguishable in practice."[24] Yet as we shall see, the conflation of parallelism and convergence causes proponents of the CFC to conclude that external constraints on design space are more pronounced than the evidence in fact warrants. If the distinction between parallelism and convergence offers a promising avenue for distinguishing Gouldian repetitions from true convergence, then crack this "old chestnut" we must try.

Parallelism could have important implications for the use of homoplasy as evidence in the contingency debate, depending on how the concept is cashed out. For example, parallelism could indicate that certain developmental parameters or "informational substrates" that lineages share due to their contingent history strongly increase the probability of certain evolutionary outcomes while dramatically reducing the probability of others. We saw this earlier with the example of the repeated evolution of flight in walking stick insects (see chapter 4). Path dependency need not entail irreversibility, and the kind of path dependency contemplated by the RCT does not entail irreversibility at all phylogenetic grains of resolution—a point that is illustrated powerfully by the positive potential of parallelism.

3.1 Parallelism as Homoplasy in Closely Related Groups

The seminal treatment of the "homoplasy" family of concepts is found in a paper by Otto Haas and G. G. Simpson published in the 1940s. According to their study, the earliest definition of "parallelism" came from W. B. Scott in 1891, who used the term to describe "the independent acquisition of similar structure in forms which are themselves nearly related," in contrast to "convergence," which referred to the acquisition of similar structures "in forms which are not closely related."[25] Scott later added that "the more nearly related any two organisms are, the more likely are they to undergo similar modifications."[26] The idea that closely related lineages will tend to undergo similar modifications because of shared development and lifeways is also one expressed by Darwin in a later edition of *On the Origin of Species*, where he writes,

> Members of the same class, although only distantly allied, have inherited so much in common in their constitution, that they are apt to vary under similar exciting causes in a similar manner; and this would obviously aid in the acquirement through natural selection of parts or organs, strikingly like each other, independently of their direct inheritance from a common progenitor.[27]

Thus, parallelism has a long history of being associated with homoplasy in closely related lineages. In addition, the greater frequency of iterations in closely related lineages (as compared to distant lineages) has historically been *attributed to shared developmental plans*.

Although Simpson read parallelism this way, his coauthor, Haas, opted for a purely geometric definition, using the term "parallelism" to describe situations in which two clades evolve along parallel trajectories but do not converge (i.e., do not come to resemble one another more than they did before the homoplasy). For instance, horses and brontotheriids (a family of extinct perissodactyl mammals) both evolved small molar cusps, but this did not result in morphological convergence because these lineages simply maintained their current levels of affinity by virtue of the homoplasy; in other words, they evolved in geometric parallel and hence did not increase their resemblance to one another. Simpson's reading of parallelism has come to be the predominant one, however.

With the growth of "evo devo" over the last two decades, there has been increasing recognition that developmental homologies are important for explaining certain homoplasies. Distinguishing between parallel and convergent homoplasy is a way of acknowledging the different causal frameworks that underlie iterative outcomes in evolution. As mentioned earlier, some accounts of homology break entirely with phylogenetic patterns of morphology to label parallelism as a type of *homology*.[28] Even if parallelism smacks of a certain "homologyness" (as Gould suggests in *The Structure of Evolutionary Theory*), in my view it is preferable to retain a pattern-based account of homology, if for no other reason than to document the distribution of iterations in the history of life and to examine causal explanations of these patterns.

This still leaves two unresolved questions. First, how can we delineate parallelism from convergence, given the universal gradations of developmental homology? Second, to what extent do parallelisms map onto Gouldian repetitions? To the first question, we could stipulate some coarse-grained relatedness metric to separate parallelisms from convergence. For instance, we could hold that iterations between taxa at or above the class level are convergences, whereas homoplasies at or below the family level are parallelisms. But this simply brings us back to the limitations of using a blunt relatedness heuristic to make inferences about developmental causes, as discussed earlier. What is needed is a more targeted causal analysis of the developmental mechanisms underlying iteration. The rest of this chapter is devoted to this philosophical task.

3.2 Causal Accounts of Parallelism

If this section had a motto, it would be, "It is not the *extent* of developmental homology, but rather the *causal type*, that counts." The last two decades have witnessed great strides in our understanding of the molecular basis of development—or what evolutionary developmental biologist Sean Carroll has called "genetic dark matter"[29]—and its role in adaptive evolution. Although organisms sharing a homoplasious trait will always share varying degrees of developmental homology merely by virtue of their relatedness, the decision to categorize a homoplasy as a parallelism need not be arbitrary or rely on coarse-grained phylogenetic affinities. The key is to identify homology in the generators that are *directly causally responsible* for the relevant iteration.

The big question, of course, is what we mean by "directly causally responsible." If by this one simply means "an actual difference-making cause" (a technical term that will be unpacked later), then iterations underwritten by deep homologs could indeed be considered parallelisms. This seems to be how Gould interpreted iterations that implicate deep homologs, which he argued are better thought of as internally constrained parallelisms than proper convergences. If this is right, then it would render many paradigmatic cases of convergence, such as the evolution of image-forming eyes (see chapter 6), consistent with the RCT—cleverly pressing the strongest cases for robust replicability into the service of radical contingency.

Take for instance *Pax6*, a so-called master control gene that is involved in the development of camera-type eyes in vertebrates and cephalopods as well as in the compound eyes of arthropods. We know from gene "knockout" and substitution experiments that *Pax6* plays a causal role in the development of eyes in each of these distant phyla. The question before us is twofold. Does this deep homology convert the iterated evolution of eyes from convergence into parallelism, as Gould contends, and, regardless, does it undercut the experimental independence of the iterations observed? I will answer "no" to both counts.

There are several ways to gloss the proposition that some developmental mechanism is "directly causally responsible" for an evolutionary iteration. One way is to hold that a homologous developmental mechanism is directly causally responsible for an iteration when it is a *proximate* cause, rather than simply *a* cause, of the phenotypic similarity. We know that the misexpression of *Pax6* transcription factors can lead to the formation of ectopic eyes (e.g., on legs, wings, etc.) in both vertebrates and arthropods. We also know, thanks to gene knockout studies, that *Pax6* is crucial for the development of healthy eyes in both animal groups. Yet neither of these important facts tell us whether *Pax6* is a proximate developmental cause of eyes.

One approach is to use a "screening-off" test to determine whether some developmental factor is a proximate cause of a given phenotypic outcome. Two basic experimental manipulations can be carried out to test whether deep homologs like *Pax6* screen off other developmental factors with respect to the production of eyes (see figure 5.3).

The first experiment, which has already been carried out by Walter Gehring and his collaborators, is to insert the arthropod version of *Pax6* into, for example, the mollusk camera-type eye development cascade or vice versa and see what type of eye develops.[30] Low and behold, if we substitute the drosophila version of *Pax6* for its homologue in the octopod eye cascade, we get a normal octopod camera-type eye, not an insect compound eye. Likewise, in a series of ectopic eye studies, Gehring and his collaborators showed that the mouse version of *Pax6* can be used to induce compound eyes in various body parts of drosophila, such as on the tips of antennae. What this shows is that while *Pax6* is acting as an early trigger of eye morphogenesis, it is not causally responsible for the specific parameters of eye development in diverse eye-bearing groups. The macro-morphological arrangements of camera and compound eyes are directed not by *Pax6* but by thousands of nonhomologous genes that lie downstream from *Pax6* in the developmental cascades that lead to the various eye types.

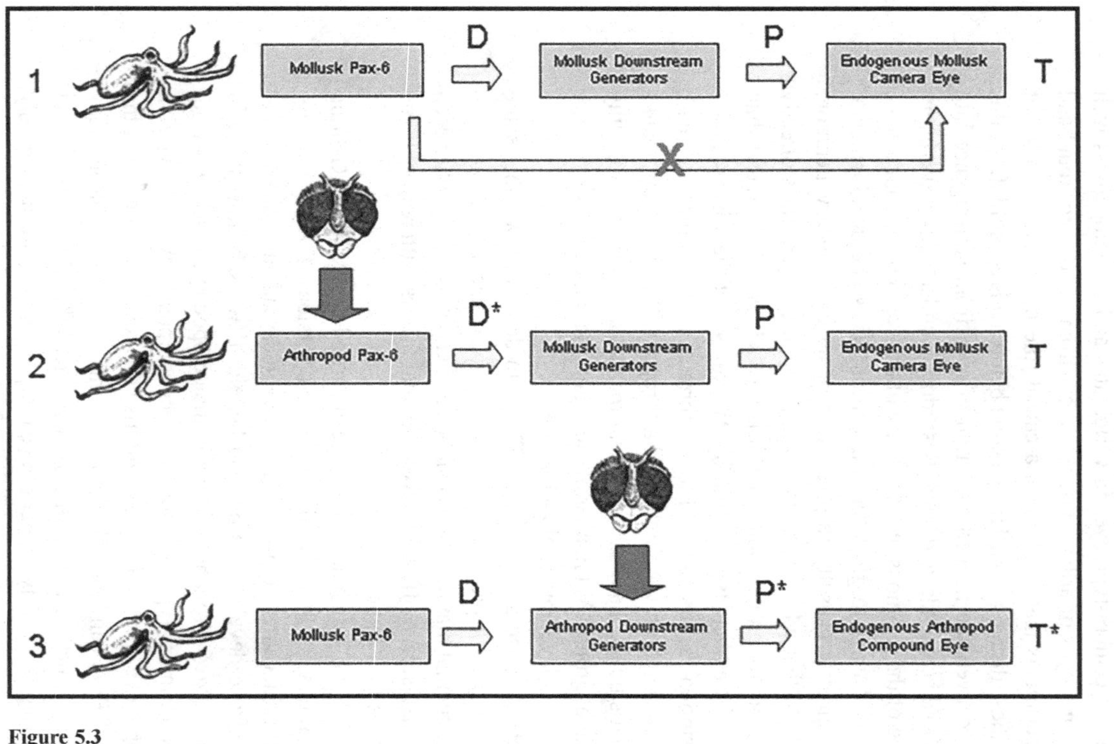

Figure 5.3
Three different manipulation scenarios show that *Pax6* is "screened off" by downstream developmental generators in the production of eyes. (1) In a normal cephalopod, *Pax6* (distal cause D) triggers downstream generators (proximate cause P), which produce a normal cephalopod camera-type eye (T). If either D or P is nonfunctional, T will not be produced, so both D and P are causes of T. (2) In the first (actualized) manipulation, an arthropod *Pax6* (D*) is inserted into the mollusk camera-type eye development cascade, resulting in the development of a normal cephalopod camera-type eye. (3) In the second (hypothetical) manipulation, normal cephalopod *Pax6* is left intact but arthropod downstream generators (P*) are substituted for their mollusk counterparts in the cephalopod developmental cascade. If the result is an arthropod compound eye (T*) rather than T, this would show that the probability of T given P is equal to the probability of T given D and P, and different from the probability of T given D; therefore, P may be said to screen off D with respect to the production of T. See text for details. From R. Powell, "Is Convergence More than an Analogy? Homoplasy and Its Implications for Macroevolutionary Predictability," *Biology and Philosophy* 22 (2007): 565–578.

This conclusion would be even more decisively confirmed by a second manipulation, which has not yet been performed and likely cannot be performed given the complex causal structure of gene-regulatory networks. This would be to replace the downstream, nonhomologous generators of the octopod camera-type eye with those of the arthropod compound eye (or vice versa), while leaving the mollusk *Pax6* intact. If in this scenario an arthropod eye develops (even if it is poorly integrated into the phenotype), then it will have demonstrated that the macroscopic arrangements of the eye are causally determined by their proximate downstream generators, which screen off *Pax6* and other upstream homologues. This second, fanciful manipulation need not be carried out, however, because ectopic eye studies in drosophila, in which the mouse *Pax6* is substituted for the insect *Pax6*, are sufficient to establish this asymmetrical causal relation, even if the downstream generators remain largely a black box. A similar causal analysis would extend to other *Pax* family genes implicated in skeletal-muscle differentiation, as well as to *Hox* genes involved in iterated appendage modifications.

Philosopher Treavor Pearce criticizes this screening-off approach to parallelism on the grounds that it only works in the case of topographically linear developmental pathways, when in fact most complex traits that are subject to evolutionary iteration will be produced by messy, nonlinear gene networks, not tidy causal chains. This seems right, although as we have seen there is an important temporal asymmetry between upstream components of the developmental cascade and their downstream sequelae. Gene networks may not behave in a simplistic linear fashion, but it is still plausible to say that certain upstream components in the cascade are screened off by certain downstream components with respect to some phenotypic outcome (e.g., the macromorphological arrangements of the eye), even if the downstream battery of causes is interactive and poorly understood.

Pearce also points out, again quite rightly, that we should not assume that the proximate cause of any given outcome is the most important cause or explanation of that outcome. Philosopher Christopher Hitchcock argues that pragmatics are an unavoidable aspect of explanation, and consequently that explanation need not track proximate causes. Hitchcock offers the following example to illustrate the point: consider the case where an individual's unprotected sexual activity leads to an infection with human immunodeficiency virus (HIV), which in turn leads to acquired immunodeficiency syndrome (AIDS).[31] Does the HIV infection (the proximate cause) provide a better explanation of why that individual has AIDS than does that individual's earlier unprotected sexual behavior (the distal cause)? The microbiologist and the public health worker might very well disagree on which explanation is best, given their differing interests. So which explanation is best will be relative to a given set of pragmatic interests and goals.

Elliott Sober is likewise skeptical that proximate causes are better explanations than distal causes, as this would imply that *synchronic* (morphological/physiological) explanations are categorically better than *diachronic* (evolutionary) explanations.[32] Yet, as Sober correctly points out, these two explanatory programs are largely orthogonal to one another, and so neither has a claim to explanatory primacy over the other. If the causal-explanatory import is relative to the research questions being posed and the pragmatic goals being pursued, then the question becomes whether the current investigation justifies prioritizing proximate developmental causes over distal ones in explaining evolutionary iterations. Perhaps the best way to appreciate the role of *Pax6* in eye development and evolution, for purposes of assessing the independence of eye iteration, is to shift the discussion away from screening-off relations and toward the different causal roles that developmental factors play in iterated evolutionary outcomes.

3.3 The Specific Causes of Parallelism

Another way of glossing the notion of a direct developmental cause is in terms of "specific causation." Recall that the problem we are confronted with in attempting to delineate parallelism is double-edged. First, all iterations will be underwritten by a certain degree of developmental homology, given the broad relatedness of life; conversely, rarely will there be homology in all of the developmental components that are necessary and sufficient to produce a given homoplasy. If there is no nonarbitrary way of privileging some developmental causes over others, then the causal conception of parallelism may be unworkable. If so, then we cannot use parallelism to parse convergence data in ways that allow it to play the evidential role that many want it to play.

However, when it comes to the pragmatic goal of explaining evolutionary iterations, not all developmental causes are of equal relevance. Building on Jim Woodward's counterfactual account of causation, philosopher Ken Waters offered a typology of causes that can be usefully pressed into service for thinking about causal explanations of development and evolution.[33] On Woodward's interventionist theory of causation, a causal relation is defined as a statistical relationship between variables as revealed by manipulations of their respective values.[34] More specifically, *A* is the cause of *B* only if some intervention with respect to *A*—that is, a change in *A*'s value while holding other variables constant—would result in some change in the value (or probability distribution) of *B*.

Waters gives the following example to illustrate how the interventionist theory of causation works. We can say that Mary's striking a match is a cause of it lighting, because if we hold all other variables constant, had Mary not struck the match, the match would not have lit. By the same token, the presence of oxygen in the room is also a cause of the match lighting, because had oxygen not been present—again, *ceteris paribus*—the match also would not have lit. To the contrary, the position of the planet Saturn is not a cause of the match lighting, because variation in Saturn's position does not affect whether the match lights, whereas Mary's striking of the match and the presence of oxygen in the room do.

It follows from this view that developmental factors are causes so long as their manipulation would have a statistical bearing on the outcome. On this view, *Pax6* is clearly a cause of eye development and therefore of eye evolution, providing the molecular substrate of the phenotypic structures on which selection has acted to shape the visual sense in disparate groups. However, not all causes of an outcome are of equal explanatory value. We can further distinguish the "actual difference-makers" from the vast set of "potential difference-makers" in explaining the variation observed across a population of outcomes. This enables us to say that Mary's striking the match is a distinct cause of it lighting in one case, and it not lighting in another. Other causes, such as the presence of oxygen, do not vary across scenarios in which there is an actual empirical difference in the outcome (though of course they could). Waters extends Woodward's counterfactual theory of causation into the realm of developmental biology, allowing us to say that a DNA sequence is the actual cause of RNA structure in a bacterium, even though RNA polymerase and other accessory proteins are necessary causes as well. This is because actual differences in DNA explain the actual variations in RNA sequence, whereas the accessory proteins do not vary. In situations where DNA and accessory proteins vary, both will be actual causes of RNA structure because both are causes that make a difference.

The distinction between potential and actual causes solves the broad developmental overlap problem for parallelism because it enables us to distinguish the actual developmental causes of the iteration from all potential developmental causes in the background, such as the structure of the DNA code, mechanisms of transcription and translation, cellular signaling machinery, patterns of embryogenesis, and so on. These causes of eye development and evolution do not vary such that they figure in explanations as to why some lineages have eyes and others do not. In contrast, *Pax6* is an actual (not merely potential) cause of eye development and evolution because it has been recruited for similar developmental functions in numerous eye-bearing animal groups. *Pax6* is certainly not sufficient for eye development, as it is present in various non–eye-bearing animal groups such as echinoderms, which suggests that it had a more primitive, nonvisual function (or, alternatively and less probably, that it lost its ancestral function in the vast majority of animal phyla—see chapter 9 for a detailed discussion). Nevertheless, even if *Pax6* is not *the* actual cause of bilaterian eyes, it is *an* actual cause of eye development and evolution, as it provides some of the shared molecular substrate on which disparate eye systems have been built.

Our causal ontology is not exhausted, however. Again following Waters, we can further distinguish between specific and general actual causes. "Specific causes" are those factors that, if subjected to a battery of interventions, tend to change the outcome in *detailed ways*. "General causes," on the other hand, merely determine *whether* or *when* an outcome will occur; they have no influence on precisely how it will do so, where "how" relates to the particular parameters of the outcome. Let us consider RNA synthesis again. On Woodward's causal regime, there is no basis by which to assign causal priority to DNA over and above accessory proteins with respect to the construction of RNA so long as both causes are necessary and actually vary. However, Waters's specific-general distinction enables us to say that DNA is a specific actual difference-maker with respect to RNA structure, because alterations in DNA engender particular changes in RNA sequence whereas interventions with respect to accessory proteins are limited to halting the synthesis process entirely or merely altering the rate at which it occurs. (The same cannot be said of RNA splicing agents, which do have specific effects on RNA sequence.) In the context of morphogenesis, alterations in the rate and timing of developmental events (heterochrony) can have profound morphological consequences and thus are not limited to Waters's "whether/when" criteria for nonspecific actual causation. The point, however, is that we can prioritize some token biological causes over others, even if we cannot a priori privilege certain classes of biological entities (e.g., genes) over others (e.g., proteins).

This philosophical machinery can also help resolve the second prong of the parallelism problem—namely, the fact that many or most cases of iteration implicate some proportion of nonhomologous developmental mechanisms in addition to homologous ones. Pressing this subtler causal typology into service, we can hold that a homoplasy is a parallelism when some of its developmental machinery is both homologous and causally specific. On this definition, an iteration is not precluded from being a parallelism merely because it is bound up with nonhomologous gene products or pathways that affect "whether/when" the iterated phenotype develops. For the same reasons, homology in regulatory regions of the genome, such as promoters, enhancers, silencers, and other factors affecting gene expression, will often (but not always) be insufficient grounds for parallelism. The more specific-causal homology that underlies an iteration, the more clearly that iteration will be a case of parallelism.

The notion of specific causation not only resolves the problems that arise from degrees of homology in the developmental machinery underlying iterations, but it helps to clarify some key cases. For instance, it explains why upstream homologues like *Pax6* fail the screening-off manipulation test: namely, they are general rather than specific causes of the iterated outcome. Deep homologs essentially act as triggers that merely allow traits like eyes or limbs to develop or not, but they do not specify *how* these traits should develop in three-dimensional space. Thus, *contra* Gould, the repeated activation of deep homologs like *Pax6* does not convert these cases of convergence into parallelisms.

Conversely, emerging examples of parallelism appear to satisfy the definition given above. Take, for instance, the iterated evolution of elongated and shortened pelvic spines in stickleback fish over the last 10,000 years in isolated North American glacial lakes—a pattern that has been interpreted by Beatty and others as in tension with the Gouldian view of life.[35] Stickleback populations have repeatedly assumed two ecomorphs in response to common selective pressures: a benthic short-spined form and a pelagic long-spined form. The former configuration reduces the chances of the fish being snagged by predatory dragonfly larvae in the shallows, while the latter increases the diameter of the fish so as to exceed the gape of many open-water predators. This adaptive feat has been accomplished independently numerous times by recurrent genetic modifications of hind-limb development—specifically, in regulatory regions of the pituitary homeobox transcription factor 1 (*Pitx1*) gene,[36] which plays a crucial role in pelvic fin development in sticklebacks and determines the specific length of the pelvic spine and girdle. The iterative evolution of *Anolis* ecomorphs (section 1.2) is also likely to implicate causally specific developmental homologs, given the close relatedness of the lineages.

There are also examples of what we might think of as "potential parallelism"—conserved genetic potential which, if activated under suitable selection regimes, could be the basis of future parallelism. Extant birds, for instance, have retained the ability to develop archosaurian teeth, which have been experimentally induced in chicken embryos.[37] Birds lost their dinosaurian teeth about 75 million years ago; similar losses of dentition occurred several times in theropod dinosaur evolution in favor of keratinized beaks. We can speculate that if in the future history of life mammals were to suffer a major setback and thereby surrender their predatory niches to birds, then dormant avian developmental programs (if they have not irreversibly deteriorated into pseudogenes) could be reactivated by natural selection and avian dinosaurs could once again express teeth. The same may be said for re-evolving claws in the place of wings, as occurs in juvenile forms of some living birds, such as the hoatzin. Given the conserved potential to produce archosaurian teeth and claws, birds could then "re-evolve" a guild of Mesozoic-grade "theropodomorphs"—not identical to theropods proper, of course, but close enough to deserve the moniker. Such iterated outcomes, as striking as they might be, would only be possible due to a shared set of specific genetic potentials and body plan constraints.

Birds may contain other dormant dinosaurian potentials, such as the ability to modify feathers into scales and vice versa. The standard view has it that feathers evolved from modified scales in dromaeosaurid dinosaurs (informally, "raptors") and were retained by birds; birds, in turn, lost their scales entirely but then re-evolved scales around their feet. Very recent findings from China, however, indicate the presence of feather-like structures in pterosaurs—archosaurs that evolved powered flight well before the dinosaurs.[38] Pterosaurs have long been thought to have a fuzz-like covering made up of simple "pycnofibers." However, the presence of bilaterally branched filaments of varying functionalities and colors in pterosaurs suggests that either feathers originated even earlier in archosaurian evolution, before the emergence of dinosaurs, or else functional feathers evolved multiple times in archosaurs (the group that includes dinosaurs, pterosaurs, and crocodylomorphs), perhaps from an ancestral pycnofiber fuzz. Either way, the developmental potential for scale-to-fuzz-to-feather transitions appears to be confined to archosaurs, with no similar feather-like structures having evolved in any other animal group. Unlike fuzzy coverings and gossamer skin wings, feathers may be a peculiar evolutionary innovation of a single clade that is only replicable, if at all, within that group.

On the other hand, some functionally constrained morphologies, broadly defined, can be achieved through multiple developmental pathways. For instance, teeth have in a sense revolved in birds: Pelagornithids are a fascinating group of large Cenozoic seabirds whose relatively fragile "teeth" lacked any mineralized dental tissue. Instead, they were bony outgrowths of the jaw covered in a hardened keratinized beak, designed more for grasping prey than for forcibly dismantling them.[39] In short, some repetitions may be due to loaded developmental dice that belie their independence and robustness. Just what proportion of repetitions can be characterized in this way is unclear.

3.4 Parallelism and Modal Robustness

Why is it that the same genes often act as specific difference-makers in the production of homoplasies, even in distant lineages? Francois Jacob famously likened natural selection to a tinkerer, working with the developmental odds and ends at hand to repurpose existing machinery for novel and unforeseen tasks.[40] But this only partly explains the repeated evolutionary cooptation of dormant developmental potential. A fuller explanation has to do with evolutionary constraints imposed by the causal topography of developmental systems (discussed in detail in chapter 2). There may be only a handful of genes that can produce major phenotypic effects, and thereby drive substantive adaptive evolution, without having deleterious consequences for other crucial features of the organism. Genes that are expressed "downstream" in the developmental cascade, or those that only affect a particular cell or tissue type, may be more amenable to selective modification. In contrast, "upstream" genes with ramifying pleiotropic effects that are not confined to a single tissue type may be limited to regulatory (rather than structural) modification, because any fundamental alteration in their function is likely to result in a nonviable organism. If this is so, then the vast majority of morphological innovations will be achieved through the regulatory activation of upstream networks and the structural modification of downstream components.

Given these topographic constraints, it is not surprising that two lineages working within a shared developmental plan will tend to arrive at similar solutions to common ecological design problems by drawing upon homologous generators. Although such cases of parallelism are evidence for a degree of modal robustness—namely, modal robustness within a given set of developmental parameters—they do not preclude the possibility that the set of developmental parameters within which selection acts to optimize form is itself radically contingent.

The final question is the extent to which parallelism, conceived roughly along the lines suggested above, maps on to the category of Gouldian repetitions. I have suggested that the more overlap there is in the causally specific developmental homology that underlies an iteration, the more that iteration resembles a clear-cut case of parallelism. And the more an iteration resembles a clear-cut case of parallelism, the less independent that evolutionary replication may be said to be, and thus the weaker its evidentiary force against the RCT. That said, parallelism is neither necessary nor sufficient to undermine the deep modal robustness of an iteration. It is not necessary because some Gouldian repetitions will not be parallelisms. Take, for instance, the evolution of ant mimicry, discussed earlier; even if this outcome were produced by different developmental pathways in different mimics (spiders, beetles, etc.), it would still be radically contingent so long as the ant phenotype itself is radically contingent.

Consider another case where some radically contingent set of developmental parameters makes a certain outcome a "Good Trick" in Dan Dennett's terminology (see chapter 4). An example might be the iterated evolution of retractable necks in turtles, which offers one of the best-known illustrations of competitive replacement.[41] The retractable neck has been achieved through at least two distinct specialized methods: flexing the neck sideways (*Pleurodira*) and flexing the neck into an S-curve (*Cryptodira*). Even if these iterations were produced by disparate developmental mechanisms, the reason why the retractable neck is an attractor in turtle morphospace is because of the shared (and quite possibly, radially contingent) parameters of the turtle body plan, combined with a common selective regime. If this is right, then such replications are Gouldian repetitions whether or not they are properly considered parallelisms. The same may be true of countless other iterations, such as the evolution of barracuda-like forms in both marine (sphyraenids) and freshwater (characins) lineages of ray-finned fish.

By the same token, parallelism is not sufficient for an iteration to be a Gouldian repetition because we can imagine an iteration that is at once a parallelism and a case of true convergence. For instance, if a particular developmental substrate is highly externally constrained and thus robustly replicable—such as the proteins involved in photoreception or light refraction—then the repeated specific cooptation of those conserved mechanisms in distant lineages would not undermine the evolutionary robustness of the iterations they underlie. This is why, in the final analysis, it is best to avoid tying our notion of Gouldian repetition too closely to the concept of parallelism, even though their theoretical motivations are closely allied.

The take-home lesson of this chapter is that many evolutionary iterations start from similar developmental conditions that bias the set of solutions that are accessible to selection. This in, turn, undermines the causal independence of the natural experimental replications observed and thus invalidates them *qua* tests of the RCT. Some of these internally constrained/facilitated iterations may constitute proper parallelisms, while others, even though they are produced by nonhomologous developmental mechanisms, are the product of broader body plan parameters that could easily have been otherwise and thus amount to Gouldian repetitions. Yet there is still much work to be done if we are to situate convergence within a broadly Gouldian view of life. The next chapter will delve deeper into this problem.

6 The Entanglement Problem

Science fiction has long been the stomping ground of grand thought experiments—a vehicle for imagining radically different biological, social, and technological worlds. Contemplating the properties of intelligent life on other planets has been a mainstay of this philosophical enterprise. Rarely, though, have depictions of extraterrestrial life broken free of our anthropocentric, and more broadly, vertebrate-centric framing of evolution. Science fiction has given us an endless parade of alien humanoids, some to copulate with (à la Captain James T. Kirk in *Star Trek*) and others to punch out (à la Captain Steven Hiller in *Independence Day*). In so doing, it has reinforced in the public mind the notion that macroevolution proceeds inexorably from "monad to man," to borrow the title of a searching book on progressivist narratives in evolutionary theory by historian of ideas Michael Ruse.[1] Few cinematic and literary works aim to disabuse us of the notion that life begins with simple microbes and reliably culminates in the grandiose complexity of humanlike bodies, minds, and societies. Works of "hard science fiction," such as the *Three Body Problem* trilogy by Liu Cixin, paint a similar "ascent-like" picture of technological evolution, which is assumed to proceed in a replicable series of stages.

Nevertheless, there are simply no good reasons for thinking that there is a law-like connection between the evolution of distinctively hominin or even vertebrate morphology on the one hand and the emergence of sophisticated forms of intelligence on the other. We know from our own example that the vertebrate body plan is *amenable* to intelligence, but we have no reason to project this body-mind trait cluster out into the universe or onto alternative deep histories of life on Earth. In fact, as we shall see in part II, the independent evolution of complex brains and cognition in several invertebrate body plans counsels against projecting the vertebrate body-mind cluster onto other life worlds. Proponents of the search for extraterrestrial intelligence (SETI), such as Carl Sagan and Milan Ćirković, as well as contingency theorists like Stephen

Jay Gould, are quick to acknowledge that complex forms of cognition could be instantiated in any number of theoretically possible body plans.

And yet there is a predilection to bundle morphological and cognitive traits together and to conceive of the entire "humanoid package" as a replicable evolutionary outcome. This "bundling problem," as we shall see, extends well beyond the humanoid fallacy and is merely one symptom, or perhaps cause, of the broader failure to disentangle radically contingent features of the living world from robustly replicable ones. Finding a solution to this "entanglement problem" is the holy grail of astrobiology, and I am under no illusions that it will be solved here. The aim of this chapter is more modest. In what follows, we will elaborate on the bundling and entanglement problems, consider how convergence data might bear on them, and then go on to sketch some potential avenues for making progress on these issues.

1. The Bundling Problem

The "entanglement problem," as I understand it here, refers to the great challenge of picking apart biological properties that are likely to be robust features of the evolutionary process from those that are likely to be radically contingent. The "bundling problem" contributes to, but is distinct from, the entanglement problem, in that bundling implicates specific cognitive biases or foibles of human psychology that motivate fallacious extrapolation from earthly evolutionary outcomes.

1.1 Cognitive Biases, Essentialism, and the Humanoid Epidemic

Before Darwin, theories of species were aligned with folk biological tendencies to essentialize the living world. In the Aristotelian tradition that dominated biology for centuries, species were conceived as immutable bundles of "essential" properties, with patterns of variation in morphology and behavior explained by these inner essences. On the Aristotelian view, variation within species is the result of environmental interference with a developmental process that would otherwise produce the essential characteristics of species. Work in cognitive psychology shows that humans are natural Aristotelians when it comes to thinking about biological species.[2] This explains why human minds are so resistant to the concepts of "population thinking,"[3] "lineage thinking,"[4] and "interactive development"[5] that are central to modern biological thought. The Modern Synthesis (Darwinian theory + Mendelian genetics) and the New Synthesis (Modern Synthesis + developmental biology) have exploded the assumptions and explanations of bioessentialism, which has been extinct in the life sciences for some time.

Yet a certain "evolutionary essentialism" still lingers, not only in the humanoid-dominated worlds of science fiction, but also in the annals of biological thought. There is a tendency among some researchers to view bundles of traits as co-occurring nonaccidentally, akin to the nomic clustering of properties in atomic elements. Complex bundles of traits that characterize familiar evolutionary outcomes, such as "vertebrates" or "mammals" or "hominins" (in quotes to denote nonhistorical kinds), are viewed as replicable features of the evolutionary process. This "neo-evolutionary essentialism" is defended or at least intimated in the work of Simon Conway Morris and George McGhee (discussed in chapter 4). This evolutionary essentialism is "neo" in that it is consistent with our modern understanding of evolutionary mechanisms like blind variation and natural selection, though as we saw in earlier chapters it is not adequately accounted for by these mechanisms.

The new evolutionary essentialism continues to find expression in the science fiction of our post-Darwinian world, which has widely (though not uniformly) continued to depict intelligent extraterrestrials as distinctively humanoid in form—as if the trait of higher intelligence is nomologically linked to a suite of hominin-like traits to which it is tethered by some unseen law. The humanoid cluster includes such traits as bipedalism, stereoscopic vision, prehensile hands, expanded cranium, and the broader tetrapodean and vertebrate body plans. This is best exemplified, perhaps, in films from *E.T.: The Extra-Terrestrial* and *Close Encounters of the Third Kind* to the recently expanded *Star Trek* and *Star Wars* universes.

The standing humanoid epidemic in science fiction is not due solely to the limitations of imagination or to the pragmatic demands of cinema. Nor can it be chalked up to an ignorance of the disparate body plans in which terrestrial intelligence has arisen. Rather, there is a predilection to infer from what did happen in evolution to what had to happen,[6] and this tendency is underwritten by the bundling bias. This predilection is properly characterized as a "bias" because, as we shall now see, it can lead to faulty inductions when traits co-occur accidentally.

1.2 From Humanoids to Heptapods

Ironically, the phenomenon of bundling is best illustrated by attempts to *avoid* the humanoid fallacy. Depictions of intelligent extraterrestrial life in books and films have been increasingly informed by evolutionary theory and animal cognition science, commendably avoiding the humanoid fallacy. In so doing, however, they have still fallen prey to the bundling bias. The depiction of aliens in the recent film *Arrival* (2016) nicely illustrates both the theoretical advances that have been made on this front, as well as the obstacles that bundling and entanglement continue to pose to imagining the nature of intelligent life on other worlds.

The bourgeoning field of animal cognition has begun to reveal complex cognitive mechanisms that have converged in distant animal lineages. Cognitive convergence has been demonstrated not only across mammals (such as between cetaceans and primates) and in vertebrates more broadly (such as between mammals and birds), but also between vertebrates and invertebrates (the topic of chapters 9–10). The movie *Arrival* was clearly influenced by some of this work, in particular research on coleoid cephalopod mollusks, whose extant forms include octopuses, squid, and cuttlefish. For reasons that will be explored in part II, coleoids are rightfully considered intelligent aliens on Earth: complex problem solvers with convergently evolved brains, the architecture of which is truly the stuff of science fiction.

The intelligent extraterrestrials portrayed in *Arrival* are called "heptapods" (Greek for "seven foot"). As their name suggests, heptapods have seven appendages, rather than eight like octopods (octopuses) or ten like decapods (squid and cuttlefish). But apart from this negligible modification, heptapods are, anatomically speaking, unmistakable coleoids. They boast several characters in the cluster of traits that are diagnostic of the coleoid clade: sucker-lined tentacles, jet-stream propulsion, camera-type eyes, and ink sacs from which they inexplicably squander a great deal of biosynthesis writing cryptic messages to humans (see figure 6.1).

The writers of *Arrival* deserve credit for declining to provide yet another vector for the cosmic humanoid epidemic, picking out a genuinely alien mind among us as a model for the evolution of extraterrestrial intelligence. For all its improvements, however, *Arrival* makes the same bundling error that gives rise to countless iterations of the humanoid. It lumps accidental traits with evolutionarily robust ones, and it treats the cluster as a single, law-like package, prompting a new bundle: the "cephalopoid."

The cephalopoid bundle is just as fallacious as the humanoid bundle and for exactly the same reasons. The error in each case is to entangle what are plausibly robust, interconnected outcomes of the evolutionary process, such as complex brains, cognition, and eyes (part II argues for law-like linkages among of these features), with accidental features of animal body plans, such as hominin-style bipedalism or coleoid-style jet propulsion. The portrayal of intelligent "insectoids," which have also cropped up from time to time in science fiction, emanates from the same bundling fallacy. In essence, what the standing humanoid epidemic and the incipient cephalopoid outbreak in science fiction depict is not the plausible result of extraterrestrial histories of life, but *alternative histories of the shape of life as we know it on Earth*—a worthwhile, but very different, philosophical pursuit.

Figure 6.1
Heptapods from the film *Arrival* (2016) represent a new but equally fallacious bundling of traits that make up the "coleoid," including sucker-lined tentacles, jet-stream propulsion, and ink sacs. Images from https://alienfandom.com and https://www.dailygrail.com, respectively.

There is an interesting parallel between the fallacy of bundling and the cross-cultural proclivity to postulate supernatural beings—or what cognitive scientists of religion call "minimally counterintuitive agents" (MCAs). MCAs are postulated beings whose ontology is on the whole familiar, but which also exhibit some properties that breach intuitive expectations for how entities in that familiar ontological domain will look or behave (such as a tree that talks).[7] In his impassioned defense of reason in *The Demon-Haunted World*, Carl Sagan remarks on the nonaccidental parallels between supernaturalism and folk depictions of alien visitations.[8] Despite more than a century-and-a-half of evolution and cognition research, the portrayal of extraterrestrials has achieved little more than a naturalized version of supernaturalism: there but for the point of an ear or a supernumerary tentacle (and a spaceship) goes intelligent life as we know it.

If any features of life have cosmic projectibility, then some of these features will be present on Earth. But which aspects, if any, among the complex outcomes of earthly evolution are the cosmically projectible ones? Calling out the bundling problem is only the first step, for it leaves us with the even more daunting problem of disentanglement.

2. Disentanglement

We can potentially make progress on the entanglement problem by consulting patterns of convergent evolution. However, convergence data will be unable to serve this evidentiary role so long as (1) they remain causally unparsed and

(2) the research questions to which they are put remain underspecified. Building on recent work that I have done with the philosopher of biology Carlos Mariscal,[9] I will discuss three ways of analyzing biological iterations that help to clarify their evidentiary relevance to the contingency debate. By focusing on the intersection of specificity, independence, and scope, we can identify iterations that are *structurally detailed* (rather than functionally disembodied), *modally robust* (rather than of shallow replicability), and *widely distributed in the living cosmos* (rather than vanishingly rare in the universe).

2.1 The Specificity of Iterations

Whether a given evolutionary outcome is radically contingent or robustly replicable will depend, first and foremost, on how the outcome is described. Functional properties that are characterized in vague terms and not tied to any specific structures will tend to be modally robust, though they will achieve this deep replicability at the price of being scientifically uninteresting. For instance, very generic phenomena like metabolism, thermoregulation, propulsion, mimicry, predation, and predator evasion are likely to be cosmically projectible properties of life, but not of the sort that undermine the radical contingency thesis (RCT). Universal generalizations on the basis of specific iterations, such as ant mimicry, would be interesting but probably false, whereas cosmic projections of mimicry *simplicter* would be true, but uninformative. Disembodied functions are of little use in making predictions about the evolution of specific phenotypes so long as there is an unmanageably large disjunction of configurations that can realize the same function. The more specific iterations are, the more they undercut Gould's claim that contingency dominates at the level of details "that have always defined the guts and soul of biology."[10]

2.1.1 Specifying the Specificity at Stake. A key point of weakness in the critique from convergence (CFC), discussed in chapter 4, is that it is unclear about the levels of specificity at stake in the contingency debate. Take, for instance, McGhee's formulation of the following regularity: "If any large, fast-swimming organisms exist in the oceans of Jupiter's moon Europa … I predict with confidence that they will have streamlined, fusiform bodies; that is, they will look very similar to a porpoise, an ichthyosaur, a swordfish, or a shark."[11] The ambiguity here lies with the phrase "look very similar." If McGhee is making the weaker, less interesting claim that fast-moving aquatic extraterrestrial "animals" will tend to assume fusiform shapes, then, as we saw in chapter 4, Gould would be unlikely to disagree. If, on the other hand, McGhee is making the stronger and more interesting claim that such organisms will tend to resemble sharks or porpoises in more detailed features *beyond their generic fusiform morphology*, then McGhee's claims will have outstripped the evidence. Squid have tapered spindle shapes but in all other respects (save for their camera eye) bear little resemblance to a shark or a porpoise. Because we only see "dolphinoid" convergence within the vertebrate body plan (i.e., in Mesozoic marine reptiles, cetaceans, and to a lesser degree fish), this tells us little about the evolutionary robustness of these specific outcomes across the whole of multicellular life. Given vertebrate body plan parameters, there may be only a subset of modifications that can achieve a fully aquatic lifeway, and thus dolphinoid iterations may amount to Gouldian repetitions.

Other robustly replicable outcomes may or may not lie at the level of granular detail that is at issue in the contingency debate. For instance, the filter-feeding apparatus of the wading flamingo is remarkably similar to that of the pelagic baleen whale,[12] suggesting that the structure of macroscopic filter feeding—which involves the passing of water through entrapment surfaces to simultaneously ensnare numerous prey items that are too small to be foraged individually—may be highly externally constrained (see figure 6.2).

Similar filter-feeding apparatuses have evolved in living invertebrates, and a grill-like facial structure for sieving plankton has been found among suspension-feeding anomalocarids in the Cambrian.[13] Physical constraints that determine the shape of suspension-feeding morphologies—such as the relation between the mesh size of the capture device and minimum prey size—allow for reliable predictions about morphology and ecology across disparate body plans. In all other respects, however, the outcome bundles that comprise whales, flamingos, and anomalocarids exhibit few convergent features.

This focus on specificity helps to avoid cross talk and bring out key points of disagreement in the contingency debate. Recall Gould's claim that universal generalizations about form and function do not speak to the details that fascinate biologists. Without weighing in on subjective matters of fascination, we can interpret the RCT as claiming that only *very generic* evolutionary iterations will be deeply modally robust, whereas we can interpret the robust replicability thesis (RRT) as contending that patterns of convergence support the existence of evolutionarily robust outcomes with *high levels of specificity*.

What makes one trait more specific than another? Generally speaking, we can say that one trait is more specific than another when it occupies a higher position in a nested multiple realization base of function. For example, blubber is more specific than insulation, which in turn is more specific than thermoregulation; vision is more specific than image formation, which in turn is more specific than sensory modality. Likewise, the antifreeze protein AFP type II is a molecule with a highly specific structure that has evolved several times for

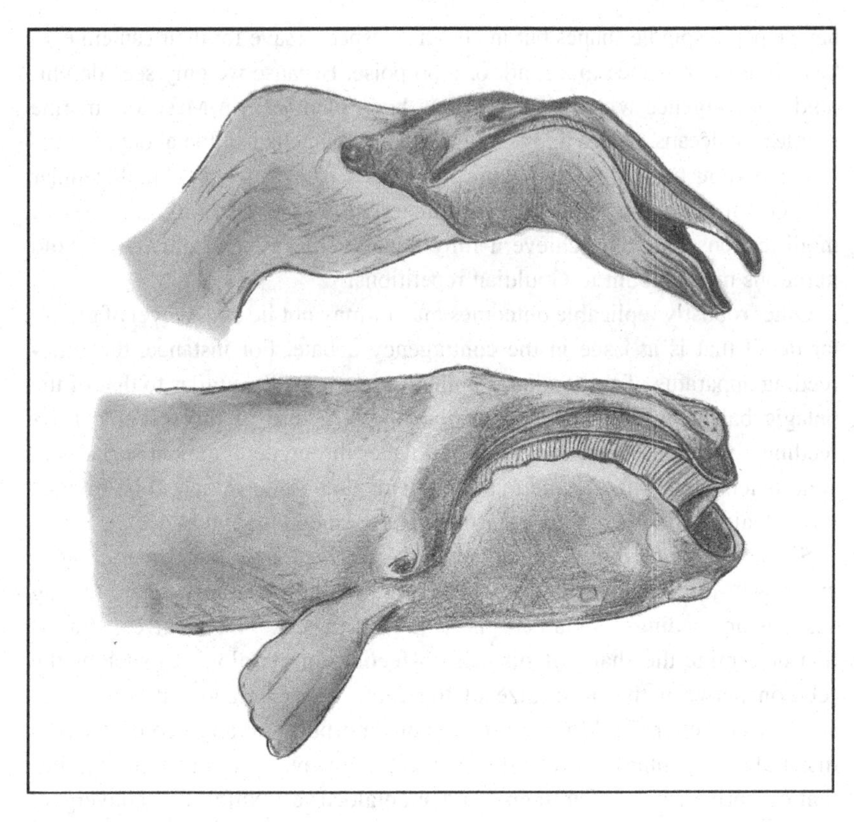

Figure 6.2
Filter-feeding morphologies of the lesser flamingo and right whale, respectively, which include a curved jaw to increase the surface area of the sieve, a narrow upper jaw and robust lower jaw, and a thick fleshy tongue. Redrawn from A. Feduccia, *The Origin and Evolution of Birds* (Yale University Press, 1999).

lowering the freezing temperature to prevent cell rupture. Because many proteins have ice-binding properties, "antifreeze" is a multiply-realizable function that has been achieved via selection on different genes and mechanisms in groups as disparate as animals, plants, fungi, and prokaryotes.[14]

All else being equal, the more generically an iteration is described, the "wider" its multiple realizability base, and hence the greater its modal robustness. Consider a sensory modality like hearing. The tympanal organ or "ear" in insects is a vibrating, drumlike structure that detects pressure waves generated by sound and converts acoustic energy into mechanical energy; it has evolved more than twenty times in the bodies, wings, and legs of various insect groups.[15] A structurally similar "ear" has evolved in vertebrates using completely different developmental machinery. However, if "hearing" is defined more broadly to include not only drumlike structures but any adaption to detect nanoscale vibrations and to transduce these stimuli into neuronal impulses that generate adaptive behavior, then the trait will be less specific but far more evolutionarily robust. Jumping spiders, for example, have long, sensitive hairs on their legs and bodies that activate auditory neurons, allowing them to "hear" predator or prey movements including airborne sounds at up to 3 meters (or 600 spider body lengths).[16] Similarly, pinnipeds (seals, sea lions, and walruses) have highly innervated whiskers that are used to detect fluid motion left behind by fish trails when foraging in low light.[17] Such vibrotactile adaptations could constitute "hearing" on very broad definitions. Likewise, the mechanosensory hair cells of the "lateral line" system in fish resemble those of the vertebrate inner ear. Bioacoustics in green plants, which initiate chemical defenses in response to vibrations caused by leaf chewing, may also meet broad definitions of "hearing."[18]

Generality, however, comes with a cost—two costs, in fact. First, as previously noted, the more generic the iterations, the less scientifically interesting they are, and the less they undercut the RCT. Compare the following three predictions:

1. Some marine-dwelling animals will evolve a subcutaneous layer of adipose-like tissue in order to conserve heat.
2. Some marine-dwelling animals will evolve a covering in order to conserve heat.
3. Some marine-dwelling animals will evolve adaptations in order to maintain viable internal temperatures.

The first description is highly specific; the second is less specific but still might admit of a limited set of realizations; the third, though by far the most evolutionarily robust of the three, is so vague that it might be realized by an indefinite number of structural, behavioral, and physiological configurations ranging from metabolic and insulatory functions to habitat choice and seasonal migration. Such generic regularities admit of no useful predictions in the abstract, and it is implausible to think that they would undermine the RCT.

Second, although massive multiple realizability can underwrite the modal robustness of functions, it also undercuts the law-like replicability of specific forms.[19] The more generically a functional outcome is described, the less reason there is to think that its realizations will be restricted to a small number of structural possibilities that could be the basis of a manageable set of laws. Establishing very generic laws of function would be little more than a pyrrhic nomological victory.

A final dimension of specificity relates to "hierarchical depth." It is widely acknowledged that convergence is level-relative: A structure can be convergent at the level of morphology while being homologous at the level of tissues, at the level of proteins that compose tissues, and/or at the level of genes that code for proteins that compose tissues. Alternatively, a trait can be convergent on all these levels simultaneously. Jeanine Donley and colleagues, for example, show that the convergence between tuna (teleosts) and lamnid sharks (e.g., makos) extends well beyond their fusiform shape to the details of mechanical design and muscle dynamics that underpin their shared force-transmission system.[20]

One might think that the "deeper" a convergence goes, the more it speaks to the power of selection to guide form toward certain evolutionary attractors, and hence the more decisively it contradicts the RCT. The fact that some iterations are more than "skin deep" is important. But if hierarchically deep convergence is due in part to shared developmental constraints operating at multiple levels simultaneously (as may be the case in teleost and lamnid fishes), then it will not be evidence of truly robust replicability. For this reason, hierarchical depth does not in itself have any clear-cut implications for the contingency or robustness of a given iteration.

Imagine that on some alien world creatures evolved that near-perfectly resembled human beings in gross morphological, cognitive, and behavioral respects. The implications of such a stunning replication for the RCT would not be undercut if we were to learn that these humanoids and their minds were constructed out of an entirely different molecular substrate. To the contrary, such multiple realizability would show just how cosmically accessible, and hence widespread, the humanoid outcome is. The point of this exercise is simply to show that whether an iteration occurs on multiple levels does not determine its evidentiary relevance to the RCT.

2.1.2 Macroevolutionary Bait and Switch. The level of specificity with which evolutionary outcomes are described can also affect one's reading of faunal turnovers, a key plank in Gould's bulwark of contingency (as discussed in chapter 2). During these transformative successions, lineages that were long subordinate are able to survive and thrive in the wake of a mass extinction perturbation, radiating into the empty niches vacated by previously dominant groups. Bats and birds filled the niches of recently extinct pterosaurs, whales and sharks filled the niches opened up by the extinction of the mosasaurs (pictured on this book's cover), late Cretaceous mosasaurs filled the vacated niches of other Mesozoic marine reptiles like pliosaurs and ichthyosaurs, and so on.

RRT proponents handle the dynamics of extinction and faunal turnover by arguing, in essence, that certain evolutionary outcomes are "overdetermined" in the sense that surviving lineages will tend to repopulate the emptied attractors in morphospace. At the extreme end of the macroevolutionary overdetermination view is the notion that deep replays will reproduce what Jonathan Losos, in a book published during the writing of this one, refers to as evolutionary "doppelgängers."[21] Losos contemplates that perhaps, following Conway Morris (but assigning more tentative credences), had the nonavian dinosaurs remained safe in their niches, the world today would be populated by humanoid doppelgängers of dinosaurian origin; or, if the dinosaurs did go extinct in the familiar way, then perhaps some other mammalian lineage would have evolved a humanoid doppelgänger.

The evidence weighs against this possibility, but we will reserve a discussion of the human(oid) case for the final chapter of this book. The general point is that describing the repopulation of niches in faunal turnovers as "iterations" runs the risk of a macroevolutionary bait and switch. In most cases, the regions of morphospace that are filled by successor lineages bear only superficial similarities to those of their predecessors. Sea lions are not all that similar in their ecomorphospace occupation to plesiosaurs, and wolves are not very similar to dromaeosaurs, even if these lineages occupy similar respective niches and share some generic morphofunctional specializations for common lifeways.

Other faunal turnovers in which morphospace repopulation is impressively specific may turn out to be Gouldian repetitions. Take the striking patterns of iteration within successive placental mammal faunas throughout the Cenozoic (discussed in more detail below). The reason why saber-toothed, mole-like, horse-like, shrew-like, hippo-like, cat-like, and wolf-like niches exist is because they represent local adaptive attractors or optimums within a mammalian body plan interacting with recurring selective environments. Local mammalian optima are illustrated by the mesonychians—a predatory group of even-toed ungulates that have been referred to as "wolves on hooves." The mesonychians emerged not long after the dinosaur extinction and were the first major mammalian predators on Earth. The "wolf" ecomorph may be a global optimum for mammals that transcends Linnaean order-level body plans, but it is a local optimum from the standpoint of tetrapods more broadly. A similar point could be made about morphological convergence in large theropod dinosaurs such as tyrannosaurids, allosaurids, and ceratosaurs, all of which were working within same theropodian ground plan.

As fewer developmental parameters are shared, however, the similarities between iterations become increasingly attenuated. Bison are only minimally reminiscent of the ceratopsian dinosaurs (such as the iconic triceratops) that crowded the plains of North America 65 million years ago, even if they occupy a similar niche and share some features and behaviors thanks to a conserved tetrapodian body plan. This raises a further issue about the nature of niches themselves: it is misleading to say that "successor lineages evolve to fill the niches that are vacated by their predecessors." For this assumes that niches are wholly external to evolving lineages and simply waiting in the world to be filled—an assumption that the phenomenon of Gouldian repetition calls into question.

2.2 The Independence of Iterations

As we saw in the last chapter, many iterations do not meet the validity conditions for natural experiments in convergent evolution because they fail to satisfy the "independence" criterion. The reason why many iterations fail to satisfy the independence criterion is that although natural experiments in convergence are isolated in space and time, they often begin from highly similar developmental starting points that "load the evolutionary dice," as it were, toward particular iterative outcomes. Such iterations are consistent with, and perhaps even corroborative of, the Gouldian view of life. How should we analyze the independence dimension?

2.2.1 A Two-Pronged Analysis. The independence of iterations can be assessed using a two-pronged analysis. First, it should aim to identify *causally specific developmental homologs* that are involved in the production of an iteration. Second, it should aim to identify features of the broader body plan that delimit the space of locally optimal solutions *even where no specific developmental homologs underwrite a given iteration.* Chapter 5 considered some of the conceptual and methodological challenges that confront these analyses and argued that these challenges must be overcome if convergence is to serve as evidence in the contingency debate.

Generally speaking, we can say that a small number of highly independent evolutionary repetitions offers stronger evidence against the RCT (and in favor of deep replicability) than does a large number of nonindependent repetitions. By the same token, iterations with a narrow phylogenetic distribution (e.g., iterations limited to taxa at or below the Linnaean class-level) are more likely to be nonindependent iterations (i.e., Gouldian repetitions) than they are to be cases of true convergence. This is because iterations among closely related taxa are more likely to deploy homologous, causally specific developmental machinery. Even where iterations among closely related taxa are produced by different developmental machinery, they will often ensue from conserved body plan parameters that are, for all we know, radically contingent.

On the other hand, if an iteration has a wide phylogenetic distribution, such as "viviparity" (the development of embryos within the body of the female), this indicates that an evolutionary outcome may transcend the peculiar body plan parameters of particular lineages. Other iterated traits with high levels of specificity are found in clades as developmentally disparate as plants and animals. An example is urticating hair, barbed bristles connected to cells that produce poisonous or acrid fluids, which are designed to break off and lodge in the skin of predators. Convergences across plants and animals—which represent two distinct origins of complex multicellularity—hints at cosmic-grade projectibility.

2.3 The Scope of Iterations

Iterative regularities also vary with respect to their "scope," or the ubiquity of the conditions under which they obtain. If convergent regularities are limited by conditions that are astronomically uncommon in the universe, then these regularities will have a very narrow scope, even if they emerge in a "law-like" way wherever these conditions obtain. We might distinguish between two broad types of limiting conditions: (1) "external" factors, which include features of both the abiotic and biotic environments, and (2) "internal" factors relating to the evolvability of the trait in question. Some external limiting conditions, like the presence of light or water or the existence of a continental crust that supports a terrestrial habitable zone, are likely to be common among life worlds and thus do little to restrict the scope of convergent regularities. Others, like the presence of particular coevolving lineages or specific internal developmental parameters, may be exceedingly rare among life worlds. Because we are not currently capable of identifying all the limiting conditions that underpin a given iteration, our pragmatic aim in assessing scope should be to identify the most restrictive limiting conditions.

For example, some iterations, such as the pollination syndromes discussed earlier, depend on the existence of both the angiosperm flowering platform and morphologically specific pollinator lineages, two very narrow limiting conditions that undermine the cosmic projectibility of these iterations. Contrast this with, say, convergence in vessels to transport water, which evolved many fewer times than pollinating morphology but which, given the invariant physics of aqueous chemistry, are likely to be universally projectible. Likewise, conditions for the evolution of the fusiform shape include active locomotion in a fluid environment and the internal structure necessary to maintain a sturdy, macroscopic shape, conditions that are probably widespread in environments that are conducive to the evolution of animal-grade multicellularity.

Developmental constraints that restrict the scope of convergent regularities are perhaps the most important class of internal limiting conditions, for two reasons. First, they explain why Gouldian repetitions are neither projectible to other life worlds nor stable across deep replays of the tape of life on the Earth.

Second, cast in in their positive role as facilitators of innovation, developmental constraints explain why many patterns of iteration exist in the first place and how they underwrite impressive degrees of replicability across "shallower" rewinds of life's tape.

2.3.1 Replay Depth. We saw in the early chapters of this book that Gould's Cambrian thought experiment presupposes the existence of multicellular animal life, and asks how the tape of animal life would unfurl were it played again. If an iteration hinges on the origin of radically contingent developmental parameters, then it will be stable across rewinds no "deeper" than the emergence and crystallization of those developmental parameters. In the case of iterations confined to lower level taxa, this stability will tend to be shallow; in the case of iterations between higher level taxa within-phyla, it will be moderate; and in the case of iterations across phyla or even the whole of multicellular life, the replay depth stability will be deeper still, verging on the cosmic.

Consider, for example, the striking set of convergences between marsupial and placental mammals, and within placentals, on a range of ecomorphologies throughout the Cenozoic (see figure 6.3).[22] Given internal limiting conditions of the mammalian body plan and certain recurring ecological conditions, a range of specific mammalian forms may be highly replicable over a 55 million year period of evolution, even if they are not deeply robust features of vertebrate life on Earth, let alone of "animals" on other potential life worlds.

The same is true of the impressive convergences in aquatic bird morphology that occurred during the post-Cretaceous diversification of birds, including between penguins and auks (flightless relatives of puffins), gulls and albatrosses, grebes and loons, diving petrels and auklets, flamingos and herons—all cases that illustrate the adaptive optimization of a common bird platform to a diverse but finite range of ecological tasks (such as foot-propelled diving, fish capture, transoceanic flight, wading, and so on).[23] The iterative evolution of stickleback and *Anolis* ecomorphs, discussed in chapter 5, are indicative of replicability across even shallower rewinds of the tape of life. In contrast, the convergent evolution of (for instance) salinity tolerance across the archeal and bacterial prokaryotic domains reflects a far deeper replay depth stability (though it scores low on specificity). In sum, a key problem with the CFC is that it has failed to distinguish iterations of plausibly universal scope from those with a narrower set of limiting conditions.

2.4 Case Study: Image-Forming Eyes

Let us apply our analysis to an iterated regularity that appears to score highly along all three dimensions: the image-forming eye. The motivation for this particular example is not merely illustrative, however. The evolution of vision

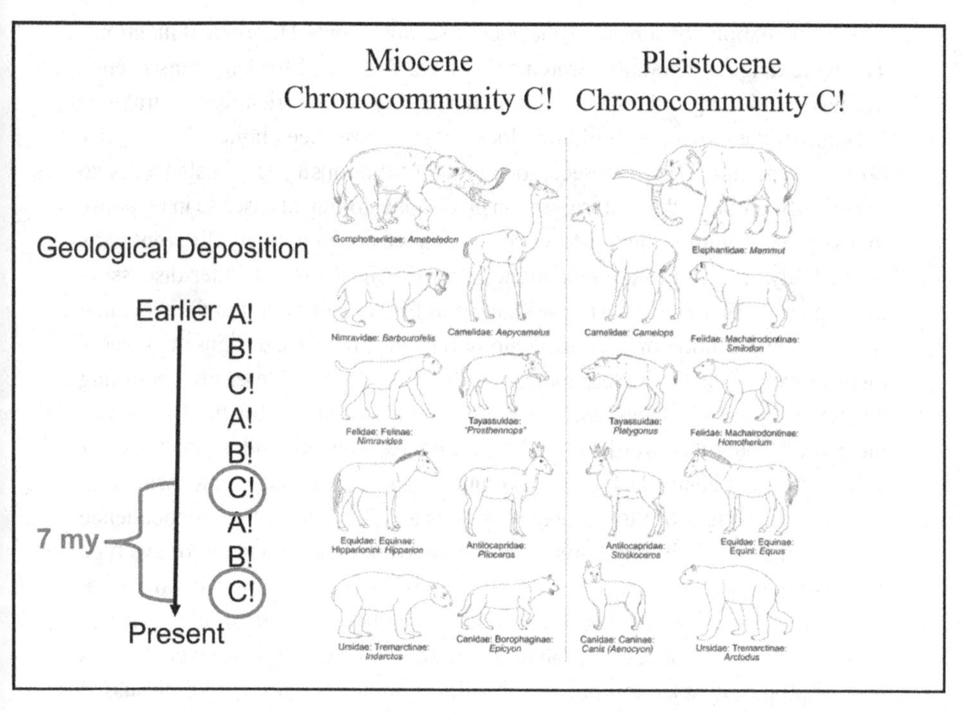

Figure 6.3
Mammalian ecomorph assemblages (or "chronocommunities") from the Miocene and Pleistocene, separated by millions of years of intervening time dominated by different faunal assemblages. Adapted from L. D. Martin and T. J. Meehan, "Extinction May Not Be Forever," *Naturwissenschaften* 92 (2005): 1–19.

will take center stage in part II, where we explore the iterated evolution of cognitive complexity and its link to image-forming perception.

First, the image-forming eye earns a high score on specificity. It includes variations on two distinct configuration types—camera and compound—which reflect the two fundamental ways that optical components can be configured so as to produce spatial vision in organisms. The first involves adding more photoreceptors to a single apparatus (resulting in the single-chambered eye), while the second involves multiplying the apparatus itself (resulting in the compound eye). Each of these eye types are structurally specific, entailing particular cornea, lens, and retina configurations.

Second, image-forming eyes are likely to have a very broad scope given the universal laws of optics, the ubiquity of the light stimulus, the fact that light waves are generally shorter than the ecologically relevant objects that bend them (permitting the extraction of various types of information about those objects), and the availability of substrates that can be readily coopted for optical functions. The wide range of crystallin proteins that have been used for lens transparency in diverse animal groups speaks to the availability of eye-friendly substrates.

Some of the molecular building blocks of eyes have been honed for globally optimal function. The photoreceptor pigment "rhodopsin," for instance, is so beautifully optimized that it triggers an electrophysiological cascade in response to a single quantum of light. Moreover, eyes have arisen in many different parts of the body, including in animals without proper heads (see further discussion in chapter 7). As biologists Michael Land and Eric Nilsson put it in their seminal work on the evolution of eyes, the "enormous range of sizes, designs, and placement of eyes … gives a clear indication that eyes can evolve easily, recruiting whatever tissue is at hand, and become superbly optimized for the lifestyle of the bearer."[24] In other words, both the external (ecological) and internal (evolvability) limiting conditions on eye evolution appear to be rather unconstrained.

Third, the weight of the evidence supports a high score on the independence dimension as well, although this issue is somewhat dicier. Each major eye type has arisen numerous times both within and between distant animal phyla, including vertebrates, mollusks, arthropods, annelids, and cnidarians.[25] Incredibly, a camera eye has even evolved in single-celled eukaryotes, where it has been sculpted by selection out of subcellular organelles, as will be discussed further in chapter 7 (see figure 6.4).[26]

Complicating the independence score, however, are two factors—deep genetic homologs and cell-type homologies—that might lead one to the conclude that eyes are either monophyletic or at best Gouldian repetitions. First, as discussed in chapter 5, certain deeply conserved developmental mechanisms, such as *Pax6*, are implicated in all known cases of eye morphogenesis. Yet, as we saw there, this does not undermine the independence of the iterations in which *Pax6* is involved. The reason why independence is preserved in crucial cases of iterative eye evolution is that convergent eye-bearing phyla do not share causally specific developmental mechanisms that specify the gross morphological contours on which judgments of convergence are based. Nor do these converging lineages share other internal constraints that shape the specific parameters of eye evolution. Indeed, *Pax6* is present in echinoderms (e.g., sea urchins), even though this phylum does not have even the simplest eye. It is true that *Pax6* betrays the signature of history, for unlike rhodopsin it is not strongly functionally constrained. However, there is nothing to suggest that *Pax6* imposes limitations on eye evolution that would convert these stunning iterations into Gouldian repetitions.

Second, there is a growing consensus that the two cell types involved in photoreception—cilia (found in vertebrates, using c-opsin photopigments) and

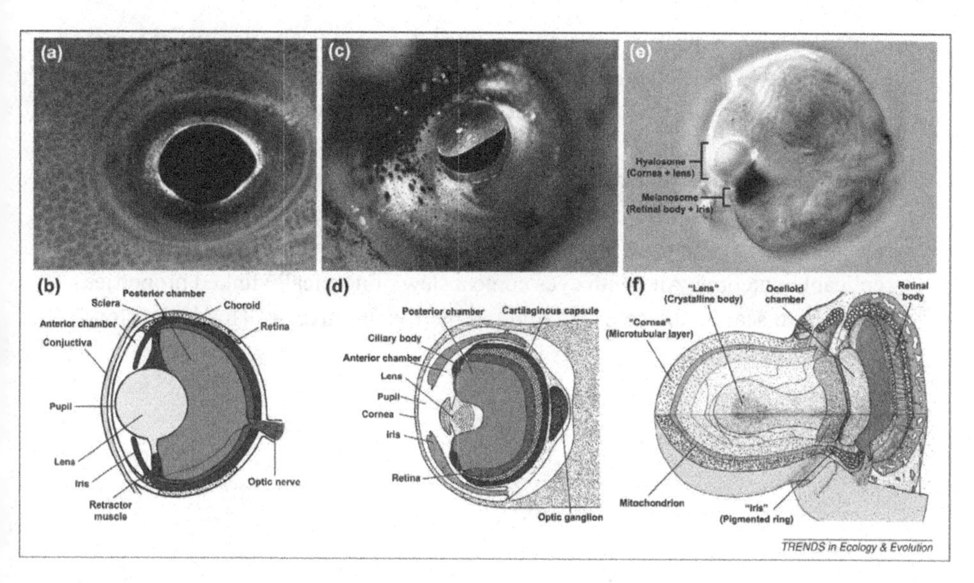

Figure 6.4
Photo and schematic pairs of the convergent multicellular camera eyes of (a and b) vertebrates and (c and d) cephalopod mollusks. (e and f) The subcellular camera eye of single-celled warnowiid dinoflagellates, composed of modified endosymbionts including a plastid (which forms the retina) and mitochondrion (which forms the cornea). From B. S. Leander, "Different Modes of Convergent Evolution Reflect Phylogenetic Distances: A Reply to Arendt and Reznick," *Trends in Ecology & Evolution* 23, no. 9 (2008): 481–482.

modified microvilli (found in invertebrates, using r-opsin photopigments)—have a single origin. It turns out that some invertebrates have both types of photoreceptors, indicating that photoreceptor cells were present in the ancestor of all bilaterians;[27] in addition, there are structural homologies between these cell types that indicate they are "sister cells,"[28] with one having been derived from the other.[29] Nevertheless, as we shall see in chapter 9, the preponderance of the evidence suggests that the last common bilaterian ancestor did not have complex eyes (or a head or brain), even if it possessed *Pax6* and both types of photoreceptors.[30] If it is correct to characterize the last biliaterian ancestor as eyeless, then describing complex eyes as monophyletic (i.e., having a single origin), as some researchers have done, is misleading.[31] For even if all eyes share some cell type homology, they would still be genuinely convergent at the level of macromorphology.

Furthermore, as with *Pax6*, there is no reason to think that conserved photoreception constrains the specific morphological parameters of eye evolution. The extent to which photoreceptor monophyly undercuts the evolutionary robustness of complex eye evolution depends on whether the origin of photoreception is itself a radically contingent event. A singular origin is not in and of itself demonstrative of radical contingency: if an innovation like photoreception occurs early in the history of a clade (such as Eukaryota), we might very well expect it to be conserved over deep time and repeatedly deployed in diverse subclades.

At the very least, we can say that if photoreception is taken among the relevant initial conditions, then the evolution of complex eyes is a robustly replicable outcome. And with eyes come a slew of nomically linked properties that help to seal the place of the mind in the living universe. This is the story to be told in part II.

Coda to Part I: Convergence at the Grandest Scales

Complex multicellularity is a precondition for nearly all of the convergent features of bodies discussed in the previous chapters. The iterated features of bodies were sculpted by selection out of highly differentiated cell types that comprise muscle, nervous, epithelial, connective, and other tissues. The evolutionary robustness of complex multicellularity in turn hinges on several earlier events in the history of life on Earth, such as the origins of the eukaryote cell and before that the simpler prokaryote cell. Several key transformations in the form and organization of life on Earth were crucial for subsequent evolutionary step-ups in hierarchical complexity; if any of these transformations are radically contingent, then all subsequent evolutionary outcomes that causally depend on them will be radically contingent as well. It would be remiss, therefore, for us not to consider the contingent nature of major transitions in the history of life.

1. Major Transitions in Evolution

In *The Major Transitions in Evolution* (1997), evolutionists John Maynard Smith and Eörs Szathmáry drew up a list of key shifts in the history of life and attempted to connect these events with a common theoretical thread.[1] Some of these transitions involve a shift from autonomous replication to cooperative group replication, in which fitness is "transferred" from lower level units to the group level, thus forging a new evolutionary individual. Other transitions involve the emergence of new systems of inheritance that allow for molecular and cellular divisions of labor, such as the separation of transcription and translation and epigenetic systems that direct cell line differentiation.

Whether Maynard Smith and Szathmáry succeeded in theoretically unifying the diverse events that they identify as major transitions is questionable.[2] In addition, some theorists have objected to their anthropocentric—and, more broadly, eukaryote-skewed—picture of macroevolution, which obscures the bacterial dominance of life on Earth from its inception to the present day.[3] To paraphrase microbiologist Ford Doolittle, living worlds will always be run by and for microbes. Nevertheless, the transition from microscopic to macroscopic life was of great ecological significance to the history of life on Earth, and it is especially crucial to the inquiries that preoccupy us here. Some events Maynard Smith and Szathmáry describe are the organizational building blocks upon which later complex morphological and functional iterations would be built.[4] The replicability of certain major transitions, especially basal ones like eukaryogenesis and multicellularity, is thus pivotal to the overall story to be told in this book.

2. Iterated Transitions

Iterated major transitions enable us to avoid inductive problems that stem from observer selection effects (discussed in chapter 1). Any observer reflecting on the prospects of extraterrestrial life need not hail from a planet on which some of the major transitions that led to the observer—such as the free-living cell, multicellularity, and society—have multiple iterations. Singular transitions, on the other hand, create problems for the robust replicability thesis (RRT) precisely because they are susceptible to observer selection effects.

In recent work, Szathmáry notes that some major transitions are subject to "recursion" (read: iteration), but he does not draw out the theoretical import of this pattern, nor does he consider the implications of the lack of recursion in other transitions for the prospects of a law-like theory of macroevolution.[5] Some major transitions in hierarchy that are perquisites for the evolution of body and mind are subject to clear iteration while others appear to be singular.

For instance, multicellularity has arisen more than two dozen times across distant branches of the eukaryote tree of life, including in animals, land plants, fungi, and various algae (brown, red, and green), with about one out of every twelve of these simple origins giving rise to complex multicellularity.[6] Basic multicellularity requires several generic innovations, including cell-to-cell adhesive properties, systems of intercellular communication, and reproductive specialization. This last innovation is especially important, for as evolutionists Rick Michod and Matthew Herron have argued, there may be a limited number of ways that transitions to complex multicellularity can occur such that fitness is transferred from lower-level to higher-level entities.[7] One such iterated mechanism is the evolution of the *germ–soma division,* wherein some cell lines specialize in reproduction while others specialize in the various nonreproductive functions of the organism. Once reproduction becomes specialized in this way, all cell lines of the body come to share a collective fate, and a new evolutionary individual is "born."

Multicellularity per se does not seem particularly hard to evolve and it appears to transcend phylogenetic constraints. It has arisen many times in eukaryotes as well as in both domains of prokaryotes to produce what might loosely be called "bacterial bodies," complete with specialized modes of adhesion, signaling, and reproduction.[8] Nevertheless, only two multicellular lineages—animals and green plants (and to a lesser extent, fungi)—have achieved levels of cellular specialization that support truly three-dimensional body plans.

Whether animal-style multicellularity, which includes a body with differentiated tissues and a digestive cavity, arose once or multiple times is less clear. If radical contingency is going to outflank the evolution of complex bodies and minds, the origin of the animal gut is a promising point of attack. The metabolic challenge for animal evolution lies in switching from a basic eukaryote mode of feeding—intracellular phagocytosis (wherein nutrients are "swallowed" and digested within the cell)—to one in which cells give up their feeding capacity in order to form specialized digestive tracks that serve the higher-level unit of selection.[9]

The lack of clarity about how many times animal-grade complexity evolved is due to two uncertainties: (1) the phylogenetic position of the enigmatic "Ediacaran fauna" (a global multicellular fauna that preceded the Cambrian explosion by some 40 million years), and (2) the uncertain phylogenetic position of ctenophores (comb jellies). Paleontologist Adolf Seilacher has argued that discrepancies between Ediacaran and Cambrian modes of body construction, behavior, and ecology point to the Ediacaran fauna as a separate origin of multicellular animal life.[10] This view remains controversial, however, and the now prevailing interpretation, based on recent morphological and molecular fossil analyses, is that some animal-like Ediacaran forms, such as *Kimberella* and *Dickinsonia*, are in fact "stem" taxa of later metazoans, harkening back to the cladistic reinterpretation of the Cambrian critters discussed in chapter 2.[11] Even so, if comb jellies turn out to be more basal than sponges, as some recent research suggests, then this would indicate multiple origins of guts as well as of nervous and motor systems (more on this in chapter 9). If, on the other hand, sponges are the most basal animal, then we are looking at a single origin of some of the key bilaterian traits on which the subsequent shape of animal life would be built. Our picture of animal evolution continues to evolve, but at present it points to a single origin of animal-grade multicellularity, even though organ-grade multicellularity per se appears to be a robustly replicable outcome.

Coloniality is another major transition listed by Maynard Smith and Szathmáry that has numerous iterations, in this case across multicellular and unicellular lineages. Among colonial organizations, caste-based eusociality comes closest to the formation of a new evolutionary individual. Eusocial societies are less physically integrated than complex multicellular organisms, but in some cases they are highly differentiated and involve reproductive specialization into sterile and reproductive castes that resemble the germ–soma differentiation in transitions to multicellularity. Although humans are capable of cooperative feats that surpass even those of the eusocial insects (see the coda to part II), though it does not seem that human societies do not rise to the level of a new individual or superorganism. In any case, the evolution of eusociality, though ecologically significant[12] and deeply iterated,[13] does not bear directly on the evolution of bodies and minds in multicellular organisms, so we will not explore it further here.

3. Singular Transitions

Let us assume for the sake of argument that some major transitions are law-like, or even "natural kinds" in the parlance of philosophy of science. If these outcomes hinge on earlier transitions that are radically contingent, then we have a situation where certain law-like generalizations only obtain given certain accidental antecedents. For reasons discussed in chapter 3, this scenario not only undermines the cosmic replicability of these transitions, but in addition it is plausibly consistent with the radical contingency thesis (RCT) depending on what the RCT takes as its initial conditions. Apart from the evolution of key animal features (which we will revisit in part II), which major events, given their lack of iteration, might undermine the cosmic projectibility of complex bodies?

Chapter 1 dispelled some concerns surrounding the apparently singular origin of the DNA code and offered reasons to think that protocells evolved more than once on Earth. The transitions to complex multicellularity and society also look to be evolutionarily robust, though there are several complications that may affect our credence in this conclusion. One is the prospect, already mentioned, of a single origin of animals. Another is the apparently singular origin of eukaryote-grade *cellular* complexity. "Eukaryogenesis" refers to the emergence of complex cells packed with distinct organelles dedicated to specialized functions. The origin of the eukaryote cell was a necessary condition—structurally, informationally, and energetically—for the emergence of all subsequent levels of hierarchy built upon it, including complex multicellular lineages like animals and green plants. Eukaryogenesis is thus a linchpin in the prospect of a cosmic biology, and yet it appears on all counts to be a nonrepeated event.

There are two features of the evolution of the eukaryote-grade cell that are hard to reconcile with the RRT. First, as just noted, it arose only once in the entire history of life.[14] Second, it took nearly 2.5 billion years to do so after life had gained a permanent foothold on the planet. There are two alternative explanations of eukaryogenesis. The first, known as the "autogenous" hypothesis, holds that the eukaryote cell was fashioned from a single prokaryote ancestor through accumulated mutations that produced invaginations of the cell membrane, which resulted in the various organelles such as the endoplasmic reticulum and golgi. This hypothesis is now largely rejected for the mitochondrion as well as for the chloroplast in plant and algal cells, which are both understood to be the descendants of once free-living bacteria that were acquired in endosymbiotic events. The second, "exogenous" account of eukaryogenesis hypothesizes that an ancestral prokaryote—probably an archaean—somehow (despite the lack of phagocytosis) engulfed and maintained a bacterium that would become the mitochondrion, and that this energetic upgrade in turn set the stage for eukaryogenesis. Proponents of this hypothesis, such as biologists Nick Lane and William Martin, reason that the increased energy production that came from the addition of mitochondrial membranes was crucial for the energetically costly expression of larger, informationally capacious genomes, which in turn supported the internal differentiation of the eukaryote cell.

What can we say about the contingency or robustness of this event, which served as the scaffolding for countless morphological convergences over the subsequent history of life? If we presuppose, as many believe, that the endosymbiotic acquisition of the mitochondrion was the chief evolvability hurdle to genomic and therefore morphological complexification, then there is some reason for optimism. For although the mitochondrion was acquired only once, the endosymbiotic acquisition of free-living prokaryotes for metabolic functions has occurred over and over again, including not only in the case of the chloroplast (mentioned earlier) but also in many secondary and tertiary plastid endosymbioses as well.

What's more, the citric acid cycle itself—the metabolic pathway through which aerobic respiration is carried out in the mitochondrion—may have evolved numerous times in prokaryotes due to its global adaptive efficiency,[15] using the smallest number of chemical steps to produce the highest energy yield.[16] The biochemical constraints on aerobic respiration, along with phylogenetic variability in citric acid cycle intermediates that appear to have been recruited from different subcellular functions, raise the possibility that aerobic respiration has arisen multiple times. This would not be surprising, given that prokaryote metabolic innovation is among the most versatile and evolvable properties in the biosphere.

What about the great delay before the emergence of eukaryotes? And why did prokaryotes not subsequently re-evolve eukaryote-grade complexity? One reason for the delay in the origin of the eukaryote cell, which could also account for its subsequent nonreplication, is an evolutionary catch-22: engulfing a fellow prokaryote (like the mitochondrion) requires sophisticated cellular capacities, but by Lane and Martin's calculations these cellular capacities could not have evolved without the added mitochondrial power. How this happened remains one of the great mysteries in biology, though there is recent evidence to suggest that phagocytosis evolved before the acquisition of the mitochondrion.[17] A much better understanding of the ecological conditions, genetic innovations, and metabolic mechanisms that precipitated these events and precluded their iteration is needed before we can draw any definitive conclusions. In the meantime, the replicability of our complex living world hangs in the balance.

4. Summary of Part I

Let us now distill the main takeaways of part I. Patterns of convergent evolution offer a promising source of evidence for adjudicating the contingency debate. They also provide a way around observer selection effects that would otherwise prevent us from projecting earthly outcomes out into the cosmos. Until they are parsed in accordance with their underlying causes, however, convergence data cannot establish the replicability of animal forms across deep rewinds of the tape of life on Earth, let alone support Simon Conway Morris's stronger assertion that "the nodes of [morphospace] occupation were effectively predetermined from the big bang."[18] This is because some iterations are consistent with, and could be read as corroborating, a broadly contingent view of life in which internal constraints set the channels for evolutionary innovation, repetition, and predictability. Once we begin the project of parsing convergent events, we find that many iterations are the result of body plan parameters that are likely to be radically contingent and thus do not speak to the replicability depths and global optimization narratives that Gould's contingency thesis denies.

At the same time, we also found that some iterations that implicate shared developmental mechanisms (like deep homologs) are nevertheless genuine cases of convergence. In disentangling robustly replicable features from radically contingent ones in "bundled" evolutionary outcomes, we can identify repetitions that are more clearly in tension with the RCT. Radical contingency remains a very real possibility, however, for some singular major transitions and key innovations out of which animal forms have been built. At this stage in the investigation, what we can say with confidence is that some specific forms (though not body plans)—and many more generic forms and functions—are robust probably across the evolution of animal-grade morphological complexity, quite possibly across the evolution of eukaryote-grade cellular complexity, and perhaps across the evolution of life itself. We have thus begun to build a case for a cosmic biology of bodies. Part II will mount a similar case for minds.

II CONVERGENT MINDS

7 Convergent Ways of Seeing

Is mind a radically contingent accident of life on Earth, or is it a replicable, law-like feature of any living world? This is the central question taken up in part II. The main conclusion will be that the RCT (or what survives of it from part I) does not preclude the evolution of complex cognition or even conscious minds. The rhetorical journey that will lead to this conclusion begins here, with the emergence of image-forming sensory modalities.

1. Image-Forming Sensory Modalities

This chapter will argue that due to the invariant physical structure of the universe, the emergence of image-forming sensory modalities (ISMs) is both robustly replicable *qua function* and highly constrained *qua form*. These constraints support detailed, law-like generalizations that transcend the peculiar features of body plans, whether on Earth or beyond. Iterated ISMs are not therefore Gouldian repetitions, despite the fact that they make use of some conserved genes and developmental machinery as might be expected for any evolutionary system that builds on itself over time. The next three chapters will extend this argument to the emergence of minds.

The chief function of sensory modalities is to tie informational inputs from a particular energy source in the environment to behavioral outputs that enhance fitness. Energy inputs are recorded by cellular receptors and then transduced into action potentials or other types of signaling cascades that culminate in behavior. Among the various forms of physical energy in the universe, only waveform energy is capable of providing the rich, real-time information necessary for organisms to construct detailed, three-dimensional "scenes" of their surrounding world—an ability that I will call "seeing."

Life has found a way to form rich scenes using all known forms of waveform energy from which ecologically useful information about the distribution of objects and their properties can be gleaned. Only a few types of waveform energy carry this information: light (within certain spectral bounds), sound, and electromagnetic fields. Other energy stimuli, such as chemical gradients that serve as the basis of olfaction, do not carry information that can underwrite a three-dimensional perceptual world that supports sophisticated locomotion, navigation, predation, and other cognition-mediated behaviors.

Olfaction can be informationally rich: just think of all that a dog can divine about a prey or conspecific from a single scent trail. But olfaction is informationally impoverished when it comes to forming real-time images of objects that comprise scenes. Scents remain even when the object that left them is long gone. Chemosensory systems are robustly convergent,[1] but they do not allow organisms to "see" in anything representationally or phenomenologically analogous to the object-based scene perception that characterizes the visual sense.[2] Chemosensory, tactile, and other nonimagistic senses do not support the spontaneous construction of a three-dimensional model of the world with the subject at its center—a model that updates in real time with a dynamically changing world, recording changing distributions of waveform energy and measuring them against internal states of the body and its position in space.

The cognitive side of this equation will be examined in the next three chapters. The focus here is on the sensory modalities themselves and how they tap into universal energy forms to enable organisms to see. Seeing, it will be argued later, is the evolutionary gateway to more complex forms of cognition and behavior, and even to consciousness itself. In what follows, the evolution of three distinct modes of image-formation will be explored—vision, echolocation, and electrolocation—each of which has been subject to iteration.

2. Vision

Of the three broad ISM types, vision is the most well-known, the most subject to iteration, the widest in scope given the ubiquity of the light stimulus and the availability of optical substrates, and the most independently derived, as indicated by the unparalleled phylogenetic breadth of its iterations. Light contains ecologically useful information in almost all habitats on Earth and is used by many organisms to fuel their metabolism. It is not surprising, therefore, that the ability to sense and respond to light is ubiquitous in the living world.

2.1 Phototaxis

All domains of life track sunlight, using signal transduction networks to link the perception of light with adaptive movement.[3] Phototrophic organisms, or those that harvest sunlight for metabolic energy, must be capable of optimally positioning their photosynthetic apparatus in relation to the sun, taking into account both the intensity and spectral quality of light as well as current physiological states of the organism. Nonphototrophic organisms may also benefit for a variety of ecological reasons from positioning themselves at some optimum distance with respect to a light source, such as to avoid damage from ultraviolet radiation, to situate themselves within hosts or symbiotic partners, or to orient toward the shoreline or ocean surface.

The diverse set of simple motility systems that are guided by the detection of light fall under the umbrella term "phototaxis." The three main components of phototactic systems are light detectors (which are molecularly diverse), signal transduction networks (involving proton gradients or electron transport chains), and motility mechanisms (such as the convergent flagella of bacteria and archaella of archaea). At the miniscule scale of prokaryotes, phototactic systems are generally (though not exclusively) limited to sensing and responding to light gradients, rather than to light vectors. Gradients, whether made up of light or chemicals, are rarely steep enough for prokaryotes to detect differing chemical concentrations at two ends of the cell, which are often no more than a couple of microns apart. This limits most prokaryotes to "temporal" rather than "spatial" perception of gradients—that is, to "two-dimensional" perception guided by whether concentrations are increasing or decreasing over time.[4]

Some of the mechanisms underlying phototaxis in prokaryotes are well established. One is the "biased random walk," in which the cell randomly tumbles or changes direction by flagellar switching, with the probability of flagellar switch reduced during movement up or down a steep light gradient depending on whether the response is photophilic or photophobic. True three-dimensional phototactic navigation—that is, orientation along a proper light vector—evolved at least eight times independently in eukaryotes, aided by the independent evolution of eyespots (known as "stigmata").[5]

In green algae, these eyespots are composed of hexagonal arrays of lipid globules—pigmented structures that provide directional shading and mirror-like focusing (refraction) of light onto adjacent photoreceptors during axial rotation; these photoreceptors trigger signaling transduction systems that link to ciliary beating, which then produce movements toward or away from the light source. All eukaryotes that independently evolved the ability to orient and move toward a light vector have done so using this same structural configuration. Some of the pigments used in stigmata, such as rhodopsins, have evolved independently as well,[6] increasing the "hierarchical depth" of phototactic convergence.

2.2 True Spatial Vision

The focus here is not basic phototactic systems or lensless pigmented eye cups, but rather focus-capable eyes which, in conjunction with sophisticated information processing capabilities, are capable of producing a "visual scene": a panoramic, dynamically updating representation of spatiotemporally distributed objects in the surrounding environment. The cognitive generators of the visual scene will be examined in the next chapter. Here, the discussion will focus on the basic principles of eye construction and the informational resources they make available to organisms that can process them.

"Spatial resolution" refers to a sensory system's ability to perceive adjacent points in space as separate points. The smaller the distance or angle between these points, the more acute the spatial resolution. Information gleaned from true spatial vision is orders of magnitude greater than that achieved in the transition from nondirectional to directional phototaxis. High-resolution spatial vision supports the effectively instantaneous detection of fine differences in light stimulus "gradients" across a sweeping spatial expanse. It does this aided by lenses, stacked photoreceptors that simultaneously monitor different elements of the scene, and screening pigments that provide directional and spectral sensitivities.

Although there are many types of eye that support some degree of spatial resolution, only variations on camera and compound eye configurations have achieved genuine panoramic vision with high angular detail that can discern and track objects over space and time. Camera-type eyes, which consist of a single optical unit with refractive elements that focus light onto a retina, have evolved in distant animal phyla, including vertebrates, mollusks, arthropods, annelids, cnidarians, and even single-cell eukaryotes. Compound eyes, which consist of numerous optical units ("ommatidia"), each with their own refractive components (lens and cornea) that focus light from a small region of space onto their corresponding photoreceptor cells, are less phylogenetically disparate than camera eyes but far more prevalent due mainly to the unparalleled success of the arthropods. The highest resolution compound eye—that of the dragonfly—consists of an array of over 30,000 individual lenses. Although simple lens eyes have evolved from lensless pigmented eye pits in numerous animal phyla, only four major animal groups (spread over three different phyla and two superphyla) have evolved high-resolution vision that supports active, dynamic lifestyles: crustaceans/insects, spiders, vertebrates, and cephalopod mollusks.

The vertebrate camera eye and the paired compound eyes situated in the head of arthropods probably arose once in each respective phylum and have been conserved in these groups ever since. Homology of the arthropod eye remains contested, however; either many groups of arthropods lost compound

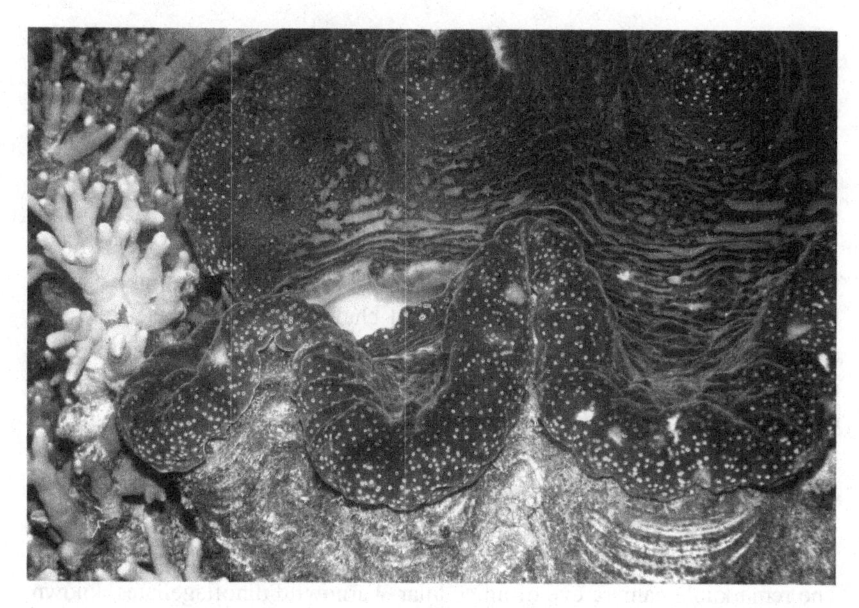

Figure 7.1
Giant clam (*Tridacna gigas*) with pinhole eyes in blue-green circles distributed throughout the mantle. Photo taken by author at Lizard Island, Great Barrier Reef, Australia.

eyes (including chelicerates, myriapods, and ancestral ostracod crustaceans), or similar compound eyes evolved in disparate arthropod groups.[7] What we know is that the arthropod compound eye arose first, in the base of the Cambrian, and the single-chambered eye evolved in craniate vertebrates not long thereafter, with cephalopod mollusks eventually following suit.

Probably the most striking example of convergence on the camera eye is that between coleoid cephalopod mollusks and vertebrates (see chapter 6), which have comparable visual acuities.[8] Vertebrates and mollusks in all likelihood share an eyeless, brainless, and relatively immobile common ancestor that lived in the Precambrian more than 550 million years ago (see chapter 9 for an extended discussion of this inference). Precisely when the cephalopod camera eye evolved is not known with certainty, though it is likely to have arisen not too long after the origin of the vertebrate eye, perhaps in the common ancestor of ammonites and belemnites in evolutionary competition with primitive fishes. Simpler eyes have evolved repeatedly in the mantles of bivalve mollusks, such as the pinhole eyes of giant clams which number in the thousands and serve as coarse-grained intruder alarms (see figure 7.1).

Spiders and other chelicerates are also an interesting case. Despite their close relation to insects and crustaceans, chelicerates generally have single-chambered eyes like vertebrates and cephalopods. The highest level of spatial resolution achieved by any chelicerate is the principal eyes of the jumping spider (salticidae), an active diurnal hunter that exhibits some of the most cognitively sophisticated behaviors observed in arthropods (discussed in detail in chapter 10). The chelicerate camera eye may have evolved through a process of simplification on the ancestral arthropod compound eye, or it may have arisen de novo. Anyone who has scoured the East Coast beaches of the United States knows that horseshoe crabs have prominent compound eyes, like those of insects; and horseshoe crabs, as it turns out, are basal chelicerates—the clade to which spiders and other arachnids belong. It is thus possible that chelicerates tweaked the existing compound eye design once or many times to produce a single-chambered eye. Convergent reduction is a far cry from convergent origin, and thus the macroscopic eye homology between insects/crustaceans and chelicerates could undercut the "independence" (*sensu* chapter 6) of this particular eye iteration.

Complex eyes are not limited to animals or even to multicellular organisms. The remarkable camera eye of unicellular Warnowiid dinoflagellates, known as the "ocelloid," has been fashioned out of endosymbiotic subcellular organelles (see figure 6.4 in chapter 6). For instance, in the green algae *Erythropsidinium*, interlocking plastids (which ordinarily carry out photosynthesis) form the retinal body, and mitochondria have been modified into a cornea-like structure.[9] Although the dinoflagellate eye has a lens and screening pigment, like the camera eyes of cubozoan cnidarians (the close relatives of jellyfish we encountered in chapter 6), the retina is placed too close to the lens to be in focus. As a result, these eyed dinoflagellates cannot direct their attention toward a detailed image or, more importantly, process any image that could be projected (they are, after all, but a single cell!).

What is the fitness advantage of such surprisingly sophisticated optics in brainless organisms like unicellular eukaryotes, cnidarians, and bivalves? The answer seems to be an increase not in resolution but in sensitivity, which suffices for the low-resolution visual tasks that these lineages face, such as coarse-grained orientation and predator evasion. The inability to resolve fine spatial details may actually be beneficial if the goal is to perceive large, stationary structures in the environment.[10] Greater resolution—often at the expense of sensitivity (as discussed later) and coupled to the increased processing power of brains—allowed for the richer forms of information extraction that underwrite the most active lineages of life.

Which then came first, the eye or the neurons necessary to process visual information? Unlike the infamous chicken-or-egg conundrum, this is not a pseudo problem because we have strong reasons to believe that neurons predated true eyes (see chapter 9), even though the evolution of eyes and the complexification of nervous, proprioceptive, and motor systems is likely to have gone hand in hand. Whereas convergent eyes abound in nature, only a handful of lineages have turned these optical devices into the input sides of sophisticated information processors, triggering runaway coevolutionary feedback between brains, bodies, behavior, and ecology. Activities like the dynamic pursuit of prey and navigation by fine-grained landmarks require vision with high spatial resolution—an angle of acuity of no more than a few degrees.[11] Only the compound eyes of arthropods and the camera eyes of vertebrates, cephalopods, and spiders have achieved this level of angular detail.

2.3 Robustness, Invariance, and Evolvability

Modeling work suggests that eyes can evolve very rapidly, geologically speaking—going from a mere patch of photosensitive cells to a high-resolution camera eye in less than half-a-million years, with each step plausibly conferring a fitness benefit by increasing sensitivity and/or resolution of the eye.[12] The evolutionary robustness of eyes was analyzed at length in the previous chapter. The argument there was that the repeated recruitment of deep homologs (like *Pax6*) in disparate eye-bearing groups does not convert eye iterations into Gouldian repetitions or parallelisms because this conserved molecular substrate does not constrain the space of gross eye morphology in specific ways. By contrast, the macromorphological arrangements of visual ISMs are highly constrained by the properties of light: all image-forming eyes must, for example, have lenses that are made of high-refractive transparent material that operate within the laws of refraction. The lenses of organisms will, of course, be constructed out of proteins rather than glass; however, as Land and Nilsson eloquently put it, "it is chemistry rather than physics that distinguishes biology from technology."[13] This bold statement is applicable not only to vision but to all three types of ISM.

Whether camera or compound, Earth-based or extraterrestrial, any visual organism must contend with a number of inexorable trade-offs imposed by the laws of optics. One such trade-off is between resolution and sensitivity: the more densely packed the photoreceptor "pixels," the better the spatial resolution; but greater density entails smaller photoreceptors, which reduces the capacity of individual photoreceptors to detect photons, and thus reduces the overall sensitivity of the eye and its ability to discern contrast. Similarly, the more photons an eye collects thanks to a wider pupil, the greater the distorting effects of spherical and chromatic aberration, and thus the blurrier the image. The evolutionary trade-off between resolution and sensitivity has been resolved in different ways, depending on the ecological context. These optical constraints and trade-offs, like the structural configuration of eyes themselves, are not contingent on the peculiar developmental parameters of lineages but rather are biological universals with which any visual life world must contend.

Vision is by far the most ubiquitous ISM on Earth, but it is not the only one. Two other image-forming sensory apparatuses have evolved, exploiting two additional waveform energies: echolocation (which uses sound) and electrolocation (which harnesses electromagnetism). These "alternative" modes of image-formation, which are explored in the remainder of this chapter, are far less common than vision, probably because they are harder to evolve. One reason they are harder to evolve is that they require high-intensity energy emissions. Active focused emissions, whether in the form of sound or electromagnetic fields, have significant energetic and anatomical requirements over and above those associated with passive reception and transduction. Not only must organisms have metabolic rates that could support persistent high-energy emissions, they must also have the anatomical structures necessary to produce these outputs, such as lungs capable of emitting concentrated bursts of sound, or electricity-generating organs capable of steady electromagnetic discharges.

That said, it is somewhat misleading to characterize vision as a "passive" sense. Constructing detailed visual scenes requires scanning movements of the eyes and adjustments of the perceiver's position relative to its surrounds so as to move objects in and out of focus (regions of acute retinal resolution) and to view them from different angles. Vision is also cognitively active, in that various properties of the scene must be stitched together by dedicated regions of the brain—cognitive activities that loom just as large, as we shall see, in echolocation and electrolocation. In contrast to alternative ISMs, however, vision is accomplished through the "passive" detection of ambient light that is generated mainly by the sun and nighttime stars and reflected by objects and terrain. A hypothetical equivalent to active emission in the case of vision would be animals in the ocean midnight zone evolving a bioluminescent apparatus that produces a flashlight-like beam of light, in combination with sophisticated optical and neural apparatuses to sense, transduce, and process reflections of that beam. The closest nature has come to this is the loosejaw, a deep-sea fish with an organ that produces red light and vision attuned to the red spectrum, which is invisible to most prey in the midnight zone. This might sound like a fantastical adaptation, but it is precisely what has been achieved by alternative ISMs like echolocation and electrolocation in which active emission capabilities evolved in tandem with detection and processing apparatuses. Let us now turn to these alternative modes of seeing.

3. Echolocation

Echolocation is a perceptually rich discrimination and navigation technique accomplished through the use of active biosonar. Echolocation is best regarded—

functionally, representationally, and phenomenologically—not as an advanced form of hearing but as an advanced form of "seeing" (see chapter 8 on the delineation of senses). It is employed as a primary image-forming sensory modality, not as a secondary sense relied upon in circumstances in which primary sensory systems are ineffective. Echoic abilities have evolved independently at least five times in the history of life, including in three orders of mammals and two orders of birds. An even greater number of animals have evolved the capacity to generate sounds and listen to their echoes in order to locate objects or sense terrain in dark environs. For instance, blind subterranean mammals, such as moles, appear to use seismic vibrations produced by striking their heads on tunnel roofs for orientation and navigation.[14] Even blind humans have been shown to successfully employ rudimentary echolocation techniques, recruiting brain regions that are normally associated with the processing of visual information.[15]

Only in bats (microchiropterans) and whales (odontocetes), however, do echoic abilities approach vision in their imagistic resolution. Because ensonified objects tend to have higher acoustic impedance (resistance to sound waves) than air or water, they absorb virtually no sound energy, which is instead reflected and scattered in much the same way that light (electromagnetic radiation) is absorbed and scattered by objects in the environment. Water has a higher acoustic impedance than air, but animal hard parts (such as skeletal structures) and soft organ systems have a higher impedance than water, and thus return echoes that make them "visible" even when cloaked under sediment. Features of the emission echo, or what is known as the "acoustic impulse response," contain information about the size, shape, velocity, and even the fine-grained textures of ensonified objects. Different reflecting points of a complex, irregular target will yield slightly different range values, creating a profile that is interpreted as shape.

Pioneering work by James Simmons in the 1970s showed that bat acoustic images are so precise that they can distinguish between targets separated by distances well under 1 millimeter in three-dimensional space—an angle of acuity that is comparable to the camera eye of vertebrates and cephalopods.[16] The acoustic impulse response is compared to the emission both in time and frequency domains in order to extract detailed spatial information about the surrounding world. Fine spatial details of objects and surrounds are resolved by detecting miniscule differences in echo return time. Bats can apparently discriminate time differences on the astonishing order of 10 nanoseconds—or 10 billionths of a second.[17]

Such fine-grained time perception is presumably done outside of conscious experience, with detailed acoustic images simply "presented" to echolocating bats and whales much as visual images are spontaneously presented to organisms through "cognitively encapsulated" visual systems that do most of their imagistic processing outside of conscious awareness. Sound objects are typically experienced as temporally (but not spatially) extended.[18] The range differences in echolocation are so infinitesimal that they leave no time for the experience of sound. Phenomenologically and representationally speaking, echolocation is a thoroughly spatial sense. It is used by bats and dolphins not only to identify objects and avoid collisions with terrain but also for short and middle-range navigation by identifying echoic landmarks and constructing acoustic maps.

3.1 Echoic Object Recognition

One of the central research questions in the initial decades of biosonar research was the extent to which echoic animals are forming true, three-dimensional acoustic images of objects. In other words, it was initially unclear whether echolocating bats and whales were "seeing" with sound or simply listening to echoes and adjusting their behavior accordingly. The results of early forced choice discrimination experiments with echolocating dolphins were called into question when it became clear that humans could discriminate between objects as successfully as dolphins on the basis of echoes alone when these were played back at the frequency range of human hearing. There is a world of a difference, cognitively and phenomenologically, between listening to echoes in order to discriminate among objects or track their movement over space and time, and spontaneously seeing three-dimensional objects through the topographic analysis of their acoustic profile in a way that is functionally analogous to vision.

It is now well established that richly echoic animals, such as bats and dolphins, are not discriminating objects on the basis of the echo alone; instead, they are carefully comparing the emission with the echo in order to generate a detailed, three-dimensional image of object shapes, textures, positions, and velocities, and relations to other objects in the environment. Bat echolocation supports activities as precise and nuanced as aerial insect hawking and other prey interception techniques, navigation through cluttered foliage, and plucking small prey items from complexly textured surfaces, all at speed and in complete darkness.[19] Bats must be capable of intercepting flying insects that are only a few millimeters in diameter while avoiding obstacles and interference by other call-emitting bats. These prey items, moreover, do not sit idly by while being captured—many deploy sophisticated antiechoic countermeasures such as jamming techniques or the initiation of erratic flying maneuvers when high-frequency ensonifications are detected. Intercepting prey on the wing in physically and acoustically cluttered environments is a daunting task that would be impossible if echolocation were not providing an acoustic analog to the integrated, object-structured visual scene.[20]

Experiments in the laboratory have shown that bats can discriminate among objects based on their detailed shapes and textures[21] irrespective of their size,[22] solely on the basis of acoustic images. Acoustic "flow fields" of reflected sound form around the bat's head as it flies, providing real-time information about the bat's movement and the orientations of objects and terrain in much the same way that visual flow fields guide the complex flight patterns of birds. Background objects are not mere clutter to be avoided but rather form part of the bat's topographically rich acoustic scene. In fact, bats form multilayered acoustic maps of their spatial world: lower resolution maps for coarse-grained navigation, which allows them to beeline back to their roost after a meandering foraging translocation (sometimes over 100 kilometers), and higher resolution maps for finer-grained spatial tasks such as cave navigation or the aerial capture of insects.[23]

Evidence for holistic object recognition in echolocating dolphins is also compelling.[24] Perhaps the strongest such evidence comes from "cross-modal recognition" studies in dolphins, whereby an object initially inspected only through echolocation is immediately and globally recognized through vision, and vice versa, with very high rates of success.[25] This strongly suggests, and perhaps decisively shows, that what a dolphin "sees" through echolocation is comparable in crucial ways to what it sees through vision. Experimental design has excluded the confounding possibility of associative learning by requiring the immediacy of discrimination and never exposing the target objects to both senses. Given these controls, the only plausible explanation of cross-modal recognition in dolphins is that they are constructing detailed, three-dimensional object percepts that are functionally comparable across both sensory modalities.

Echolocation in bats and dolphins allows for spontaneous object recognition in much the same way that vision does—probably by comparing observed patterns with three-dimensional "templates," "prototypes," "categories," or other imagistic representations stored in memory. To succeed at shape-based object recognition tasks, an organism must be capable of categorizing objects based on their shapes independently of their size, with shape representations scaled up and down to larger and smaller objects. For example, small and large triangles should be classified as the same type of object and distinguished from other shapes based on their three-dimensional profile.

Such echoic discriminations among real and virtual objects, which have been observed in laboratory experiments with bats and dolphins, cannot be explained by temporal acoustic properties alone. Rather, they implicate a complex form of neural processing that faithfully reconstructs the spatial dimensions of stimuli, which then appear to the echolocating organism as "bound" three-dimensional representations. The fact that echoic animals recognize objects and track them over space and time shows that the paradigmatic "binding" problem of vision has been solved convergently in an ISM that exploits an entirely different waveform energy, an important result that we will revisit in chapter 8.

3.2 Passive Listening and Active Feeling

It is by carefully comparing the active emission with the echo that bats and dolphins are able to construct an acoustic analog to the visual scene. However, it is possible to achieve a significant degree of spatial perception on the basis of passive audition alone. Barn owls, for instance, can localize and track prey items (e.g., mice) along a three-dimensional grid by detecting small acoustic arrival delays between their left and right ears, which are then processed by a network of neurons that form a spatial map. Cells of this spatially specific cluster of neurons are organized into a topographic array that represents the horizontal and vertical positions of the target object. Because low-frequency sounds provide general information about horizontal position and high-frequency sounds provide more precise information about elevation, the owl must analyze a spectrum of frequencies in order to construct a three-dimensional "image" of the prey item's position and velocity. It accomplishes this, as Eric Knudsen and colleagues have shown, thanks to a specialized facial anatomy: the densely packed feathers of the facial ruff that amplifies high-frequency sounds, making the right ear more sensitive to high-frequency sounds from above, and the left ear more sensitive to high-frequency sounds from below.[26] This enables the barn owl to locate prey along a three-dimensional grid and capture it in total darkness.

Scare quotes are placed around the word "image" above because it is unclear whether the barn owl "sees" its prey in a vision-like sense of the word. Even in its most spatially sensitive forms, the "passive" nature of audition severely limits the topographic information and textured detail that can be gleaned from the surrounding world. Living objects can generate sounds—such as a vole scurrying in the underbrush, a cricket chirping, or a songbird vocalizing—but they do not have sounds emanating from across their three-dimensional surfaces unless they are actively ensonified at high frequencies. When the world is dark and quiet, animals with even the most sophisticated audioreception are blind.

Moreover, pinpointing a sound made by an object in space—even the beating heart at its center—does not entail a representation of the object itself. We can infer the location of a particular object from the precise localization of a particular sound. But the individuation conditions for three-dimensional material objects are different from the individuation conditions for the objects of sound, even if the latter can be spatially located. Passive audition cannot, therefore, support imagistic object recognition, let alone the creation of a visual scene.

A certain degree of spatial imaging can be achieved by tactile senses as well. The elaborate nasal organ of the star-nosed mole, for instance, is less of a "nose" and more of a sophisticated tactile device; upon making active contact with objects, it is capable of the near-spontaneous discerning of shape and movement, which allows the mole to locate prey items in the complete absence of light.[27] The somatosensory systems of many fossorial (subterranean) mammals, some of which are blind, are connected up with brain regions that process visual information, suggesting that a degree of spatial information is afforded by the tactile sense.

The transduction of mechanical signals, such as pressure and vibration, has produced spatial cognition in other groups of mammals as well, such as in pinnipeds (seals, sea lions, and walruses). The harbor seal, for example, has highly specialized facial vibrissae (whiskers) that compliment or even substitute for vision during foraging in dark or turbid environments, in a way that is analogous to the "lateral line" of fishes and some amphibians. Unlike the lateral line, however, the vibrissae of pinnipeds are so sensitive to water disturbances that they can discern the three-dimensional vortices left by hydrodynamic fish trails, which contain information about the body shape, size, and swimming style of the organism that left them.[28] It is unclear if the somatosensory images that pinnipeds form are properly imagistic and if so how detailed these images may be. Are pinnipeds gleaning information from fish trails in the same way that dogs glean information from scent trails? In any case, tactile sensory systems are restricted in the informational detail they provide and the speed at which images can be formed, updated, and acted upon. The spatial details of hydrodynamic vortices left by moving objects can prove useful for tracking, but they do not support real-time scene reconstruction.

3.3 Universal Constraints on Acoustic ISMs

The shape of acoustic ISMs is determined by general laws that transcend the body plans of the particular lineages in which these ISMs are found. Active sonar, whether biological or human-made, is governed by universal physical constraints described by the quantitative laws and models of wave propagation (such as principles of diffraction, the inverse square law, Doppler effects, atmospheric attenuation, etc.). Biosonar is subject to trade-offs between frequency, resolution, intensity, and range that are resolved in ways that are amenable to prediction given adequate knowledge of local ecology. Bat echolocation dynamics and call structures can be predicted on the basis of niche information alone, irrespective of bat phylogeny.

A classic problem for vision science concerns how a three-dimensional representation of the world is constructed from a two-dimensional retinal array and other two-dimensional neuronal maps. Vision solves this problem by using pictorial, oculomotor, binocular, and motion-related cues to reliably infer depth. Many of these solutions are convergent among visual organisms, such as between camera-eyed vertebrates and compound-eyed arthropods.[29] Although this three-dimensional construction problem applies as much to the acoustic images formed by bats and cetaceans as it does to visual images, echolocating animals can take advantage of unambiguous spatiotemporal information encoded in the echoes of ensonified objects (information that is not available in reflected light). This enables them to construct reliable three-dimensional representations of their surrounding world. As we saw earlier, bats and dolphins can discern differences in arrival time and frequency between their ears, giving them the location of an object in two-dimensional space; the third coordinate—distance—is computed unambiguously by measuring the time delay between the emission and echo. By monitoring changes in echo delay and other parameters over time, echoic animals can construct a dynamic, rapidly updating, topographic model of the external world: a true acoustic scene.

Because successful echolocation hinges on the universal properties of sound, solutions to echolocation problems are highly constrained by waveform physics. One constraint, for instance, is atmospheric attenuation. The energy of acoustic emissions dissipates as it travels as a sound wave through the atmosphere. As a result, echolocation will always be a relatively short-range detection system as compared with vision because light propagates much more effectively than sound in air. A decrease in bandwidth increases the range at which echolocation is effective but decreases resolution; a decrease in call frequency means that small objects will not be detected but reduces atmospheric attenuation.[30]

Because of these physical constraints, certain convergent call sequence structures have emerged. For example, bat echolocation begins with a prey search phase that sweeps a wider area, using long-duration calls of narrow bandwidth and low frequency, which are less subject to attenuation and thus reach farther and wider into space. Once a potential prey item is detected, the bat switches to shorter calls of broader bandwidth, which permit greater resolution and localization of the item. If the item is determined to be a suitable prey, the bat then switches to the final “feeding buzz” phase, during which call rates increase and call duration decreases to create a buzz-like sound that optimizes prey interception in three-dimensional space.[31] Similar call structures and distance compensation dynamics are known in echolocating whales.[32]

Another universal constraint on echolocation relates to Doppler effects—shifts in waveform frequency due to movement of the sound source relative to the receiver. During flight, Doppler shifts present opportunities for overcoming certain constraints of echolocation, such as self-deafening. Because echolocation involves comparing pulse-echo pairs, it raises the specter of self-deafening, wherein an emission interferes with the return (echo) of a previous emission. This is more severe for microchiropterans than it is for other bats and echolocating animals, because they must capture small insect prey that reflect weaker echoes. One way that bats avoid self-deafening is by separating emissions in time, using the quiet interval to evaluate the echo; another method is to separate pulse emission and echo in the frequency domain, exploiting the Doppler effect to ensure that returning echoes arrive at a frequency that is optimal for hearing.

In order to avoid the deafening of conspecifics, some bats employ a jamming avoidance response, rapidly shifting frequencies or flying silent when foraging near conspecifics.[33] Because jamming is a problem facing any active emission sensory system, it is perhaps not surprising (though no less amazing) that similar jamming avoidance responses are deployed by weakly electric fish (see section 4). The speed of sound is so fast in water that it makes it difficult for echolocating whales to exploit similar Doppler effects. However, the fact that acoustic emissions propagate much farther and faster in the water medium means that there is less attenuation of ultrasound in water, and thus that echolocation can be used for broader-scale "visual" sweeping of the undersea environment.

These constraints and trade-offs must be resolved by all acoustic ISMs, on Earth and beyond. There are equally universal anatomical and metabolic constraints on the evolvability of echolocation that explain why it is "harder" to evolve than vision. First, as noted earlier, a powerful sound-production capacity, such as the lungs of tetrapods, is required to produce high-frequency emissions capable of supporting high-resolution acoustic imaging. Second, the costs of echolocation are high, which may limit acoustic imaging to organisms with high-metabolisms, such as mammals and birds. The metabolic rates of bats during echolocation, for instance, are up to five times greater than they are at rest. These costs have been offset in bats through the evolutionarily ingenious coupling of sound emission to wing-beat cycle, which function as a single unit of biomechanical and metabolic efficiency.[34] Sound emission is coupled with the upstroke phase of the wing-beat cycle, coinciding with contraction of abdominal muscles and pressure on the diaphragm. This significantly reduces the price of high-intensity pulse emission, making it nearly costless.[35] It is also why, as any careful crepuscular observer may have noticed, bats spend hardly any time gliding (which is otherwise a more efficient means of flight).

Universal constraints and internal limiting conditions can help explain cases of nonrepetition. For example, low-metabolic rates may explain why biosonar does not seem to have evolved in fully aquatic marine reptiles such as ichthyosaurs, which had active foraging strategies that parallel those of toothed whales. But why have pinnipeds not developed the ability to echolocate, given that they have the requisite anatomic and metabolic capacities and tend to forage in dark and murky coastal waters, hunting prey items that are similar to those sought by dolphins? The absence of echolocation in pinnipeds is likely due less to internal constraints and more to the physics of echolocation itself.

Pinnipeds have an obligatory amphibious lifestyle, which requires that they spend a significant portion of their time on land. Ear structures evolve to match impedance, and secondarily aquatic mammals have to modify ears adapted to air in order to match the impedance of water. Because tooted whales are fully adapted for aquatic life, they have undergone ear restructuring for a wholly maritime existence. Pinnipeds, on the other hand, have retained the ability to hear well in the air medium; they give birth terrestrially (on land or ice) and thus must spend significant portions of their time on land. These demands have constrained their ability to evolve the exceptional underwater hearing apparatus necessary for sophisticated biosonar. This could change at some time in the evolutionary future should pinnipeds take fully to the sea.

4. Electrolocation

Biosonar is not an evolutionary option for animals living or foraging in darkness if they lack either sound emission capability or the metabolic rates necessary to sustain high energy acoustic emissions. This is the case for certain fish living in dark and turbid waters, which have evolved an alternative mode of image-formation: electrolocation.

4.1 Passive Electroreception versus Active Electrolocation

The passive perception of electromagnetic fields is common in the animal world. Many animals use information about the Earth's magnetic field for migration, orientation, and navigation. Magnetoreception has evolved convergently in all groups of vertebrates as well as in arthropods and mollusks, though the precise mechanisms of magnetoinduction are poorly understood and probably variable across animal phyla. Geomagnetic fields provide coarse-grained navigational information that affords organisms with something akin to a magnetic compass. And because geomagnetic fields vary over space, they can also provide finer-grained (though still low-resolution) information about position relative a specific place or object. However, geomagnetic fields contain very limited spatial information. Although magnetoreception can support directional navigation, it does not allow for the imaging of objects and terrain in an organism's environment because geomagnetic fields (unlike light and sound waves) do not interact energetically with ecologically relevant objects.

True ISM-grade perception in the electromagnetic domain has been achieved by electrolocation. Unlike magnetoreception, electrolocation involves the active production of electric currents that generate magnetic fields. The rudimentary production of electricity has been achieved in several fish groups, some of which are capable of emitting high-voltage electric discharges to stun prey or deter predators. Passive electroreceptive organs, on the other hand, enable a wide range of taxa, from fish and birds to insects, to detect weak electric signals that enable them to zero in on visually hidden foraging items. These simple uses of electric fields are informationally depauperate and do not support the construction of "electric scenes."

Just as achieving a full-fledged acoustic ISM requires active sound emissions, achieving a full-blown ISM in the domain of electromagnetism requires active electromagnetic emissions. Electrolocation has arisen at least twice in populations of weakly electric freshwater fish—in African mormyrids and South American gymnotiformes—in each case produced by structures that were co-opted from muscle tissue. We will consider later whether this co-optation of a conserved tissue substrate undermines the independence of these iterations.

Electrical image-formation and scene reconstruction in weakly electric fish is achieved through active electric organ discharges from the tail region that create a stable, three-dimensional electromagnetic field around the signaling fish. The fish sense their own electric discharge with electroreceptor cells that specialize in detecting either the frequency or amplitude of electric signals. These electroreceptive cells, called mormyromasts, are distributed over much of their skin, and an "electric fovea" is positioned in the front of their head. An "electric flow field" around the fish's head and body provides information about movement and relative position, much as it does in vision and echolocation.

Mormyrids, which are the better studied of the two groups of weakly electric fish, actually have two specialized electric foveae: the first is in the nasal region, and is dedicated to longer-range electrolocation for navigation and object detection; the second is in the *schnauzenorgan*, a long and flexible chin-like appendage covered with densely packed mormyromast electroreceptive cells, which is used primarily for shorter-range prey detection and discrimination. This is akin to the dual visual foveae of certain birds (e.g., pigeons) which are specialized for flight control and foraging, respectively.

Electric organ discharges are not used solely for the purpose of imaging objects and the environment. They also serve to coordinate social behavior, from mating in gymnotiformes[36] to pack hunting in cichlid-eating mormyrids.[37]

Weakly electric fish are commonly kept as shy inhabitants of tropical fish aquariums. Though the electric outputs of these fish are well known, few aquarium enthusiasts appreciate just how remarkable are their sensory worlds.

4.2 Electric Object Recognition

How does electrolocation work, and what can we say about its representational and phenomenological qualities? Constant electric organ discharges emanating from the caudal region maintain a stable spatial voltage pattern over the skin surface. This voltage pattern changes when objects that have a resistance different from the surrounding water come within range of the signal and distort the field, resulting in changes of local electric voltages at particular skin loci. Objects can alter the stable electric discharge field in waveform and/or in amplitude, and weakly electric fish can detect both types of disruptions. These changes in local transepidermal electric current flow are recorded by the skin electroreceptors, which act as a "retina" upon which an electric image of the object is projected. This image is then transduced, and the information is fed to regions of the brain that process higher-order features of objects. Whereas in humans the processing of higher-order features of objects takes place in the cerebral cortex, in electrolocating fish these cognitive tasks are carried out in their hypertrophied cerebellum. The "mormyrocerebellum" is so oversized that it accounts for the vast majority of the organism's total oxygen consumption, with metabolic expenditures exceeding that of any vertebrate.[38] This, in turn, speaks to the great functional utility of electrolocation: all that brain stuff must be doing something computationally demanding and ecologically important.

Two decades of pioneering work by Gerhard von der Emde and his collaborators has shown that weakly electric fish can perceive numerous properties of objects, including the passive and resistive components of the object's electrical impedance, as well as its size, shape, depth, spatial orientation, and distance.[39] Like bats and cetaceans, weakly electric fish use active electrolocation not only to detect and classify objects, but also to evaluate and memorize distances, shapes, textures, patterns, and configurations—allowing them to solve object recognition tasks and construct electrical scenes.

Mormyrids can, solely with their electrical sense, discriminate between real and virtual electrical objects based on their three-dimensional orientations and configurations. Like the virtual acoustic stimuli that are used in bat and dolphin echolocation research, virtual electrical objects allow experimenters to control for visual, chemical, and mechanical means of perception. Mormyrids have been shown to electrically discriminate between objects on the basis of their shape, which is apparently more salient for them than other properties such as material composition, size, or spatial orientation.

For instance, where electrolocating fish were trained to receive a positive reward (in the form of a conspecific electric organ discharge) by learning and remembering to choose a metal cube, they later preferred a plastic cube to a metal cylinder, despite the difference in material; similar results have been obtained with spheres, pyramids, ellipsoids, and crosses.[40] Electrical object identification and discrimination is immediate and global, suggesting that the various parametric assessments that go into these discriminations are "bound" into a single percept. In this respect, electrolocation is similar to acoustic image-formation in echolocating animals. Although electrolocation is not focus-capable, when inspecting an object weakly electric fish engage in various "motor probing" behaviors to achieve better three-dimensional electrical resolution. By bending its body and factoring in proprioceptive information, the weakly electric fish alters field direction and skin receptor position, allowing it to glean additional information about its surrounding landscape—movements that are analogous to the "acoustic angling" behaviors observed in echolocating cetaceans.

Mormyrids have also been taught to discriminate between objects on the basis of distance irrespective of the size, shape, or electrical properties of the targets, showing that they have a true sense of depth perception.[41] They work out distance up to a maximum of 10 centimeters, apparently by measuring the normalized maximal slope of the electric image, which is affected only by object distance and not by object size. By comparing the intensity at the electrical image center with the "fuzziness" of the image boundaries, weakly electric fish can compute distance unequivocally: the farther away an object is, the more "out of focus" or fuzzy the image becomes (see figure 7.2).[42]

Thus, like echolocation, electrolocation does not need to rely on probabilistic cues to establish the depth dimension, as vision is forced to do. The degree of focus, in conjunction with maximum amplitude, is the primary means by which mormyrids compute depth. Once distance is computed, the size of the object can be ascertained unambiguously by comparing image width with the distance measure. And once distance and size are determined, the fish can then use this information to calibrate waveform and amplitude measures and hence determine the object's complex impedance. As with echolocation, this is all performed outside of experiential awareness, with holistic bound object percepts spontaneously "presented" to the fish for comparison with affectively valenced representations stored in memory, thus guiding behavior.

Even more incredibly, just as color vision measures the wavelengths of light reflected by an object, so too are electrolocating fish able to detect an object's capacitance properties through waveform distortion.[43] Just as color perception measures the color (frequency) and brightness (amplitude) of light independently of one another, electrolocating fish perceive resistance and capacitance

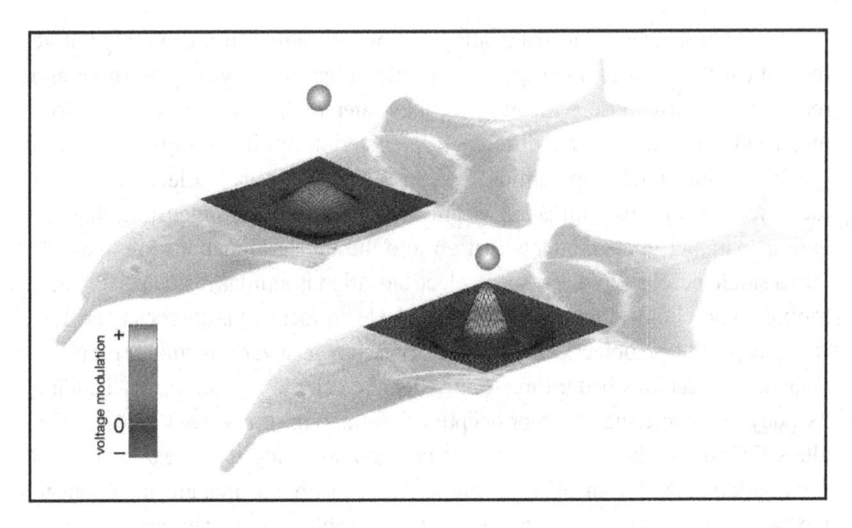

Figure 7.2
Electric image of a spherical object projected onto the skin of a mormyrid (*Gnathonemus petersii*). Amplitude changes correspond to shape and distance of the object (which is closer in the bottom image), with an increase in distance resulting in a larger electric image. By comparing the fuzziness at the image borders to the maximum amplitude change at the image center, the fish is able to compute distance unequivocally. From G. von der Emde and S. Schwarz, "Imaging of Objects through Active Electrolocation in Gnathonemus petersii," *Journal of Physiology–Paris* 96, no. 5–6 (2002): 431–444.

properties independently of one another. In this way, the perception of capacitance adds "color" to an otherwise "black-and-white" electrical world.[44] Because animate objects (such as plants and animals) tend to have capacitance properties, living things brilliantly pop out as colorful objects against a "gray" inanimate background. The result is the perceptual construction of a rich three-dimensional scene that permits sophisticated foraging and navigation in the complete absence of light and sound.

Just as striking, von der Emde and his collaborators have recently demonstrated that weakly electric fish are capable of the spontaneous recognition of objects across the visual and electric domains and vice versa.[45] In cross-modal recognition, an object is recognized as the same object based on one or more of its global features (such as shape), even though information about these global features is provided by distinct sensory modalities exploiting different energy forms in each case. To do this, the representation of one ISM (e.g., vision) must be encoded, stored in memory, and accessible for comparison against the information flowing in from an entirely different ISM (e.g., electrolocation or echolocation). It seems likely that this stored representation is encoded in a generic "imagistic" format that allows cross-modal recognition to occur spontaneously without associative learning. Mormyrid fish trained to discriminate objects using vision performed identically during modal transfer tests in which only the electric sense was available for discrimination, and they did so without any previous training.

As with cross-modal task transfer in echolocating dolphins (discussed earlier), spontaneous cross-modal recognition in weakly electric fish strongly suggests that electrolocated objects are being perceived holistically in three dimensions with a representational format and/or phenomenological quality that is analogous in fundamental ways to vision. Object recognition across ISMs is thus a robustly replicable phenomenon and is indicative of both the common representational formats of ISM percepts and their global access. Further, as von der Emde and colleagues point out, cross-modal recognition is not a quirk of experimental artifice. Rather, it is a crucial adaptive functionality that ensures reliable perception in complex environments in which information flowing in from different senses must be weighted and adjusted in accordance with fluctuating conditions, such as changes in turbidity, lighting conditions, and so forth.

4.3 Phenomenology of the Electric Sense

What is it like to "see" as the weakly electric fish does? Some would deny that there is anything "it is like" to be a weakly electric fish at all, and thus maintain that there is no electric phenomenology to speak of. Let us set aside questions of phenomenal consciousness until the next chapter and instead use representation as a fill-in for phenomenology. Electrolocation is a sensory modality that is qualitatively alien to us, and it has some notably weird representational properties that have no analog in vision or echolocation. For instance, larger electric images may be nonsymmetrical even if the objects themselves are symmetrical (such as cubes). This is because the electric discharge organ is located in the tail, and larger objects will cross areas of the electric field that bend at different angles along the fish's body and head, resulting in "cubist"-esque percepts.

Furthermore, the electric image of a multiobject scene is not the mere addition or patchwork of individual object images (as in vision), because the presence of proximate objects may distort the image of others and the images of different objects can superimpose. Given all of these differences, are we still justified in calling electrolocation a mode of "seeing"? We will return to the relevance of physics, representation, and phenomenology to the delineation of sensory modalities in the next chapter. The main lesson for now is that electrolocation is a full-fledged ISM that draws upon an entirely different energy source than vision and echolocation.

5. Evolutionary Robustness of the ISM

Let us bring the various lines of discussion together in considering the evolutionary replicability of the ISM. The visual ISM is likely to evolve earliest and most often, given its high evolvability and wide ecological applicability. This is precisely what we have seen: vision arose repeatedly in the earliest phases of animal evolution, and as visual acuity increased, so did the perceptual and cognitive machinery necessary to capitalize on the rich flow of information that it provided, resulting in the convergent evolution of brains with minds (see chapters 9 and 10). The emergence of visual ISMs—not one, but several—so soon after complex multicellular life arose speaks strongly in favor of its law-like replicability (see chapter 1), at least taking animal-grade multicellularity among the initial conditions of the replay.

But if it is so evolutionarily robust, then why did true vision and its neural-cognitive accoutrements never arise again in more than 400 million years, even while optics have evolved myriad times? The answer is unclear. We can surmise that it has to do with a combination of packed visual niches (with incumbent advantage precluding additional transitions to visual morphologies and lifeways) and the "congealing" of body plans that became recalcitrant to the developmental transformations that would be entailed by later origins of the visual ISM.

Alternative ISMs arose well after all major origins of the visual ISM, which is consistent with their narrower evolvability conditions. Once the neural structures dedicated to spatial perception, and their accompanying motor and proprioceptive systems, were up and running, the evolution of alternative ISMs would be likely to draw upon preexisting neurocognitive architecture even while they exploit a different waveform energy through a new transduction apparatus. Echolocating bats and dolphins, for instance, have co-opted regions of the mammalian brain that subserve spatial perception, such as place cells of the hippocampus. And recent work on dolphin brains shows that auditory information is fed into both the temporal lobe (associated with audition) and primary visual regions that process imagistic information.[46] In contrast, in electrolocating fish, the higher order features of images are processed in nonhomologous regions of the cerebellum.

We should also expect to see the recruitment of common molecular substrates in the evolutionary development of alternative ISMs. Convergence on echoic abilities between bats and dolphins, for example, has occurred not only at the level of functional morphology but also at the genomic level, with numerous convergent amino acid substitutions in genes linked to hearing.[47] As with the role of *Pax6* in the evolution of vision (see chapters 5 and 6), the fact that generic gene-regulatory resources associated with hearing are implicated in the evolution of echolocation in distant mammalian groups does not convert these iterations into parallelisms or Gouldian repetitions. This would only be so if we had reason to believe that the coopted resources were themselves radically contingent—which we do not, given the robust replicability of vision, hearing, and the brains that process their inputs (see chapter 9).

In sum, we have identified convergent ISMs in many living lineages, and it seems likely that many more of them occurred in the deep evolutionary past without leaving any trace of their existence in the fossil record (or leaving traces that have yet to be found). The iterated evolution of three distinct classes of ISMs shows that complex forms of perception have been realized through multiple, externally constrained routes by exploiting the limited set of energy forms that can support image-formation—presenting numerous evolutionary gateways to mind. Each type of ISM has its own distinct advantages, limitations, and associated phenomenology. Common to each of them, however, is that they enable organisms to construct a panorama of identifiable objects, bundled with properties, that are distributed in space and time relative to the perceiver—a phenomenal world that is fundamentally familiar to the one that we experience. The next three chapters will examine these convergent solutions to the problem of mind and its evolution.

8 Convergent Evolution of the Umwelt

The world as we experience it is strikingly ordered. Features of the surrounding environment do not insensibly grade into one another, nor are they jumbled in a chaotic maelstrom of incoming sensory data. Rather, we find ourselves situated at the center of a field of discrete objects. Each object is bound with various properties (such as shape, color, and motion) and retains its identity over space and time. This object-structured phenomenal field is somehow seamlessly stitched together in the brain and "presented" to the subject of experience, which has little choice but to take it all in. Our unified model of the external world is updated in real time as we navigate a meaningful landscape, incorporating feedback and filling in perceptual gaps to maintain a stable scene, which in turn guides active, purposeful behavior.

This is how we, and probably many other animals, experience the world at very coarse grains of resolution: namely, as a unified field of property-bound objects distributed over space and time. This chapter and the next two chapters will argue that the core structure of experience—the unified phenomenal field—is evolutionarily primitive, convergently derived, and likely to be a cosmically projectible property of the living universe. Precisely why the structure of experience is constrained in this way is unknown. Possible explanations range from constraints imposed by the external world to computational limitations that are inherent to cognition itself. There is clearly much philosophical territory to be explored; yet when all is said and done, most of the landscape will remain uncharted, and the mystery of mind will remain.

1. The Deep Structure of Experience

The phenomenal scene is unified in several respects. First, it is experienced as a single whole that cannot be decomposed into its phenomenal constituents. For example, the scene cannot be parsed into separate experiences of shape, color, or motion, or even into separate experiences of distinct objects. The components of conscious experience have what the philosopher Tim Bayne calls a "conjoint phenomenal character": they are contained within and subsumed by the larger phenomenal scene, which presents to the subject as a single global tapestry bound in time.[1]

Second, object percepts within the scene are individually unified in that they exhibit the same phenomenal holism or decomposability that can be attributed to global experience. Our experience of an apple, for instance, cannot be decomposed into experiences of shape and texture and color. Borrowing a term from modern aviation, we can call the simultaneous experience of numerous bound objects of the phenomenal scene "situational awareness": the ability to situate the self (or one's body) in the midst of the dynamic, fluctuating flow of the phenomenal field by simultaneously keeping track of multiple object variables and interrelations. The more entities that can be experienced synchronically and tracked diachronically, the richer the situational awareness of the subject.

Third, consciousness is unified in the sense that the phenomenal scene cannot be divided into multiple separate points of view. For instance, there are some indications that patients whose corpus callosum (the band of nerve fibers that functionally integrates the two hemispheres of the brain) is severed experience the world as two separate, unified subjects of experience, each with their own exclusive first-person perspective (more on this later). Splitting a unified subject of experience appears to result in multiple unified experiencers, each with an inner world that is inaccessible to the other.

A fourth aspect of experiential unity is stability: the phenomenal scene is remarkably coherent despite incongruities among incoming streams of sensory data. For instance, even though each eye has a slightly different view of the world, and even though objects may be partially occluded by clutter in the environment, the scene as we experience it is smooth and coherent. The phenomenon of bistable percepts (such as the "duck-rabbit"), which applies also to auditory and olfactory experiences, lends additional credence to the notion that experience is not decomposable into its basic elements. We can only experience the world in one way at a time; this experience is immediate, and its character is accessible only to the embodied subject.

The main contention of part II is that the unified structure of experience is not a radically contingent accident of human, primate, or even vertebrate evolution. Rather, it is a law-like feature of the evolutionary process that is intimately connected to the emergence of image-forming sensory modalities (ISMs). We saw in the previous chapter that the ISM transcends both phylogenetic constraints and the boundaries of sensory modalities traditionally conceived. The present chapter will make a similar case for the unified phenomenal world—or what I will call the "Umwelt" (a term coined by Jakob von Uexküll that I will adapt to the evolutionary context)—whether this concept is cashed out phenomenologically or in more tractable terms like representation. This chapter will focus on the conceptual dimensions of Umweltian cognition and consciousness. Empirical evidence for the convergent evolution of Umweltian minds will be explored over the next two chapters.

2. Umweltian Cognition and Consciousness

The term "Umwelt," which was coined by theoretical biologist Jakob von Uexküll in the early 1900s, literally means "the world around."[2] The Umwelt is a first-personal portal on the world, an internal model of meaningful objects in the surrounding environment as interpreted through an organism's peculiar set of neurocognitive and sensory apparatuses. The notion of Umwelt that will be deployed here differs in a number of respects from the term as it was originally conceived, though I hope to retain some of the elegance and utility of the original.

2.1 Adapting a Nonevolutionary Concept

At the core of the original Umwelt theory is the notion that there is a "semiotic" relationship between the subject of experience and its surrounding environment. That is to say, each species bears its own relationship of meaning and interpretation to the external world, and this constructed relationship of meaning is the primary means through which species interact with the world around them. The Umwelt is essentially an inner model of the external world which, according to von Uexküll, remains in itself unknowable. Von Uexküll conceives of the Umwelt as a broad-based biological phenomenon, one that does not require linguistic or other high-level cognitive capacities. Umwelten vary in content from species to species, depending on the "meaning organs" (senses and processing powers) they possess, which in turn are teleologically designed to detect aspects of the external world that are relevant to an organism's particular lifeways. Animals are thus attuned to ecologically meaningful features of their surrounds, and their ontologies are populated by, and limited to, this relevant subset. As von Uexküll puts it, the earthworm's world is made up only of earthworm things, and the dragonfly's world is composed only of dragonfly things.[3] As we shall see, this simple statement contains rich philosophical and methodological lessons for the evolution of perception and cognition.

According to von Uexküll, brainless animals are limited to reacting to relatively simple sensory inputs, whereas more behaviorally sophisticated animals are capable of synthesizing numerous aspects of the stimulus field in their central nervous system and directing motor movements in a top-down way. Thus, he says, "when a dog runs, the animal moves its legs—when a sea urchin moves, the legs move the animal."[4] He describes the sea urchin and other critters lacking a central nervous system as a "republic of reflexes" in which goal-directed behavior is built into the body plan itself. A qualitatively different kind of Umwelt arises with the capacity for object recognition, which involves integrating numerous stimuli into a single, holistic representation. In the next section, we will see how information integration is crucial not only for complex forms of representation but also perhaps for the emergence of phenomenal consciousness. But perhaps the chief contribution of Umwelt theory to modern biology has been to evolutionarily informed theories of cognition. The Umwelt lays the groundwork for a fruitful research strategy in comparative cognition: if we are to understand an animal's behavior, we should attempt to see the world through its eyes (so to speak), or as interpreted through its interconnected suite of sensory, nervous, cognitive, and motor systems.

However, there are serious problems with the original Umwelt theory that must be addressed if the concept is to prove useful in a modern biotheoretic context. These problems come from essentially three sources. The first is von Uexküll's general nonengagement with, or even outright rejection of, the two central contributions of Darwinian theory: the inference of common ancestry and natural selection as a mechanistic explanation of adaptive match. As conceived by von Uexküll, the Umwelt presupposes the immutability of species, as well as a perfect harmony between organism and environment that is permanently recalcitrant to mechanistic explanation. The dismissal of Darwinian principles, for reasons we shall see, significantly limits the explanatory power of the original Umwelt theory.

A second and related problem with the "Umwelt" as it was originally conceived stems from its entrenchment in transcendental Kantian metaphysics and epistemology. This heavy Kantian baggage leads von Uexküll not only to disregard the possibility of mechanistic explanations of biological teleology, but also to postulate mysterious vital forces that present in the form of a "Bauplan" (body plan). On his view, *Baupläne*—or, more precisely, the inscrutable forces that we attempt to reconstruct in the form of a Bauplan model—guide the teleological unfolding of an organism–environment match. Some of the reluctance to embrace the Gouldian notion of internal constraint comes from Gould's attempt to reinvigorate the scientific study of the Bauplan while jettisoning its vitalistic connotations. In its original von Uexküllian formulation, the Bauplan is unabashedly vitalistic, antimechanistic, and nonevolutionary. Von Uexküll's commitment to Kantian metaphysics also has the odd (and unhelpful) result that meaning organs do not provide information about the external environment because the external world remains permanently unknowable. Hence there is no meaningful sense in which sensory-nervous systems can be said to generate information or representational content that can be acted upon by the organism. For von Uexküll, the surrounding world is part of the animal itself.

A third source of difficulty for original Umwelt theory is its self-defeating rejection of animal psychology (*Tierpsychologie*). Von Uexküll forwent psychological categories like memory, perception, and representation, as well as attributions of subjective experience—all of which he deemed objectionably anthropomorphic as applied to nonhuman animals. This has the strange and counterintuitive consequence that von Uexküll's inner world is a purely "physiological" notion for all animals other than humans. On this view, the inner world of animals is unmediated by cognitive processes and hence is "dark" through and through, even if a central nervous system is critical to producing an appropriate behavioral response in the "effector organs." By today's lights, and indeed by the lights of von Uexküll's time (compare, for instance, his view to that of his continental contemporaries like Konrad Lorenz and Nikolaas Tinbergen), it is hard to imagine successful explanations of many animal behaviors without adverting to distinctively cognitive categories. Thus, as students of von Uexküll have been quick to point out, his third-personal model of the inner world of animals cannot plausibly be construed as a purely physiological or behavioral notion, despite his proclamations to that effect.[5]

A final feature of Umwelt theory that many contemporary evolutionists should be reluctant to take on board is its very expansive notion of biosemiotics, or a theory of signs in the living world. I suspect that most biological theorists would take the position that although "meaning"—or what we might now call "information"—is pervasive in an organism's world (and more richly present for some than it is for others), not all information should be thought of as a "signal" or "sign" properly conceived. A more restrictive but still broad-based account of biological signs, such as that proposed by philosopher Peter Godfrey-Smith,[6] might pick out the coevolution of sign production and interpretation as occurs in cases of mimicry, costly signals, and other forms of signaling that carry semantic (rather than merely causal) information.

Although the field of biosemiotics has made a comeback as of late, it has yet to formulate empirically tractable hypotheses, and some of it is as epistemically impenetrable as Kant's "thing-in-itself." To the extent that modern biosemiotics offers a cogent theory, it may plausibly be assimilated into more rigorously developed information-theoretic and teleosemantic frameworks. Biosemiotics has largely spurned information theory because of what it rightfully perceives as the inability of causal or correlational accounts of information to adequately capture the "normativity" of biological representation. Umweltian objects do seem to carry semantic information in a way that reliably co-occurring physical variables, which carry mere causal information, do not. However, there has been a recent wave of attempts to ground the "aboutness" of biological information in evolutionary function,[7] and this may be the more natural move, given that teleosemantic accounts of cognitive content are well-developed and widely deployed in biological and cognitive science.

Problems with the original conception of the Umwelt are analyzed in depth by Carlo Brentari in his impressive historical and philosophical treatment of the concept,[8] as well as by Morten Tønnessen in his own work on the topic.[9] I shall not explore these problems further. Instead, my aim will be to extract a thoroughly evolutionary, psychological, and phenomenological conception of the Umwelt that can be more readily assimilated into our contemporary scientific understanding of the living world. On this account, Umwelten, like other complex traits of the organism, are shaped by natural selection, and natural selection is guided in part by the objective structure of the external world. From a contemporary evolutionary perspective, we have strong reasons to think that perception is largely a veridical, truth-tracking process. As theoretical biologist Stanley Salthe, who is otherwise sympathetic to biosemiotics, points out, the material reality that lies beyond (and behind) Umweltian objects is a purveyor of true signs, lest the Umwelt never evolve at all.[10]

Recognizing that there are both internal and external constraints on the shape of Umwelten can go some ways toward explaining why Umweltian ontologies are "filtered" in the ways that they are, and why in other respects—such as with regard to the common core of consciousness—they are deeply convergent. Umwelten will vary as a function of the ISMs that lineages possess and the particular subset of environmental signals that a lineage has evolved to detect, interpret, represent, attach affective markers to, and act upon. Because only a subset of available information will be useful to an organism, natural selection will ensure that the lion's share of an organism's perceptual and cognitive resources is devoted to processing information that *matters*. The Umwelt will thus employ adaptive filters that sift informational environments for meaning; and the more information that a lineage can process, package, and act upon, the richer and more meaningful its Umwelt becomes.

Von Uexküll, for his part, was loathe to cede causal powers of the organism to the environment, for he feared that doing so would be to surrender to an environmental determinism of development (recall that Von Uexküll's account is developmental, not evolutionary). Yet we can acknowledge that the external environment shapes perception and cognition in meaningful ways without presupposing any sort of environmental determinism for ontogeny, and without rejecting the idea that organisms actively construct inner models of the external world rather than passively perceive their surrounds. Acknowledging that features of the external world constrain adaptive evolution, including the evolution of the Umwelt, does not commit us to an objectionably strong "externalism" about selection or to what Lewontin has critically dubbed the "lock and key" model of adaptation.

The final bit of von Uexküllian baggage that needs to be jettisoned before we can proceed unencumbered with an evolutionary notion of the Umwelt is his eschewing of psychological and phenomenological attributions in favor of exclusively physiological ones. As noted earlier, von Uexküll avoided mental concepts in characterizing the Umwelt of nonhuman animals, even those that have central nervous systems similar to our own. Yet it is impossible to make sense of the Umwelt and how it underwrites sophisticated goal-oriented behavior without adverting to distinctively cognitive categories, such as representation, memory, and other forms of information packaging and processing. In addition, subjective experience—actually, *subjectivity of a specific sort*—is also built into the deep structure of the Umwelt. Or so I will argue.

2.2 *Sentio Ergo Sum*

What do sunsets, symphonies, sewage, and sadness all have in common? They each have a unique experiential quality that cannot be captured, perhaps even in principle, by third-personal descriptions, no matter how complete these descriptions might be. To trot out the standard primate-centered example, even a complete understanding of the neurocognitive mechanisms that give rise to our experience of the redness of a sunset would not give us access to the experience of redness itself—they would not tell us, as the philosopher Tom Nagel famously put it, "what it's like" to see red.[11] The central task of the science of consciousness, according to philosopher David Chalmers, is to relate and integrate these first-personal (internalist) and third-personal (externalist) perspectives on the world.[12] The Umwelt construct, as conceived here, spans this internalist–externalist divide.

The inner world of subjective experience may not only be epistemically impenetrable to external observers, it may also be epistemically privileged among sources of knowledge available to the subject. Descartes's famous insight *cogito ergo sum* (canonically formulated, "I think therefore I am") can be modified to make an equally foundational epistemological claim about experience: I know first and foremost that I am an *experiencing* thing. Whether these experiences correspond to features of the world is a secondary question, and one that can never be answered as self-evidently, or with as great a certainty, as the first. As psychologist William James noted, what we conclude from introspection is not that experiences exist out there in the world, but rather that these experiences are *our own*. Some of these experiences involve thinking in the rich Cartesian sense, but many, perhaps most, involve brute perceptions of the world. These perceptions may not be veridical: I could be dreaming, delusional, or a virtual cog in a grand electronic simulation run by superintelligent aliens, and yet the bare fact that I am an experiencing thing is unassailable.

I may attribute similar conscious states to other similarly situated beings—and the better part of the next two chapters will be spent fleshing out what "similarly situated" here might mean. But any such third-personal attributions will be grounded in empirical *inferences*, not self-evident propositions. My own existence as a subject of experience would seem to entail the existence of an appropriately configured physical substrate, such as a brain, out of which my subjectivity emerges. But this, too, is not an epistemically self-evident proposition, but rather an inference made on the basis of third-personal observations of the causal relations between behaviors (including conscious reports) and brain states.

Importantly, we can accept the central Cartesian insight about subjective experience without signing on to Descartes's metaphysical commitment to "substance dualism"—the notion that mind and body are composed of distinct substances—or to his claim that the mind can outlive the body. We now know from careful studies of anesthesia and brain-damaged patients that subjective experience is intimately connected to brain activity and, in particular, to brain activity in specific regions of the brain. We know, for instance, that damage to certain functional brain areas, such as the cerebellum, does not affect consciousness, whereas lesions to other brain areas, such as the cortex, can have profound, content-specific effects on consciousness and its character. Other neurological structures, like the reticular formation of the brain stem, appear to be crucial for conscious activity but do not influence the specific contents of conscious experience. It is also clear that consciousness can be present in varying degrees in the same human subject, depending on the subject's underlying brain activity: we are more fully conscious while downhill skiing and less so when we are asleep or under a twilight anesthesia. And even though sensory inputs influence the content of conscious, conscious states can be present despite the complete absence of sensory inputs, as in the case of dreams.

At the same time, consciousness does not seem to be *equivalent* to any brain states or processes. Neurocognitive functions could in theory be described completely and entirely in mechanistic (or processual) terms without any reference to experience. Furthermore, we can logically conceive of a being that possesses all of the same brain and behavioral states that we have but nevertheless lacks the subjective qualities that, in our nomic universe, accompany them. The logical possibility of philosophical "zombies," as such beings are called, is one of many arguments intended to establish "property dualism" about mind, according to which conscious states are not identical to brain states. Property dualism rejects Cartesian dualism insofar as it accepts that conscious minds cannot exist without brains (or some functionally equivalent material realization base), but it maintains that states of subjective experience are not identical to states of the brain. On this view, the ontology of mind is not coextensive with the ontology of the brain, even if the former "supervenes" on the latter. Some theorists subscribe to Marvin Minsky's view that "minds are simply what brains do"; but on the property dualist view, one thing that minds do is generate conscious states that are not equivalent to brain states.

The deep mystery of consciousness is thus threefold. First, why is the internal perspective of subjective experience epistemically inaccessible to external observers? Second, why should brain states feel like anything at all to the embodied system? And third, why do particular instances of consciousness have the contentful qualities that they do? Why, for example, is our experience of blueness not like our experience of redness or, for that matter, like our experience of the rising sound of a malleted gong?

It could be that phenomenal consciousness is simply a brute set of necessary relations that can adequately be described but not explained in mechanistic terms. It could also be that in the future, consciousness will be shown to reduce to physical states or be exposed as an artifact of asking the wrong questions. I have no decisive sense of which view is more likely to be right. I wake to find myself skeptical of philosophical pronouncements of in-principle limitations of third-personal explanation; but by the end of the day I become convinced that the problem of subjective experience is a profound one, leaving me to suspect that consciousness may be built into the fabric of the universe.

This chapter will ultimately give the Umwelt an experiential gloss. However, we can understand the evolution of Umwelten and how they transformed cognition, behavior, and ecology without making any conscious commitments, so to speak. That is to say, we can talk about how independent lineages of animals came to represent the world as a unified field of discrete, spatiotemporally distributed, property-bound objects with a representation of the self (loosely conceived[13]) at the center of this field—and we can begin to sketch how this cognitive innovation spurred further behavioral, anatomical, and neurological evolution—all with nary a mention of the 500-pound gorilla in the room: subjective experience.

And yet this would do great epistemic injustice to the gorilla. We do not know *why* we are conscious, but we know, perhaps above all else, *that* we are.

The third-personal inference from our own subjective experience to the existence of other subjective experiencers seems if not self-evident, then unproblematic. Why should the laws of nature be different for me than for other nearly identically configured beings whose behavior is remarkably similar to my own? When it comes to many nonhuman animals, however, this inference enters choppier epistemic waters. It is an open question whether nonhuman animals, which differ to varying degrees both neuroanatomically and behaviorally from human beings, are suitably configured for conscious experience. Similar inferential problems arise in relation to less metaphysically problematic mental categories, such as cognitive mechanisms, which also cannot be directly observed.

Some theorists, going back to Descartes, have argued that consciousness did not spring into being until the origin of language and its attendant capacities for symbolism and logical recursion. However, there is neuroscientific evidence that subjectivity (phenomenal consciousness) can be physically disassociated from both self-awareness and linguistic capacities. The latter two higher-level properties are linked to functions of the cortex, whereas subjectivity appears to be generated by more evolutionarily primitive areas of the midbrain—a finding that is consistent with so-called "first-order" representational theories of consciousness.[14] Thus, humans can be phenomenally conscious even when their cortical functions are blocked. There is, moreover, nothing inherent to linguistic capacities that would suggest they are a precondition for bare bones subjectivity, or even for subjectivity of the sort that we experience during much of our waking (and dreaming) lives. The strong language-subjectivity link thesis would imply, for example, that if archaic *Homo sapiens* lacked language and other symbolic recursive abilities, then they were mere zombie automata—a most implausible result.

It seems far more likely that language and other recursive or metacognitive additions to the mental repertoire added additional layers of phenomenal quality on top of a more basic platform of subjectivity. This view jibes with a distinction drawn by some leading researchers of consciousness, such as David Edelman and his collaborators, between "primary consciousness"—defined as the ability to construct a unified, multimodal phenomenal scene—and "higher-order consciousness"—defined as the ability to recall past scenes and project future scenes and to have a narrativistic mental representation of self.[15]

The claim that consciousness arose in tandem with language is plausible only if "consciousness" is taken to refer to something very cognitively rich—such as the awareness of self (in a thick sense) or some other meta-representational ability picked out by higher-order theories of consciousness.[16] Even this more limited claim may be a stretch, however, given that a number of animals appear to be self-aware and yet lack natural language abilities. Because our best operational measures of self-awareness, such as the mirror self-recognition test, are met by a wide range of animals that are not natural (or even artificial) language users, this suggests that the link between language and consciousness-as-self-awareness is tenuous at best.

The restrictive view that phenomenal consciousness arose in the most recent eye blink of a >600 million-year history of animal evolution can be contrasted with maximally promiscuous views, such as panpsychism and biopsychism, which hold that mind is in all matter and in all living things, respectively. The position advanced here is that consciousness in any meaningful form did not arise until the emergence of the Umwelt, an event that occurred several times independently in the early phases of animal evolution (see chapters 9 and 10). The qualification "in any meaningful form" is intended to leave open the question of proto-subjectivity in non-Umweltian life: for instance, whether there is "something it feels like" to be a brainless multicellular or unicellular organism, even if that feeling is very minimal. As Peter Godfrey-Smith has noted, the difficulty with extreme continuity views such as biopsychism lies in "thinking about the difference between a complete absence of subjective experience and a minimal but nonzero scrap of it."[17] This difficulty compels Godfrey-Smith to reintroduce an older distinction between qualia and consciousness, wherein "consciousness" refers to a thicker, richer notion of subjective experience while "qualia" refers to a diffuse feeling that need not be associated with any particular form of cognition.

Similar work is done here by the Umwelt, which in broad strokes describes the kind of consciousness with which we are intimately acquainted—a form of experience that we can, in effect, imagine. Drawing inferences about the phylogenetic distribution of Umweltian consciousness requires that we commit to substantive claims about (1) what conscious experience consists in, (2) how it is related to brain structure and functionality, and (3) how it is reflected, if at all, in measurable neuroanatomical and behavioral data. In the remainder of this chapter, we will tangle with the first two elements of this problem, and the next two chapters will wrestle with the last.

3. The Umwelt Experience

Meaning in the universe arises not with the origins of life *simplicter*, but with the origins of *conscious* life. Meaning is *felt*, not merely represented. A universe teeming with zombie animals, though rich in representation, would be bankrupt of meaning. Is consciousness a freak accident of earthly evolution, or is there a law-like necessity to the way in which, as David Edelman and

Giulio Tononi elegantly put it, “matter becomes imagination”?[18] How can we determine whether the light of consciousness is part of the nomic structure of the universe?

Here again we are encumbered by an observer selection bias. Any being contemplating the contingency of consciousness must necessarily hail from a planet, and from a universe more broadly, in which consciousness arose at least once. All we are permitted to conclude from the existence of our own mind is that consciousness is *not prohibited* by the laws that govern our universe; this tells us nothing, however, about the frequency distribution of consciousness across the cosmos. Once again, convergence offers a way out of this epistemic quandary: any observer contemplating the contingency of consciousness need not hail from a history of life in which consciousness arose *multiple times.*

In order to identify evolutionary replications of consciousness, we must first establish the distribution of consciousness in the tree of life. Before we can do that, however, we must address several big ticket philosophical questions. First, what is phenomenal consciousness, what brain functions are implicated in subjective experience, and how is consciousness related to cognition in general and Umweltian cognition in particular? Second, how can we make sense of the notion that phenomenal consciousness admits of degrees in richness or complexity? An answer to the latter question will prove important if consciousness, like cognition, turns out to be a basal animal trait with a continuous ontological distribution. Finally, what do our answers to these questions reveal about the adaptive function of consciousness or its underlying substrate?

3.1 The Evolutionary Replicability of Cognition

Let us begin by considering the relation between consciousness and cognition. On the view defended here, cognition is a broader phenomenon than consciousness, in that only some cognitive functions generate, or are associated with, subjective experience, but no subjective experience can exist in the absence of cognition. What then is cognition, and do we have reason to think that it is a universally projectible feature of living worlds?

As with evolutionary outcomes in general (see part I), the extent to which cognition is evolutionarily replicable will depend in part on how the trait is described. For example, if we presuppose an account that conceives of cognition not merely *in terms of* information processing, but *as* mere information processing,[19] then it will encompass the quite sophisticated information processing and attendant behavioral-response capacities of plants.[20] Indeed, on such inclusive accounts, prokaryotes and other unicellular microbes, and perhaps all lifeforms of any appreciable complexity, could be said to exhibit minimal cognition. This is because cognition will be instantiated in bare metabolic and reproductive processes, and because metabolism and replication are probably universal features of life in the universe, this would imply that cognition is a universal feature of life as well. Homeostatic mechanisms, for example, involve processing and responding to information about internal and external states of the organism in order to maintain energetic nonequilibrium (i.e., to avoid death). As Godfrey-Smith has explained, metabolism actually entails quite a bit more than information processing; it involves sensing and responding to the world (such as via gene-regulation) in order to maintain the integrity of the organism.[21] As a result, metabolic capacities may be thought of as "proto-cognitive."

Other more restrictive but still biologically broad-based accounts aim to more cleanly distinguish minimal cognition from metabolism. They do this, for example, by holding that minimal cognition consists in sensorimotor mechanisms that are decoupled from metabolic processes and thus support faster information flows—a function that is paradigmatically realized by nervous systems. Yet even on these accounts there are functionally analogous molecular signaling systems in plants, in unicellular eukaryotes, and even in bacteria—all of which exhibit functional analogs to memory and learning that are decoupled from metabolism.[22] Brainless organisms, and even organism without neurons, have been shown to possess molecular signaling mechanisms that are designed specifically for information transfer.[23] Thus, even the "sensory-motor" account of cognition is not limited to organisms with brains and can be expected to be a universal feature of life.

Notice that on these big-tent accounts, there is nothing inherently *mental* about cognition. Thus, cognitive capacities broadly conceived are not sufficient for the emergence of experience in anything like the Umweltian sense. The Umwelt requires specialized modes of information processing and integration that are only realized in suitably configured brains.

3.2 Information and Integration

To understand how information processing centers might be configured for consciousness, we will help ourselves to the work of theoretical neuroscientist Giulio Tononi and his collaborators.[24] In a series of papers that are as stunning in their clarity as they are impressive in their quantitative rigor, Tononi argues that conscious experience consists of the integration of information, and he provides a theoretical framework for measuring this property. Although there are many working definitions of consciousness on offer in the literature, Tononi's account dovetails with the kind of information processing capacities that underpin the Umwelt. It also helps to explain how richer forms of experience might arise from more basal, simpler forms of subjectivity over the course of evolution.

Unlike many neuroscientists and some skeptical philosophers, Tononi takes consciousness—first-personal subjectivity—as a central explanandum of mind science. He presents the problem of consciousness and motivates a solution by way of several key thought experiments. In the first, he invites us to consider a photodiode, or a simple light-sensitive device, placed in front of a screen; the photodiode is set up to either initiate a beep when the screen lights up or to remain silent when the screen is dark. A human observer can be asked to perform the same task as the photodiode, but in the human case the observer "sees" the screen light up whereas the photodiode presumably has no such experience.

According to Tononi's integrated information theory of consciousness (IIT), there are two key differences between the human and photodiode systems that result in conscious experience in the first case but not in the second. The first difference concerns the sheer amount of information that is generated. On classic interpretations of information, information just is the reduction of uncertainty in an unpredictable world.[25] Unpredictable worlds are worlds in which there are many possible outcomes, moment to moment, with more or less equal probabilities. For instance, in unpredictable environments, whether a particular light, shape, color, conspecific, predator, foraging item, obstacle, chemical gradient, or danger is present in the surrounding environment may be more or less a crapshoot. The more alternative possible outcomes that a system can detect, the more it is capable of reducing uncertainty and thus the more information that it generates or processes. When a human registers the screen light up, the observed state is one of a vast number of possible alternative states of light, color, shape, duration, or motion corresponding to a vast amount of information. In contrast, for the photodiode, the detection possibility space is maximally depauperate: the system either detects or fails to detect a single state.

Now modify Tononi's thought experiment to include a photodiode that is set up so that it initiates a beep only when light waves in the blue spectrum are detected. The result would presumably be the same as the simple photodiode described above: there is no sense in which the quale "blue" would be generated or experienced by this binary detection system. The ability of a system to detect blue-wavelength light is clearly not a sufficient condition for generating the quale blue, nor for that matter any qualia at all. Why might this be so? One reason may be that there is no contrasting set of color possibilities—no informational color space, as it were—with respect to which the discrimination "blue" could be made.

A biological analog to the photodiode set up to detect a narrow band of electromagnetic radiation is the thermal imaging organs of pit viper snakes.

Pit vipers have a series of deep pits on each side of their head that are packed with cells that are highly sensitive to infrared wavelength light (or heat). The pits, which essentially act as lensless eyes that allow the snake to detect the heat signatures of prey and predators, are capable of detecting heat differentials as small as 0.01°C. Whether there are any visual qualia associated with these organs—whether there is something it is like for pit vipers to see heat—depends not merely on detection but also and crucially on what is done with that information after detection. In particular, it depends on whether and how that information is integrated into other aspects of the pit viper's visual scene. Given that thermal pits lack lenses, it is likely that infrared light is not bound into visual objects of the pit viper's Umwelt. Precisely what it feels like for the pit viper to sense heat, if it feels like anything at all, is unclear.

Further lessons about the nature of visual qualia can be gleaned from rare but revealing brain disorders in humans. The neuroscientist Oliver Sacks recounts that patients with total cerebral achromatopsia—the complete loss of color vision—report that they do not experience a world in shades of grey, as one might expect under such limited sensory conditions.[26] Rather, they report that their deficit is more fundamental and indescribable. This suggests that our experience of grey depends on the informational color space that is available to us. If this is right, then perhaps it is wrong to say that my Australian Shepherd, who has dichromatic (blue-yellow) vision, sees the red outline of her beloved flying disc as "brown." And perhaps bees experience colors differently than we do because they have an expanded repertoire of color perception that includes ultraviolet light which paints tiny landing pads on nectar-bearing flowers.

What if instead we set up a photodiode system so that it lights up red if light of red wavelengths is detected, blue if light of blue wavelengths is detected, green if light of green wavelengths is detected, and so on down the spectrum. Does this more informationally discriminating system have experiences of color? Though more sophisticated, the discrimination is still discrete, whereas colors are experienced as having continuous or spectral qualities. But suppose that we configure a more sophisticated photodiode array so that it can detect nuanced spectral differences in visible light arranged in virtually an unlimited number of spatial combinations, thereby generating vastly more information than the binary photodiode. Does this informational complexity endow the system with the experience of color?

Tononi considers just such a system—a digital camera—in order to show that subjective experience does not fall out of information processing alone. He replaces the photodiode in his thought experiment with a digital camera that comprises over 1 million photodiodes and hence can record a vast number of possible configurations of the world. Even though the camera system can process a huge amount of information, there is probably nothing it is like to be the camera—no private first-personal perspective, no inner camera world. Why is this so?

Tononi's answer is that the camera is a collection of low-information photodiodes that is decomposable into its elements, not an integrated system where the actions of one component are causally related to the actions of another component. In contrast, in human brains, a vast amount of information is not only generated but also *integrated*, with the states of some nervous system components influenced by the states of others at global scales. Integration is realized in nervous systems by virtue of what Gerald Edelman dubbed "reentry": recursive interchanges or coordinated mutual stimulations of dispersed neuronal groups. In ways that are poorly understood, reentry allows information to be packaged—and thus experienced—in more complex forms, as exemplified by the bound object percepts that crowd the Umwelt experience.

The full spectrum photodiode is just as much of an automaton as the binary photodiode, even though it processes vastly more information. More important than the total quantity of information processed by a system is how that information is packaged and combined with other bits and made available to the wider system. Why should we think that information integration is crucial for conscious experience? Tononi argues that the strongest evidence for the role of integration in consciousness comes from the phenomenological unity of consciousness. Consciousness cannot be parsed or decomposed into separate experiences of shape, color, motion, or space, even though these features are processed by different specialized brain areas dispersed throughout the cortex. As far as we can tell, there is no common cortical area where integration takes place, yet somehow properties are integrated or "bound" into unified object percepts (see section 3), which in turn are woven into the phenomenal fabric of the Umwelt.

Earlier we spoke of the Umwelt as a unified model of the external world that is "presented" to the subject. This characterization might seem to presuppose the existence of what Daniel Dennett has derisively called the "Cartesian Theater"—the intuitive but empirically unsupported notion that there must be a physical location in the brain where it all comes together and percepts are presented to the conscious subject. Yet as Dennett points out, the Cartesian theater leads to an infinite regress of homuncular minds: an infinite, nested set of experiencers within experiencers, with no explanatory purchase to be had. Several decades of neuroscientific research support Dennett's contention that there is no physical place in the brain where it all comes together. It does not follow from this, however, that the unity of conscious experience is itself an illusion. We know with great certainty that it does all come together *for the perceiver*. And this phenomenal coming together is a central feature of consciousness as we understand it. The coordinated firing of dispersed neuronal subsystems is likely to play a crucial role in this unifying process, but most of the integration story remains to be told.

Somewhat harder to situate within the Tononian framework is the nontrivial amount of integration that goes on in the cortex outside of conscious awareness (such as in certain streams of visual processing), which could lead one to question whether integration is in fact sufficient for subjective experience. Perhaps these subsystems are weakly and privately conscious in ways that we cannot assess or comprehend; or perhaps they are miniature zombies that simply feed information into the larger conscious system that we identify with and report on. At the most minimally conceivable level, perhaps there is something it is like to be two integrated neurons.[27] It is hard to know how we could adjudicate these questions, given that these modules are not wired-up for consciousness-indicative behaviors.

Tononi's response to the "nested experiencers" problem, it seems, is to argue that only *maximally* integrated information structures are the loci of subjective experience. On this view, there is nothing it is like to be the internet, or, presumably, an ant colony; although these systems exhibit minimal integration, they contain maximally integrated loci within them (e.g., individual ants) that effectively divide or decompose any subjective experience that might exist at the level of the whole. On this view, damage to subsystems that are weakly integrated with the main system, such as "encapsulated" modules, may affect the quality or richness of conscious experience (as with patients who have lost partial color vision), but these subsystems do not themselves give rise to subjective experience—they simply feed into the main system that does. This is the case for structures like the brain stem and cerebellum as well as for sensory input apparatuses (like the retina) and behavioral output devices (like the motor system) that stream information into or out of the main system.

If this is right, it would explain why a severed corpus callosum (mentioned earlier) results in the collapse of a single unified subject into two separate independently embodied loci of experience, but no more.[28] Yet even if patients with a severed corpus callosum are in fact harboring two distinct loci of experience, this does not show that they or individuals with normal brains do not harbor many more experiencers within them, because the behaviors and reports elicited in experiments do not speak to that question. As Thomas Nagel points out in one of the first philosophical treatments of the split-brain phenomenon,[29] there is no reason to think that verbalizability, or, we might add, any motor output capacity, is a necessary condition for subjective experience. Because measurable outputs will generally occur at the level of maximal integration—

that is, at the level of the organism as a whole—they say rather little about the presence of subjective experience in subsystems that are nested within the main system, whether these subsystems are neuronal networks or even neurons themselves. The notion of nested experiencers may be counterintuitive, but if we have learned any lesson from modern science, it is that the range of things that exist and the range of things that are intuitively plausible often fail to overlap. It is probably best, therefore, to remain agnostic as to whether there are nested experiencers within maximally integrated conscious systems.

Thankfully, the explanatory aims of the present project permit us to avoid taking a stand on the question of nested experiencers within a single animal. Our goal here is to zero in on the evolutionary function of Umweltian consciousness by linking it (or its neurocognitive generators) to more sophisticated forms of adaptive behavior. Because the evolutionary functions of traits are generally assessed at the level of the organism, we can safely elide the question of nested experiencers whose functions are causally screened off by the maximally integrated information structures that power the behavior of organisms upon which selection can act. Informational integration at the organismic level is therefore the appropriate level of analysis for the present study, whether or not it exhausts the universe of subjective experience.

Tononi insists that phenomenal consciousness just is the integration of information as expressed in mathematical terms. On his view, phenomenal states are metaphysically identical to integrative informational states. And if experiences just are maximally integrated information structures, then experiences cannot be said to *arise from* these structures. Conceived in this way, information integration is not a correlate, generator, or signature of consciousness, but rather what consciousness consists in. Edelman and Tononi thus claim to solve the first-personal/third-personal explanatory divide,[30] but they do so by essentially defining the problem away: consciousness just is informational integration.

Nevertheless, it is clear that the IIT does not solve what Chalmers understatedly called the "hard problem" of consciousness: the challenge of explaining why integrated informational processes, or for that matter any functional cognitive architectures, should generate private experiences at all, and why these experiences are only accessible from the inside. If we can reasonably ask whether integrative informational structures do or do not generate subjective experiences, then this suggests that informational structures are not metaphysically identical to conscious states, even if they are nomically linked to them. In fact, the IIT can be *glossed entirely in third-personal terms*. That is, it could characterize integrated information structures in a way that successfully explains cognition and behavior without making any reference whatsoever to subjective experience. If this is right, then the subjectivity element does no explanatory work in the model—and this lack of explanatory work is precisely what compels some leading philosophers of mind to defend an epiphenomenalism about consciousness.

3.3 The Binding Problem(s)

The role of informational integration is perhaps most salient in the case of binding. The "binding problem" refers to the puzzle of how sensory information about different features of objects (such as shape, color, and motion), which are processed in different specialized regions of the brain, are somehow integrated or brought together to form the array of complex object percepts that make up the Umweltian scene. In fact, there are two distinct versions of the binding problem. There is the binding problem in cognitive neuroscience, which is addressed to a representational puzzle; and there is the binding problem of subjective experience, which is addressed to the unity of consciousness.

The first problem is entirely third-personal and queries the neural mechanisms through which objects are "bound" with a constellation of properties that move with them over space and time. Redness and roundness, for instance, are properties that are bound to the apple and travel with it as it retains its perceived identity across temporal frames. There has been much work in cognitive neuroscience attempting to understand the mechanisms that are used to carve up the world of objects at its joints. The cognitive psychologist Anne Treisman provides a typology of binding according to which objects are bound *in properties* (such as color, shape, and motion); objects are bound *in space* (placed in specific locations) and object properties are spatially localized within objects; objects are bound *in time* such that they are recognized as the same object at different temporal intervals, even though they assume different orientations and occupy different locations in a cluttered scene; and objects are bound *in their parts* such that all of their parts are recognized as features of the same object, which "pops" out of the visual scene. Although such representations are imagistic, they are not purely "iconic" (in the technical sense that different aspects of the representation correspond to different aspects of the object or scene); additional representational formats are required for objects bound with numerous features to pop out and maintain a permanent identity over space and time.[31] This "temporal thickness" of cognition, as Zohar Bronfman and colleagues call it,[32] is the basis of working memory and figures into sophisticated goal-oriented behavior (see chapter 10).

We can add to this typology of binding the notion that objects are bound *in meaning*: that is, they are seen "as an obstacle" or "as food" or "as a predator" or "as noise"—categorizations that are attended by corresponding affective valences (positive, negative, or neutral) encouraging adaptive approach, avoidance, or ignoring behaviors. Some valences attached to complexly bound objects or action sequences may be innately specified while many others will be learned through experience via mechanisms of associative learning. When meaning is attached to a bound representation through categorization and the attachment of an affective marker, this amounts to "semantic-conceptual binding," as philosopher Antti Revonsuo calls it.[33]

One reason to think that different aspects of binding are handled by distinct subsystems of the brain is that different "agnosias," or sensory deficits caused by brain damage, are associated with specific binding failures. Oliver Sacks offers striking depictions of patients who are unable to bind colors to objects, others who can bind properties to objects but cannot bind objects to locations, and still others who can bind properties to objects and objects to locations but cannot bind meaning to objects.[34] Patients with these agnosias experience visual stimuli, but their visual experience is so fractured that they appear by their actions to be blind or incoherent. More common and less catastrophic disorders of binding include various types of synesthesia in which patients not only associate but actually experience colors in response to sounds or shapes in response to tastes.[35]

Visual stimulus binding is not the end of the sensory integration story, however. For animals that have multiple modalities, information can be integrated cross-modally so that smells and sounds are attributed to visual objects and vice versa, allowing distinct sensory systems to inform one another in working toward the best perceptual guess about states of the external world. Mundane cross-modal illusions,[36] like the ventriloquist effect, show how one sensory modality—such as a visual percept of the "chattering" ventriloquist dummy—can influence another modality—such as the perceived location of audible speech, which is erroneously bound to the marionette.

Any full-bodied solution to the binding problem must describe not only the mechanisms that underlie stimulus and semantic property binding but also the neural-cognitive generators of the final bound representation—the one that we "see" and that is made available for thinking and action. In addition, it must explain how the constellation of final bound representations are woven together into a single, unified Umweltian scene. This last aspect of the binding problem is probably the least understood. There is growing evidence that neural synchrony and reentry are both important mechanisms underlying the information integration that occurs in binding.[37] Described thusly, however, the binding problem is a wholly third-personal puzzle: it concerns how complexly integrated representations are formed in the brain and made available to memory and reasoning in order to guide action.

The relation between binding and consciousness is less clear. Neurocognitive binding surely affects the character of Umweltian consciousness and may be a necessary condition for it. Whether binding provides sufficient conditions for subjectivity in Umweltian or even more basic forms is not evident. In theory, neurocognitive binding and phenomenological binding could come apart—this would be implied, for example, by evidence that bound representations are sometimes formed outside of subjective experience. One possibility is that binding may not be sufficient for phenomenal consciousness unless it also includes affective content that projects meaning onto objects and events, thereby creating an Umweltian subject (see chapter 10 for more details). If so, then bound representations that lack affective content would fail to generate conscious experience. It could be, for instance, that bound representations are formed by the synchronous firing of distributed neural systems that specialize in different features of the scene, but that this bound representation must then be made accessible to the wider system for memory, categorization, and affective response if it is to become part of the stream of consciousness. And if this is so, then information integration would not be sufficient to generate subjectivity. Although much of this picture remains opaque, work on binding is providing the first glimpses of how the Umwelt was made.

4. Evolutionary Gateways to the Umwelt

One might argue that the best way to understand the causal structure of mind is to understand its evolution. This was perhaps Charles Darwin's deep insight when he famously jotted down in his notebook that "He who understands baboon would do more toward metaphysics than Locke."[38] Conceiving of consciousness in terms of the information integration that is implicated in the various forms of binding we have discussed renders the Umwelt an epistemically accessible target of evolutionary explanation.

4.1 Consciousness as Evolutionary Explanandum

There are two broad causal frameworks of evolutionary explanation that could account for the emergence of Umweltian experience: an adaptationist account and a by-product account. The adaptationist account takes consciousness to have causal properties that are relevant to fitness—properties that explain its origin, molding, and proliferation under the forces of natural selection. The by-product account, on the other hand, takes consciousness to be a collateral consequence—an incidental side effect—of selection for other neurocognitive properties. "Exaptationist" accounts of consciousness, meanwhile, hold that consciousness arose as a by-product of other neurocognitive adaptations but was subsequently shaped by selection for its own fitness-relevant properties. For our present purposes, exaptationist and adaptationist accounts are aligned because both attribute causal properties to subjectivity, whereas the by-product account is consistent with denying that subjectivity has any causal properties at all (though it need not entail this).

Conceived as global information integration, consciousness is not only causal, but causally predominant—for in typical cases, global integration screens off the causal inputs of modular subsystems with respect to organismic behavior. An exception is the cognitive phenomenon of "blind-sight,"[39] which shows that some behavior-affecting visual processing and attention can exist outside of conscious awareness—which, for some authors, raises the very real prospect that nonhuman animals are totally blind-sighted.[40] But if consciousness is, in general, a predominant cause of organismic behavior, then this opens the door to its having an evolutionary function; or, more precisely, it opens the door to its *being* an evolutionary function of suitably configured brains. The capacity to integrate information is presumably an evolutionary function of specialized neural anatomies, such as reentrant pathways, neuronal synchrony, and the central processing centers that serve as what theoretical neuroscientist Bernard Baars has called a "global workspace" for the flow of information, in which different modular inputs compete for the "spotlight" of selective attention.[41] As we shall see in the next two chapters, such integrative neural architectures look to be convergent across vertebrates and invertebrates, and their functions are corroborated by comparative studies of animal behavior.

Whether the adaptationist account takes consciousness-as-information-integration to be a *trait* (like the ichthyosaur dorsal fin) or the *function* of a trait (like stabilization during swimming), information-integrative structures are astronomically unlikely to have evolved by sheer accident—for they are as superbly matched to complex cognitive tasks like binding as ichthyosaur fins are to swimming at speed. And just as the convergent evolution of dorsal fins in fish, ichthyosaurs, and dolphins is indicative of external constraints on form, so too is deep convergence on information-integrative architectures evidence of external constraints on cognition that transcend the body plans of the particular animal groups in which they are found.

Alternatively, one might want to resist Tononi's stipulation that consciousness equals information integration. In this case, one might hold that integrated informational structures are the function of the adaptive neurocognitive configurations mentioned earlier but that consciousness itself is a by-product of selection for that function. On this by-product view, first-personal perspectives arise under the nomic conditions approximated by IIT (or some other successful third-personal theory), but they are not equivalent to those conditions. The by-product theorist might further insist that consciousness is a *noncausal* side effect, in that it exerts no influences on any events (neurocognitive or behavioral) that could be "seen" by natural selection. For present purposes, we can remain neutral as to whether the adaptationist or by-product account of consciousness is more plausible. Although each view conceives of the nature of consciousness very differently, both take the gorilla seriously as a legitimate target of evolutionary explanation.

With consciousness now firmly within the ambit of evolutionary explanation, let us consider how ISMs are connected to the origins and character of subjective experience. If a broadly biopsychist account of mind is correct, then minimal subjectivity would have preceded the evolution of the Umwelt, even if we have little imaginative sense of what minimal subjectivity might feel like. But consciousness *as we know it*—the construction of an Umweltian scene with the subject of experience at the center—arose only with the evolution of centralized brains and the informational integrative functions they achieved. The evolution of centralized brains, in turn, arose in coevolutionary feedback with ISMs as well as motor and proprioceptive systems, with the ISM and embodied Umwelt elements of a single, law-like coevolutionary package. As primitive ISMs increased in resolution and provided access to vastly more information, the nervous systems of ISM-bearing lineages underwent selection for neurocognitive architectures that could bundle that information into more complex, bound representations, which in turn could be made available for decision making and action.

Binding is implicated in the two types of phenomenological unification that form the Umwelt: local unification, in the form of bound object percepts, and global unification, in the form of a seamless panoramic field of objects with the subject at the center. Each of these types of unification increase situational awareness, or the simultaneous tracking of numerous bound features of the environment, in order to guide the organism in real time through a complex, three-dimensional, and meaningful world. Information about nonimagistic (e.g., olfactory, auditory) stimuli can be integrated into and inform the construction of the phenomenal scene, but it is properly imagistic perception afforded by an embodied ISM that renders the Umwelt.

4.2 Phenomenology of the ISM

What are the physical, cognitive, and phenomenological features that make a sensory modality an ISM? In chapter 7, we saw that vision, echolocation, and electrolocation all permit holistic object recognition and the construction of a unified phenomenal scene. One justification for this typology of "seeing" is cognitive: what makes these sensory modalities types of ISM, notwithstanding their very different energetic bases (electromagnetic radiation, sound, and electromagnetism, respectively), is their representational contents—the features of the world that they represent. These features include things like shape, texture, color, motion, and relative position, which are bound (along with affective valences) into spatiotemporally differentiated objects that are in turn integrated into a meaningful Umweltian scene. Add to this the not implausible assumption that the character of experience is roughly determined by representational content—such that there can be no change in phenomenal character without a change in representational content, and such that many if not all changes in representational content will result in changes in phenomenal character. It follows that since ISMs overlap to a significant degree in representational content, they also overlap to a significant degree in the character of experience. Thus, both cognitive and phenomenological similarities point to a common sensory system type: the ISM.

Alternative ISMs, such as echolocation and electrolocation, are representationally and experientially more like vision—which exploits a different waveform energy—than they are like other sensory modalities that use the same waveform energetics and homologous sensory components but differ greatly in their representational and phenomenal content. For example, vision is more like echolocation and electrolocation than it is like phototaxis or heat sensing, even though vision exploits light and uses photoreceptors; echolocation is more like vision and electrolocation than it is like audition, even though it exploits sound and uses the ear; and electrolocation is more like echolocation and vision than it is like magnetoreception, even though it exploits electromagnetic fields and uses electroreceptors. Another way of putting the point is that if we were able to construct a "sensory system phase space" along representational and experiential dimensions, then ISMs would all be closer to one another in that map than they would be to other modalities like phototaxis, heat sensing, magnetoreception, and audition.[42] Nagel may be right that we can never know what it is like to be a bat without being bat-embodied; but however alien the phenomenology of the echoic scene might be, the bat Umwelt is a coherent way of relating to the world, one that is fundamentally intelligible to visual organisms like ourselves.

There is an ongoing philosophical discussion over whether representational content, subjective experience, energetic stimulus, selectively shaped sense organs, or some combination of these features should be the criteria used to construct a typology of biological senses. We will not delve into this debate here. Instead, we can simply make the pragmatic observation that how one thinks sensory modalities ought to be individuated depends on what explanatory role the concept is playing in the particular streams of scientific thought in which it is deployed.

Psychology-oriented accounts of sensory modalities are typically built around mental categories like representation and experience, whereas biology-oriented approaches have tended to eschew psychological factors in favor of physical and biochemical criteria. A contemporary defender of the biological approach is the philosopher Brian Keeley, who argues that "to possess a genuine sensory modality is to possess an appropriately wired up sense organ that is historically dedicated to facilitating behaviour with respect to an identifiable physical class of energy."[43] On this view, two token sensory modalities cannot be of the same type—or even in the same sensory space vicinity—if they process information that is produced by different classes of physical energy. Echolocation would thus be considered closer to hearing, and electrolocation closer to magnetoreception, than either are to vision.

There is a serious problem with this view, however, even if we presuppose the biological perspective. Selection molds traits for particular functions, and as we saw in part I, functional traits are multiply realizable. Wings, eyes, and filter-feeding devices can be, and have been, produced out of entirely different structures and molecular substrates. The same is true for sensory modalities, which are best delineated in virtue of their *selected function*, not the structures, substrates, or energies that realize them. Moreover, using energetic bases to delineate senses may be useful in the context of some scientific questions, but less so for others. As previously defined, ISMs are modes of imaging the surrounding world that overlap significantly in representational and phenomenal content. This psychological overlap is crucial, moreover, for explaining the evolution of complex brains and sophisticated behavior (see chapters 9 and 10). It is not vision qua vision, but *vision qua ISM*, that explains the coevolutionary feedback process that results in the neural, cognitive, motor, and behavioral complexification associated with the evolution of active intelligent animals. Understanding this complexification process requires delineating sensory modalities in terms of psychological and informational predicates that are multiply energetically realizable.

4.3 The Cosmic Umwelt

ISMs have been dominated by vision ever since their origin in the Cambrian. The same integration and unification problems that had to be solved (and solved repeatedly) for vision also had to be solved (and solved repeatedly) for alternative ISMs, such as echolocation and electrolocation. As we saw in chapter 7, these alternative ISMs support holistic object recognition and scene construction using entirely different waveform energetics as their informational basis. Contours, textures, motions, positions, affective valences, and even analogs of color are integrated into acoustic and electric object percepts, which retain their meaningful identities over space and time.

If Umweltian consciousness consists in, or arises from, the integrative informational structures that seamlessly stitch together the phenomenal scene, then the fact that this unification has been achieved convergently in alternative modes of image formation should make us even more confident in the cosmic nature of the Umwelt. No matter how bizarre and unimaginable the mental life of intelligent aliens might be, their first-personal portal on the world is likely to be fundamentally familiar. Consciousness binds minds and connects the cosmos.

9 Finding Minds: Evidence from Neuroanatomy

Where can minds be found? This question is crucial for establishing the replicability of Umweltian cognition and consciousness. Yet it is exceedingly difficult to answer. We cannot observe minds in the way that we can observe other features of organisms, such as morphologies, behaviors, and even proteins and genetic sequences with the aid of a sufficiently powerful microscope. Mind is not a substance that reflects light and bends gravity; nor are mental states identical to brain states, given their multiple neural realization: there is no reason to think that the same biochemical brain state will reflect the same mental state in the brains of different animals or extraterrestrials.[1]

The most sensible view is that mind is a functional realization of the organization of bodies, including but not necessarily limited to nervous systems. However, the nature of this functional realization remains contested.[2] Even if we managed to agree on the relevant functional relations and devised reliable ways of probing for them in the world, there is still the problem, discussed in the previous chapter, that minds, or at least the feelings that accompany them, cannot be accessed from the third-personal perspective and hence cannot be observed directly with any scientific apparatus. Thankfully, direct perception, whether naked or technologically enhanced, is not the only means by which to generate reliable inferences about the presence of entities that we cannot see, so long as these entities have causal powers. Much as physicists can detect the traces of subatomic particles by observing the aftermath of particle collisions in high-speed accelerators, so too can we detect the causal signatures of mind in the world.

This signature is composed of three mutually informing lines of evidence: neuroanatomy, behavior, and evolution. Precisely how these lines of evidence work together to establish the existence of minds of particular types will be explored in the next chapter. The task that will occupy us here is to review the evidence from neuroanatomy and explain how it can be used to construct a provisional phylogenetic distribution of mind from which inferences about its evolutionary replicability can be drawn.

1. Convergent Bilaterian Brains

We saw in the previous chapter that minimal cognition is likely to be a cosmic feature of living worlds. Even if cognition is conceived in somewhat more demanding terms than bare information processing, such as in terms of sensorimotor mechanisms that produce faster information flows, cross-kingdom convergence on such mechanisms suggests that cognition has wide scope and is relatively easy to evolve from disparate developmental starting points. Yet cognition in this minimal sense does not rise to the level of what many would want to call "mind," at least not in anything like the rich Umweltian sense contemplated in the previous chapter. What we are in search of is the kind of mind that comes with the rich information-integrative capacities of *brains*. Brains are hierarchically organized centers of parallel processing that receive, interpret, and integrate information from sensory modalities and other peripheral modular systems, packaging complex representations and making them available for executive decision making and action. Can we glean anything from the way brains are distributed in the living world about whether and how intelligence is likely to arise elsewhere in the universe?

The nervous systems of extant animal groups can be divided into three basic categories: diffuse neural nets, cerebral ganglion, and brains. As with much else in the biological world, there is no clean dividing line between these grades of neural complexity, and there is considerable variation within them. At precisely what point, for example, cephalic ganglionic complexity rises to the level of a brain is essentially arbitrary. Notwithstanding this continuum in neural complexity, the above coarse-grained categories are useful: there are stark morphological and functional differences between full-blown central processing centers, on the one hand, and primitive cephalic ganglia, on the other. And as we shall see in chapter 10, these neural complexity grades map reasonably well onto the behavioral sophistication of lineages and the lifeways they lead.

The discussion that follows will pivot around the provocative claim made by Simon Conway Morris that centralized nervous systems are an inevitable outcome of evolution.[3] In order to answer this question, we need to know how many times centralized nervous systems evolved and what sorts of contingencies these iterations (if they exist) may have relied upon. But learning the distribution of neural complexity among extant animals does not, by itself, reveal how many times neural complexity grades evolved, so it tells us little about the evolutionary robustness of such outcomes. To shed light on this question, information about the existing distribution of neural complexity must be superimposed onto our scientific understanding of the relatedness of animal groups. In essence, resolving the evolutionary history of brains is the first step toward constructing a phylogeny of minds.

1.1 Observer Selection Effects and Contingency Credences

Let us say, for the sake of argument, that brains evolved only once in the history of life on Earth, and that this trait was transmitted continuously via descent from a common ancestor to all existing animals with brains. This single origin scenario for brains should increase our credence in the contingency hypothesis about mind because (1) the existence of minds depends on the existence of functional brains and (2) observer selection biases (see chapter 1) guarantee that brains will evolve in any history of life in which a species comes to ponder the prospects of intelligent life in the cosmos, no matter how radically contingent and infinitesimally rare such an evolutionary outcome might be. The alternative scenario is that brains evolved multiple times such that some of the animal brains we observe descend from separate origin points in the history of life on Earth. This multiple origins scenario should increase our credence in the robust replicability thesis as it applies to brains and, by a somewhat more tentative extension, to minds. For any intelligent observer posing these cosmic questions need not hail from a history of life in which brains and their attendant minds evolved multiple times. The multiple origins of brain/mind complexes gestures at law-like constraint, rather than cosmic accident.

If the multiple origins scenario is correct, then we have the beginnings of a case for the law-like expectability of mind—but only the beginning. Whether this pattern of iteration supports the the robust replicability thesis as it relates to brain–mind complexes would depend on three things. First, it would depend on what the relevant initial conditions are taken to be and how evolutionarily robust we think they are. For instance, does our question about the robustness of brain–mind complexes presuppose the origin of eukaryote-grade complexity, complex multicellularity, functional neurons, or none of these things? Second, it would depend on the extent to which the co-optation of conserved substrates, such as homologous cell types or gene regulatory networks, undermines the independence of specific brain iterations. Third, it would depend on whether we are licensed to make an inference from the existence of brains to the existence of minds. Before addressing these points, let us begin with the basic homology-convergence question first, which, as it turns out, is surprisingly difficult to answer.

1.2 How Many Times Have Brains Evolved?

To work out how many times brains have evolved in the history of life on Earth, we need an animal cladogram—a working reconstruction of evolutionary relationships among animal groups on which we can place the extant nodes of neural complexity. Animals are grouped into a number of clades that reflect high-level taxonomic categories in the Linnaean system of classification, like phyla and superphyla. Whereas nervous systems of varying levels of complexity can be found in virtually all animal groups (save for a few basal clades, as we will discuss), true brains are only present in a handful of phyla. Specifically, centralized nervous systems are found in chordates, arthropods, annelids, and mollusks, which in turn span the two bilaterian superphyla known as *protostomes* and *deuterostomes*, respectively.

Many readers will recall being taught in high school biology that insects—arthropods nested within *Protostomia*—have decentralized ganglionic nervous systems. This view is outdated and now known to be wrong. Like the vertebrate brain, the *mushroom bodies* of arthropods (analogs of which can be found in polychaete annelids) are centralized neural structures characterized by reentrant pathways that permit an ongoing exchange of signals to and from all major brain areas and sensory systems, supporting attentional processes, learning, and the executive control of the body and its movements.[4] Insect mushroom bodies have significant dendritic ramifications and take inputs from all critical regions of the brain—a signature of higher-order information integration and, potentially, of sophisticated forms of cognition like learning and memory.[5]

Working out how many evolutionary origins of brains are reflected in the extant distribution of centralized nervous systems among bilaterian phyla requires that we accurately reconstruct the evolutionary relationships between major bilaterian groups. Unfortunately, resolving evolutionary relations among phyla has proved difficult—much more so than discerning the phyla themselves. This because bilaterian phyla arose in the geological eye-blink that was the Cambrian explosion, and they have remained morphologically stable ever since (see chapter 2). During this Big Bang of animal evolution, nearly all bilaterian clades split from their last common ancestor in as few as tens of millions of years—an incomprehensibly vast span of time from the human perspective, but too brief for evolutionary descent to leave an unmistakable trace in the geological and genomic records.

We face three possible scenarios for the evolution of bilaterian brains. In the first scenario, the last common ancestor to all bilaterians not only had a through-gut (a crucial animal innovation that allowed for the processing of ingested sediment) but also a complex head encasing a centralized nervous system with a tripartite organization (composed of a forebrain, midbrain, and hindbrain), which was transmitted continuously through common descent to all bilaterian lineages that have brains today. In the second scenario, the last common ancestor to all bilaterians had a through-gut and neural nets (and perhaps even neural ganglion), but true brains evolved from these homologous ancestral nervous systems at least three or four separate times in bilaterians. In the third scenario, the last common ancestor to all bilaterians had no nervous system (and perhaps no gut) at all, with the implication being that neurons, ganglia, and central nervous systems all arose multiple times within *Bilateria.*

Which of these scenarios is most likely to be correct? The third scenario—the repeated evolution of neurons and basic nervous systems in *Bilateria*—may be the least plausible for two reasons. First, the fossil record of the late Ediacaran—which, it may be recalled, is the period immediately preceding the Cambrian explosion, characterized by an enigmatic multicellular fauna with opaque affinities to metazoans—shows trace fossils of burrowing and other avoidance behaviors, which indicate that basic sensory-motor systems predate and hence were probably present in the first bilaterians.

Second, cnidarians—the radially symmetric phylum that includes jellyfish, corals, sea anemones, hydrozoans, and cubozoans—have decentralized nerve nets of varying complexities. And cnidarians are widely thought to be the sister group to *Bilateria*, from which they are estimated to have split more than 700 million years ago—nearly 200 million years before the Cambrian explosion. If this is right, then a parsimonious reading suggests that the last common ancestor of cnidarians and bilaterians had a primitive nervous system that was transmitted to bilaterians.

The alternative, of course, is that cnidarians and bilaterians evolved neurons independently (or perhaps from a conserved genetic potential). But this flies in the face of a guiding epistemic assumption of cladistics—namely, that a hypothesis that postulates the single origin of a complex character state followed by ancestral transmission is more parsimonious (all else being equal), and hence a preferable explanation of observed data, than a hypothesis that postulates multiple origins.[6] Moreover, genetic and developmental affinities between cnidarian and bilaterian neurons and nervous systems generally (but not exclusively) corroborate a single origin.[7] This does not rule out the possibility that neurons as a functional cell type evolved more than once, as we shall see in the next section. But it does speak quite clearly in favor of the proposition that the last common ancestor of all bilaterians had ancestral neurons and a primitive nervous system.

Having dispensed with the third scenario, we are left to the trickier task of adjudicating the first two. Given the general theoretical preference for single-origin hypotheses, one might think that, all else being equal, the first scenario (brain homology) would be preferred over the second (brain convergence). But in fact a stronger cladistic case can be made for the multiple origins of brains given the sheer number of major character state changes that are entailed by the brain homology hypothesis. The crux of the matter boils down to which set of evolutionary outcomes is deemed more likely, given all of the available evidence and background theory: a *small* number of brain/head/eye *gains* or a *very substantial* number of brain/head/eye losses.

The brain homology hypothesis entails losses across the board, with more than 75 percent of existing animal phyla having quite literally lost their heads and exhibiting secondarily degenerated nervous systems. In contrast, the convergence scenario does not require any brain/head/eye losses in the bilaterian phyla that lack these features, because in that hypothesis these groups never had them in the first place. Instead, the heavy lifting for the convergence hypothesis comes in through its remarkable postulation that brains originated from primeval nervous systems *at least three or four times within Bilateria.* Both the homology scenario and the convergence scenario seem improbable, and yet one of them must be true.

Given that both scenarios are consistent with the extant phylogenetic distribution of nervous systems, which hypothesis offers the better explanation of the observed data? The answer that one gives to this question will depend on whether one accords greater evidential weight to phylogenetic patterns of morphology or to the genetic factors that underlie the development of nervous systems in distant groups. In truth, this is an oversimplification because morphological and molecular evidence point to some extent in both directions. There is, in my view, a clear frontrunner, but the scientific jury is still deliberating and far from a verdict. We are therefore venturing into the frontiers of scientific knowledge.

1.3 Gains, Losses, and Evolutionary Parsimony

The question before us is whether the brains observed in distant bilaterians represent phylogenetically primitive or convergently derived character states of nervous systems. Any answer to this question will turn on the features of the last common ancestor of *Bilateria,* about which little is currently known. The received view, at least among paleobiologists, is that the last common bilaterian ancestor was comparably simple: an eyeless, headless, and brainless animal with very limited motor capabilities.[8] Not all biologists agree with this assessment, however, for reasons that will soon be clear.

On standard approaches, "out-group analysis" is used to determine the "polarity" of a character in a cladogram—that is, to determine which character state is likely to be ancestral and which is likely to be derived among the alternative character states exhibited by a given set of taxa (the "in-group"). Out-group comparison looks at the character state in an "out-group" (typically a sister taxon to minimize evolutionary distance); if there is only one out-group (or if all out-groups exhibit the same character state), then the character state of the out-group is taken to be the most likely ancestral condition. Once trait polarity is determined, the most likely evolutionary reconstruction of the trait is deemed to be the one with the fewest character state changes. Unlikely events do occur, of course, and so the findings of out-group analysis are to be taken as *prima facie* but defeasible evidence for a particular evolutionary hypothesis.

If we use out-group analysis to map nervous system character states—broadly partitioned into neural nets, ganglia, and brains—onto our best current reconstructions of metazoan evolutionary relationships, we find that brains are likely to have evolved multiple times in *Bilateria*. Figure 9.1 illustrates current scientific thinking about the evolutionary relations among metazoans in light of both fossil and molecular data, together with what is arguably the most parsimonious evolutionary history of nervous systems in light of these clade relations.

Not surprisingly, there are some important wrinkles and fluid elements in these reconstructions that we will consider below. But if we assume that sponges (which lack neurons, digestive tracks, and circulatory systems) are the most basal animal clade, then the presence of neural nets but not true brains in nonbilaterian animals like cnidarians and ctenophores, which branched off after the divergence of sponges but before the origins of *Bilateria*, suggests that neural nets are the ancestral character state of bilaterian nervous systems and brains are the derived state. And if this is so, then it implies that brains arose several times in *Bilateria*. Lending further weight to the multiple origins scenario are more taxonomically fine-grained out-group analyses that indicate convergence on varying degrees of nervous system centralization *within* phyla, such as within mollusks. The hypothesis that the ancestral mollusk had a complex brain that was repeatedly lost within the clade entails many more character state changes than does the intra-clade convergence scenario.

Having said all this, there is a large phylogenetic gulf between cnidarians and bilaterians. Were we to discover an extinct taxon in between whose nervous system structure could be ascertained, this would substantially affect our credence in the previous analysis of trait polarity. Another thing to take into account is that losses may be evolutionarily more accessible than gains when it comes to complex characters like brains; if so, then we should not assume that a tree entailing fewer character state changes but more brain gains is more likely than one entailing more character states changes but fewer brain grains. Still, there are sound reasons to prefer the convergence hypothesis beyond an epistemic preference for fewer character state changes.

If the bilaterian brain homology hypothesis is true—that is, if arthropods, vertebrates, mollusks, and annelids all have robust centralized nervous systems because they descend from a common ancestor that had a similarly complex nervous system—this would entail the loss of brains and a massive reduction

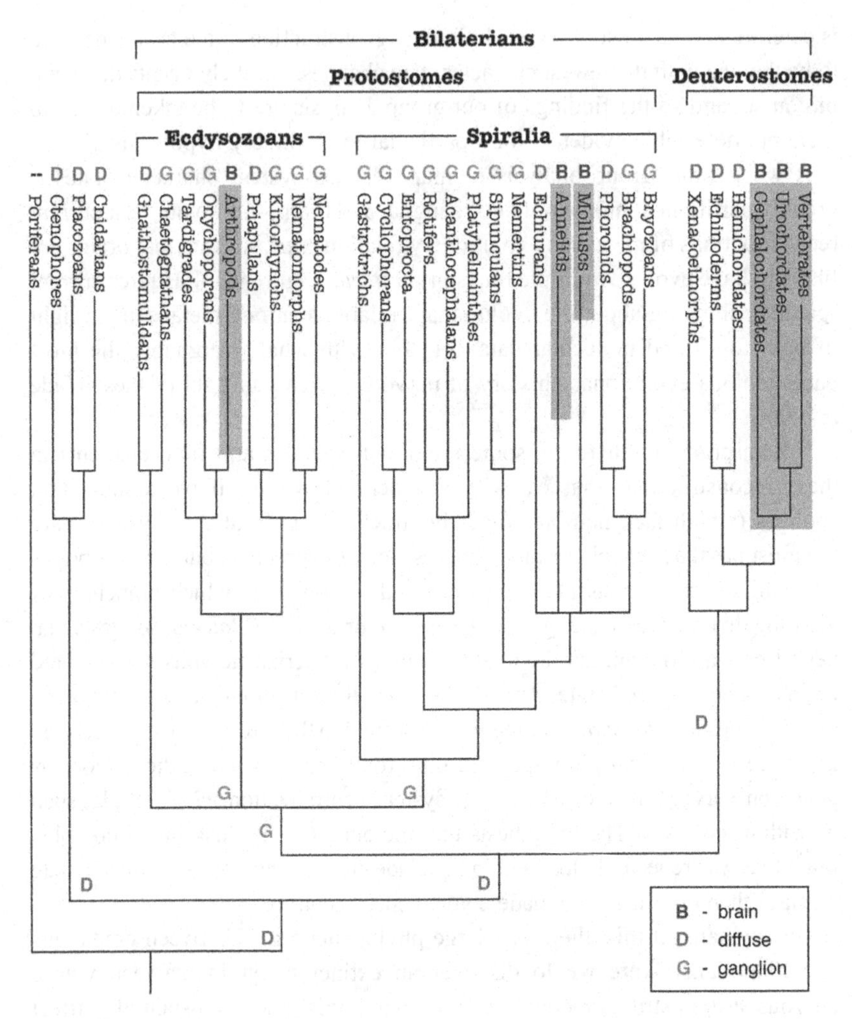

Figure 9.1
Out-group analysis of nervous system phylogeny across extant metazoans indicating that brains independently evolved at least four times in bilaterians. From R. G. Northcutt, "Evolution of Centralized Nervous Systems: Two Schools of Evolutionary Thought," *Proceedings of the National Academy of Sciences of the United States of America* 109, Suppl. 1 (2012): 10626–10633.

of nervous systems in more than a dozen bilaterian phyla quite rapidly in the early Phanerozoic, with the implication that twenty-three of the thirty existing animal phyla have secondarily degenerated brains and entirely vanished heads.[9] In fact, the situation is even worse for the bilaterian brain homology hypothesis because there is substantial nervous system variation *within* phyla that would (*ex hypothesi*) have retained the ancestral executive brain, such as within the mollusk and annelid clades. Taking into account this intra-clade variation would imply an even greater total number of brain and head losses—such as in bivalves, chitons, and snails within *Mollusca*; and in earthworms and clitellates within *Annelida*.

We are right to be wary of progressivist penchants that might cause us to downplay the likelihood of reductions and simplifications in macroevolution. Still, the loss of heads/brains/eyes in 75 percent of bilaterian phyla as well as the radical functional and ecological reorganization of the same proportion of bilaterian body plans seems on its face to be implausible, particularly given the developmental stability of animal body plans for the duration of the Phanerozoic.

It is true that neural tissue imposes significant metabolic demands on organisms that natural selection will tend to shed if doing so is beneficial. It is also true that brain *size* has been reduced in many animal lineages for whom the metabolic costs of cognitive substrate outweigh the benefits of enhanced cognition. This is poignantly illustrated by secondarily herbivorous vertebrates (like pandas) whose calorie-frugal diet can no longer sustain their carnivorous clade's historical brain tissue expenditures. It is the case as well for lineages whose ecology calls for the reduction of neurologically demanding somatosensory functions, such as "cavefish"—several groups of freshwater fish adapted to lightless underground habitats that have repeatedly lost portions of the cortex dedicated to visual processing. The loss of a complex head is thus not totally inconceivable.

However, as evolutionary neuroscientist Leonid Moroz and his collaborators note, apart from parasites, there are no instances of major clades radically altering their lifeways—for example, switching from active predation to sessile filter-feeding—and, we might note, losing their heads and brains in the process. Even barnacles—arthropod crustacean filter-feeders with sessile, parasitic lifestyles—have retained minimal brains and compound eyes. The same is true of tiny copepod crustaceans and other meiofauna (minute interstitial animals that live in marine and terrestrial sediments), some of which are parasitic and lack circulatory systems but nevertheless have retained centralized nervous systems and eyes.

It is possible that there were weaker developmental constraints on early animals before bilaterian anatomies and lifeways congealed. Perhaps, for instance, there was less prohibitive pleiotropy, and hence more "quasi-independence," in upstream components of the bilaterian developmental plan before it evolved into the highly specialized forms that are reflected in modern phyla. The trouble with this is that, as we have seen, the brain homology hypothesis requires that the last common ancestor of *Bilateria* was equipped with a highly complex body, brain, and visual system, presumably with an active lifestyle to match—precisely the sort of body plan that we would expect to be entrenched due to the causal topography of development (described in chapter 2).

Without being able to point to a single clear example of total brain/head/eye reduction in a nonparasitic lineage in the history of life on Earth, we should be loath to believe that it occurred on more (and perhaps many more) than a dozen occasions in the base of the Cambrian. The bilaterian brain homology hypothesis thus paints a decidedly unlikely picture of the evolution of bilaterian body plans. We would therefore need overwhelming evidence from other biological sources before it could be accepted. Enter developmental genetics.

1.4 When *Ceteris* Is Not *Paribus*

One reason why the question of brain convergence remains unresolved is that there is an apparent conflict between phylogenetic (out-group) analysis and molecular-developmental data. In particular, there are compelling structural similarities between arthropod and vertebrate brains, as well as orthologous genes and neural patterning mechanisms shared between these groups, that appear to support the brain homology hypothesis over the convergence hypothesis.

For instance, some of the genes that are implicated in the patterning of the nervous system into forebrain, midbrain, and hindbrain are shared between arthropods and vertebrates. This deep nervous system homology has been established in much the same way that deep homology in eye morphogenesis has been established, namely by swapping genes involved in the construction of brains between arthropods (drosophila) and vertebrates (mice) and showing that this fails to derail or otherwise affect healthy nervous system development in these groups.[10] In addition, gene knockout studies in drosophila and mice that deliberately interfere with the expression of molecules involved in memory and learning (such as protein kinase A) suggest that similar gene-regulatory networks underlie brain development in these disparate phyla.[11]

Nevertheless, there are several reasons to be cautious about the genetic-developmental data as they pertain to the brain homology-convergence question. First, to reiterate a point made earlier, the loss of the head/brain/eye complex and radical body plan reorganization in at least a dozen major animal phyla and many more classes within phyla is a difficult phylogenetic pill to swallow—though this goes to how we should balance conflicting sources of evidence, not to the merits of the evidence itself. Second, and more to the point, the fact that there are conserved genes and regulatory networks that underwrite the development of nervous systems across brained bilaterians does not imply that those genes functioned similarly in the last common ancestor and hence that they lost their function in most animal phyla.

Some have argued that the presence of conserved genes and homologous cell types in the development of eyes in distant phyla indicates that the last common ancestor of eye-bearing bilaterians possessed vision, and hence that the visual adaptation is homologous (monophyletic) rather than convergent (polyphyletic). But as we saw in chapter 6, this inference is problematic for a number of reasons that carry over into ruminations on the deep homology of nervous systems. The lesson there was that although the deep homolog *Pax6* is involved in the development of all known eyes, it is also present in non-eye-bearing phyla, such as echinoderms, that do not have eyes or photoreceptive cells of any kind; moreover, *Pax6* does not direct the specific contours of eye morphogenesis—instead it acts more like an on-off switch. This suggests that *Pax6* was repeatedly co-opted in the convergent construction of eyes, not that eyes are monophyletic *at the level of macroscopic organization.* When viewed in the light of out-group analysis, this speaks in favor of the convergent cooptation of *Pax6* for visual functions.

The same lessons can be brought to bear on the evolution of heads and nervous systems. We should not assume from the presence of deep homologs in the patterning of vertebrate and arthropod nervous systems that this is reflective of the ancestral function of the underlying orthologous genes—particularly when these genes are responsible for very coarse-grained neuroanatomical outcomes, rather than specifying the particular contours of brain development. The fact that the head-to-tail patterning of animals is directed by homologous transcription switches does not make arthropod, vertebrate, and cephalopod heads and mouth parts homologous. As zoologist Thomas Cavalier-Smith has aptly put it, "thinking human and grasshopper heads structurally homologous is as bad as calling a vacuum cleaner and light bulb homologous, because identical switches can turn both on."[12]

This brings us to the evidence of gross structural homology. It has long been recognized that the mushroom bodies of insects and stomatopod crustaceans—the centralized processing lobes responsible for learning, memory, and spatial cognition in these groups—appear to have organizational similarities to the mammalian hippocampus. For instance, brain regions associated with specific behaviors in vertebrates and arthropods, such as locomotion, are located in the same relative position with respect to other neural subsystems, such as those associated with light sensing. Why should this be so, if not a contingent quirk of history transmitted faithfully through common ancestry? Other structural similarities between arthropod and vertebrate brains include the parallel arrangement of neuronal fibers recently reported by Wolff and Strausfeld,[13] as well as reentrant connections that produce a bidirectional flow of information to and from disparate regions of the brain, allowing for the integration of sensory data and the executive control of bodily movement.

To bear the clear stamp of history, however, it is crucial that the organizational similarities identified between arthropod and vertebrate brains are not functionally constrained features of nervous systems in general, lest similarities due to convergence be mistaken for similarities due to common ancestry. The key question, therefore, is whether the structural resemblances that have been identified in the ground plan of insect mushroom bodies and the vertebrate hippocampus reflect the quirks of common ancestry or broad external constraints on nervous system form.

Some "vertebrate-like" features of invertebrate nervous systems—especially in the mushroom bodies of insects and the vertical lobe of coleoids—are probably due to the convergent evolution of cognitive functions that support activity-dependent and long-term plasticity mechanisms that underwrite memory, associative learning, categorization, and the flexible behavior these capacities underwrite.[14] Thus, at least some of the neuroanatomical similarities across brain-bearing phyla could indicate functional constraints rather than a single macro-morphological origin of the brain.

In focusing on the compelling organizational similarities between vertebrate and arthropod brains, it is easy to overlook the fact that the complex brains of coleoid cephalopod mollusks do not share any of these features, which in turn suggests that it evolved independently. Mollusk brains thus represent an important anomaly facing the bilaterian brain homology hypothesis. Let us now examine this anomaly more closely.

1.5 The Curious Case of the Coleoids

Even if one takes developmental similarities to be more probative than outgroup analyses in determining the polarity of a trait, a further wrench is thrown into the bilaterian brain homology hypothesis by the complex brains of coleoid mollusks, which do not fit neatly into the homology picture. Coleoids are a subclass of cephalopod mollusks whose extant members include octopuses, squid, and cuttlefish, and whose extinct members include the belemnoids (whose conical calcite guards pervade Mesozoic marine rocks). Coleoids are closely related to other cephalopods, such as the extinct ammonites and the living nautilids; and as mollusks, they are more closely related to all brainless bivalves and ganglionated snails than they are to other animal phyla with executive brains, such as vertebrates and arthropods. We saw in chapter 7 that the convergent camera eye of coleoids is strikingly similar to the vertebrate eye in terms of its gross morphology and functionality. If complex eyes are the evolutionary gateway to the mind, then we should expect to find the high-resolution, focus-capable eyes of coleoids connected to a coevolved neuronal mass that rises to the level of a brain.

The coleoid camera eye is indeed complimented by a behemoth brain—one that is stunningly complex by molluscan standards, but also not too shabby by vertebrate standards, weighing in at 140 million neurons (which is significantly larger than the brain of a mouse). This massive expansion of the ancestral mollusk nervous system includes not only formidable optic lobes to process visual information, but also the vertical lobes: centers of higher-order integration that are associated with memory and learning (though long-term storage occurs in a different area of the coleoid brain).[15] Octopod nervous systems are more distributed than those of vertebrates and arthropods, however, with nearly three-fifths of their neuronal mass dedicated to the operation of semi-autonomous sucker-lined arms and chromotaphores (color pigment cells used in crypsis, mimicry, and hunting displays), both of which are capable of autonomous activity. The extent of executive autonomy in octopuses, and in coleoids more generally, remains an area of active research.[16] Nevertheless, we know from coleoid lesioning studies—in which researchers execute carefully controlled damage to the vertical lobe and then infer from behavioral observations of the lesioned animal that certain cognitive changes have occurred—that visual, chemosensory, tactile, and proprioceptive system inputs are integrated, complex representations are formed, and behaviors are executively controlled by the coleoid central nervous system.

Whereas all vertebrates have executive brains that are presumably descended from a single common vertebrate ancestor, mollusks show tremendous diversity in their nervous system organization. If the last common ancestor to *Bilateria* had a complex centralized nervous system, this would imply (among other things) that noncephalopod mollusks, such as bivalves, snails, and chitons, lost (or massively reduced) the ancestral bilaterian brain. As noted earlier, this significantly multiplies the number of brain losses and body plan reconfigurations that are entailed by the bilaterian brain homology hypothesis.

The brains of some coleoid mollusks, such as the common octopus, are encased in a cartilaginous cranium between the eyes and contain far more neurons than many mammals. Yet they are organized in a fundamentally different way than the brains of vertebrates and arthropods. There is currently no evidence that the vertical lobe of coleoid mollusks is homologous to the vertebrate hippocampus or to the arthropodian mushroom bodies, even if it is similar at a cellular and functional level, including the reentrant pathways and long-term potentiation capacities that are characteristic of an integrative and associative brain structure.

The alien nature of coleoid brain anatomy is complemented by the unique developmental patterning of coleoid nervous systems. The molecular and developmental signatures of brain homology that have been identified for arthropods and vertebrates also appear to be lacking in coleoids.[17] For instance, the deep homologs that are implicated in the development of vertebrate and arthropodian nervous systems do not appear to have similar functions in the development of coleoid mollusk brains. Furthermore, although cephalopod neurons are similar to those of vertebrates, the molecular basis of learning and neuromodulation in cephalopods draws upon a family of proteins that are uniquely conserved in mollusks.[18] Finally, analyses of octopod genomes show massive parallel expansions in gene families associated with the specification of nervous systems in vertebrates, which indicates multilevel convergence on enhanced cognitive functions via the repeated modifications of conserved genes.[19] In short, the developmental data bolster out-group analyses that indicate a separate origin of brains in mollusks, with some convergent co-optation of conserved molecules. This, in turn, bolsters the out-group analysis of brain evolution in bilaterians more broadly.

How can the bilaterian brain homology hypothesis make sense of the coleoid case? One response would be to propose that vertebrates and arthropods share a common brainy ancestor that mollusks do not also share. But this would entail a massive overhaul of bilaterian phylogeny, requiring that we reject the protostome-deuterostome division—a grouping that few would dispute and that is itself premised on the profound developmental differences between these clades (which are distinguished in part by whether the mouth is generated from earlier or later invaginations in embryonic development).

A second, more plausible response is to propose that the ancestral bilaterian head/brain/eye complex was lost in early mollusks but then regained in cephalopods. In that case, however, there would be at least two origins of brains, which refutes the bilaterian brain homology hypothesis. Moreover, the loss-followed-by-gain scenario in mollusks proposes two fairly complex character state changes, which is less parsimonious than the convergence hypothesis, which proposes only one. Even if one thinks that brain losses are more probable than gains, this is of little help here because a gain is presupposed in addition to a loss, though this burden would even out if arthropod and vertebrate brains were deemed homologous. In short, there are no clear advantages to this hypothesis although, of course, it could still be true.

There is a meaningful sense in which coleoids are "intelligent aliens on Earth." That phrase is not meant to be taken literally to imply that coleoids originated on another planet and were transported to Earth on a soft-landing meteor (though at least one scientific paper makes that eccentric argument![20]). Rather, it is to say that coleoids are among the best approximations we have of intelligent extraterrestrial life because they represent a clearly separate origin of brains boasting mammalian levels of neural complexity, giving us a rare window into the possibility space of alien minds.

1.6 The Gateway Revisited

Let us consider one final piece of the puzzle that comes on the heels of the foregoing discussion of eyes and brains. The "evolutionary gateway" argument, laid out in previous chapters, contends that image-forming sensory modalities, especially vision, are not only robustly replicable in their own right but also tied to the evolution of sophisticated cognition and behavior through a coevolutionary feedback process that resulted in neural-motor-proprioceptive complexification in several bilaterian phyla. Once the Umwelt platform was established and organisms found their place in the world, more complex cognitive functions, such as flexible learning capacities, could then be folded in.

If this story is right, then the prospect of bilaterian brain homology hinges on the prospect of bilaterian eye homology—and the latter scenario, as we saw in earlier chapters, is implausible (though not impossible). The ontogeny of the camera eye in cephalopod mollusks is completely different than that of the vertebrate camera eye, resulting in a blind spot in the latter but not in the former. The cephalopod camera eye also employs radically different mechanisms of focus: cephalopod eyes focus by moving the lens closer to or further from the retina, whereas vertebrate eyes focus by changing the shape of the lens itself. Cephalopod camera eyes do not seem to be merely arthropod compound eyes that have been reduced to a single chamber and afforded a few bells and whistles. Consistent with the profound developmental and structural differences between arthropod, vertebrate, and cephalopod eyes, there are no known genes that direct the specific gross morphological arrangements of all eye types. Eyes are thus likely to be polyphyletic both at the phenotypic level and at the level of genes. This, in turn, suggests that brains are probably polyphyletic as well.

Fossil evidence bolsters this conclusion. Proto-vertebrates found among the early Cambrian fauna, such as *Pikaia* (the critter we encountered in chapter 2), lack true eyes and a brain. It is hard not to interpret such animals as basal or "stem" vertebrates that branched off before the vertebrate lineage had developed complex eyes and cranium-encased central nervous systems. One strains to see *Pikaia* and its early vertebrate ilk as a secondarily reduced vertebrate condition, which despite a pelagic swimming ecology, lost all of the major neurological and optical accoutrements of its ghostly bilaterian ancestor.

Moreover, cephalopod brains clearly postdate the Cambrian explosion, perhaps by tens of millions of years; the same may be true, though to a lesser extent, of vertebrate brains (see discussion of Cambrian craniates in chapter 2). All this suggests that image-forming eyes—and with them, heads, brains, and mobile bodies—either were absent in the last bilaterian ancestor (the more parsimonious reading) or else were present in the last bilaterian ancestor but were completely lost and then regained in cephalopods and vertebrates (the less parsimonious reading). Either way, we are looking at the multiple origins of eyes, cephalic brains, and the active lifestyles these structures support.

In sum, the totality of the evidence at present speaks to the multiple origins of brains via the repeated cooptation of conserved genetic, regulatory, and cellular elements. Even if brains were constructed out of homologous cell types, patterning mechanisms, and secretory signaling molecules, their information-integrative functions would have been realized independently in several bilaterian lineages. Central nervous systems are thus likely to be robustly replicable outcomes, at least given the origins of neurons and synapses—a problem to which we now turn.

2. The Perplexing Phylogeny of the Neuron

If the evolutionary history of animal brains is uncertain, the phylogeny of the neuron and synaptic signaling systems is shrouded in even greater mystery. As with brains, the evidence for neuron cell type homology is mixed. There are currently no known genes or regulatory networks that produce neurons in all groups of animals that have them, which is contrary to what we should expect if the metazoan neuron homology hypothesis were true. And just as importantly, neurons do not arise from the same germ layer in embryonic development in all animals that have nervous systems. In some animals, such as vertebrates and arthropods, neurons arise from the ectoderm (the outer embryonic layer) while in others, such as cnidarians, neurons arise from the endoderm (the middle embryonic germ layer). These differences in development could imply separate epigenetic origins of nervous tissue in metazoans, leading some neuroevolutionary biologists to treat the neuron cell type as a de facto functional kind so as to leave open the question of common ancestry.[21]

Arthropods and vertebrates have similar electrical synaptic connections and use the same low-weight neurotransmitters to regulate voltage-gated ion channels between neurons; but the gap junctions that establish these intercellular channels are formed by entirely different molecules in each phylum.[22] Although there are molecules involved in synaptic signaling in most animal neurons (but see the discussion of ctenophores in the next section), these molecules are also used in non-neuronal cells and are present in unicellular eukaryotes and prokaryotes—and thus could easily have been co-opted on separate occasions for neuronal functions.[23] But just as genetic and developmental similarities do not necessarily indicate homology, neither are genetic and developmental differences dispositive on the issue of convergence. Perhaps neurons and synaptic structures originated once but were genetically and developmentally reorganized in early metazoan evolution. Out-group comparisons add the much-needed paleontological context in which to interpret the comparative genetics and development of nervous systems.

2.1 A View from the Abyss

A lynchpin to the neuron evolution debate is the phylogenetic position of a group of animals called "ctenophores." Ctenophores are beautiful, active predators that superficially resemble jellyfish; they have surprisingly complex sensorimotor systems (including high neural concentrations and true muscular systems), which underwrite a range of hunting behaviors that have been compared in their diversity to the predatory repertoire of spiders. The deep-sea "aliens" of James Cameron's film *The Abyss* (1989) seem to have been loosely—and, in hindsight, serendipitously—modeled on ctenophores, though they still bear the scars of the humanoid epidemic.

Ctenophore nervous and muscular systems are produced by nonhomologous genes and germ layers and deploy synaptic signaling mechanisms that differ from those of other metazoans. It is true that homology in a trait can be maintained despite a significant turnover in its genetic underpinnings.[24] However, the sheer turnover that is contemplated for the evolution of ctenophore nervous systems is so comprehensive, and on so many levels, that a homology-preserving transition seems on its face unlikely.

This leaves us with two options: either (1) the genes and low-molecular-weight signaling mechanisms (such as serotonin, dopamine, adrenaline, acetylcholine, and histamine) associated with nervous system ontogeny and function have been entirely lost and replaced in ctenophores, despite being deeply conserved in cnidarians/bilaterians and despite ctenophores maintaining an active predatory (rather than sessile or parasitic) lifestyle; or (2) functional neurons and nervous systems arose at least twice in metazoans.

Which of these scenarios is likely to be correct depends, in no small part, on the phylogenetic position of ctenophores. If ctenophores are the sister taxa to all other animals and hence more basal than sponges, as some recent research suggests,[25] then this militates in favor of the multiple origins scenario, especially when taken in conjunction with the molecular-developmental data. Alternatively, if sponges, which are sessile and have no neurons or nervous systems, are the most basal animal group—which has long been suspected to

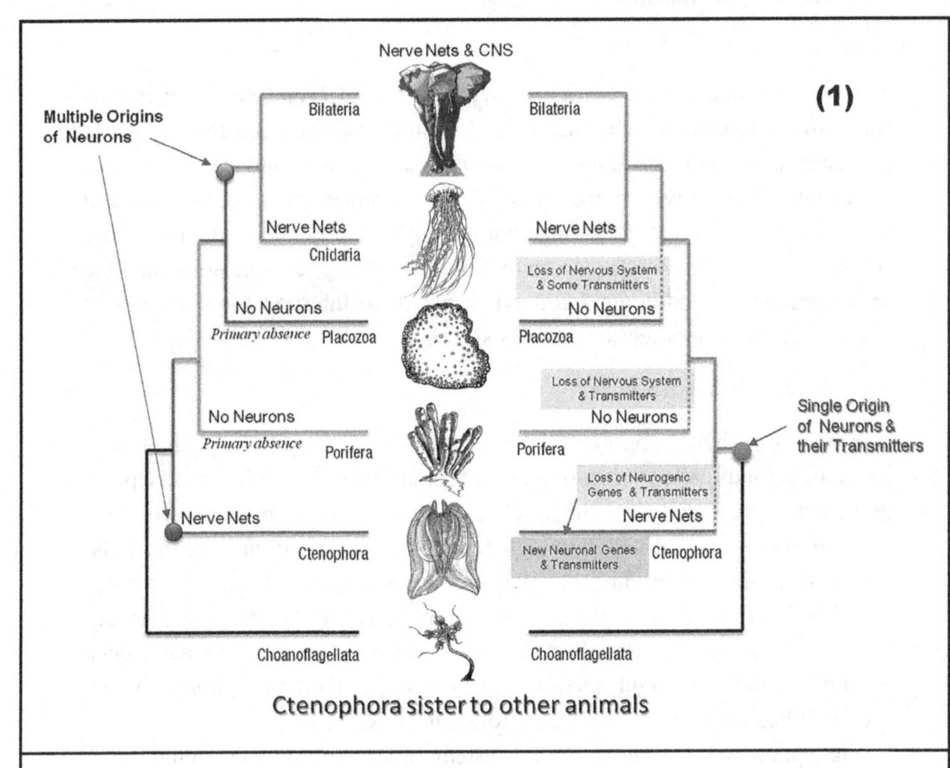
Nerve Nets & CNS
(1)
Multiple Origins of Neurons
Bilateria
Bilateria
Nerve Nets
Cnidaria
Nerve Nets
Loss of Nervous System & Some Transmitters
No Neurons
Primary absence
Placozoa
No Neurons
Placozoa
Loss of Nervous System & Transmitters
No Neurons
Primary absence
Porifera
No Neurons
Porifera
Single Origin of Neurons & their Transmitters
Loss of Neurogenic Genes & Transmitters
Nerve Nets
Ctenophora
Nerve Nets
New Neuronal Genes & Transmitters
Ctenophora
Choanoflagellata
Choanoflagellata
Ctenophora sister to other animals

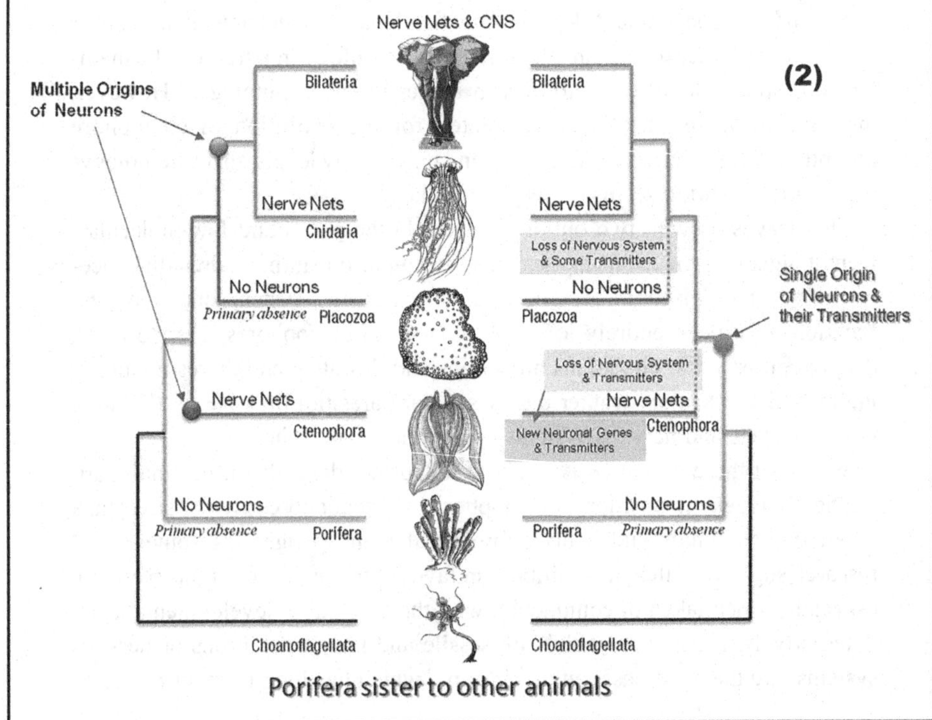
Nerve Nets & CNS
(2)
Multiple Origins of Neurons
Bilateria
Bilateria
Nerve Nets
Cnidaria
Nerve Nets
Loss of Nervous System & Some Transmitters
No Neurons
Primary absence
Placozoa
No Neurons
Placozoa
Single Origin of Neurons & their Transmitters
Loss of Nervous System & Transmitters
Nerve Nets
Ctenophora
Nerve Nets
New Neuronal Genes & Transmitters
Ctenophora
No Neurons
Primary absence
Porifera
No Neurons
Porifera
Primary absence
Choanoflagellata
Choanoflagellata
Porifera sister to other animals

be the case and also garners recent support[26]—then this militates in favor of the comprehensive neuronal loss/replacement scenario in ctenophores. Within that picture, two possible scenarios emerge: (1) nervous systems were maintained in ctenophores through an unbroken chain of common decent despite a comprehensive turnover in the molecular and genetic mechanisms underlying them; or (2) nervous systems were lost and then regained in ctenophores, an event that would effectively constitute a novel origin.

As Moroz and colleagues point out,[27] however, even if sponges turn out to be the most basal animal, this is consistent with a multiple origins scenario for neurons, synaptic morphology, and primitive nervous systems, so long as ctenophores are the second lineage to branch off from sponges, followed by placozoans (the simplest known nonparasitic animal, but according to molecular analysis not the most basal), cnidarians, and bilaterians, respectively (see figure 9.2). Alternatively, in the single-origin scenario, neurons could have been present in the last common ancestor of ctenophores/cnidarians/placozoans but subsequently lost in placozoans and completely turned over in ctenophores. It is not clear which of these scenarios is most probable.

In short, the same dilemmas that impede the evolutionary reconstruction of bilaterian brains—such as how we ought to weigh the relative probabilities of character state gains and losses, the prospects of developmental co-optations and turnovers, the relative weight accorded to out-group versus developmental analyses, and so forth—also encumber the phylogenetic construction of the *building blocks* of brains. If the neuron cell type did turn out to be an iterated functional kind, this might explain some of the genuinely perplexing aspects of neuron phylogeny. As biologist William Kristan concludes in reviewing the current scientific picture of neuronal evolution, "the multiple origins of neurons may, in fact, be why defining 'neuron' is so difficult, and why defining the origin of neurons is so complex."[28]

Figure 9.2
Two scenarios for the evolution of neurons. In the first scenario (*top*), ctenophores are taken to be the most basal animal, and neurons either arose twice (once in ctenophores and once in the last common ancestor of cnidarians and bilaterians—see left side); or else they arose once (in the common metazoan ancestor), neuronal genes and neurotransmitters were completely turned over in ctenophores, and nervous systems and neurotransmitters were entirely lost in sponges and placozoans (see right side). In the second scenario (*bottom*), sponges are taken to be the most basal animal, and neurons either arose twice—once in ctenophores and once in the last common ancestor of cnidarians and bilaterians, so long as ctenophores are more basal than placozoans (see left side); or neurons arose once and neuronal genes, neurotransmitters were completely turned over in ctenophores, and nervous systems were lost in placozoans (assuming, again, that ctenophores are more basal than placozoans). Redrawn from L. L. Moroz, "The Genealogy of Genealogy of Neurons," *Communicative and Integrative Biology* 7, no. 6 (2014): e993269.

2.2 The Evolvability of Functional Neurons

There appears to be no shortage of ways to realize neuron-like potentials in complex multicellular organisms. Many animal cells other than neurons are capable of generating action potentials that facilitate communication across cell populations, so it is not difficult to imagine secretory cells being repeatedly co-opted for neuron-like functions. Neurons can be created in vitro by modifying the epigenesis of cnidarian epithelial cells, which suggests that the repeated evolution of functional neurons from non-neuronal cell lines cannot be too difficult to achieve. The evolvability of functional neurons is further supported by convergence on action potentials and information-transfer mechanisms in lineages for whom rapid sensory-motor mechanisms are either inaccessible or not required.

For instance, action potentials have evolved in the first major origin of complex multicellularity: the green plants, some of whom, such as the carnivorous Venus flytrap, are capable of limited rapid movements. Such "real-time" plant behaviors are made possible by action potentials that are analogous in certain ways to animal nervous systems. Mechanosensory stimuli trigger sensory hairs, which then generate a propagating action potential that initiates a rapid motor response—such as the snapping shut of two leaf lobes, resulting in the imprisonment of hapless insect prey. Though the precise biochemical mechanisms of this snapping mechanism are poorly understood,[29] it is likely achieved by gated ion channels, which produce a flow of water or acid molecules that cause cells in the lobes to change shape, causing the lobes, which are held under tension, to snap shut. A basic memory system is also employed: to avoid snapping shut due to noise (such as raindrops), the snapping mechanism is only initiated when two stimuli separated in time by a few seconds are detected.

Even plants that lack action potentials have cognition-like links between stimulus and behavior, albeit on a timescale that is not intuitively obvious to humans. Like sessile animals, most plants have lifestyles that do not require active movement. As a result, they can afford to rely on more sluggish, protein-based molecular signaling mechanisms that have "on-off" (Boolean) switch-like capabilities and exhibit feedback and frequency modulation effects. These complex and poorly understood metabolic transduction networks, which are functionally analogous in certain ways to neural nets, enable plants to learn, remember, navigate stimulus fields and behave "intelligently" to vagaries of the external environment.[30]

Hundreds of intercellular signaling molecules are probably deployed by these systems, including but not limited to hormones. The information integrational capacities of hormone-style signaling systems may have outer limits.

For example, due to their low signaling speed, such systems are probably unable to achieve the synchrony and reentry necessary for stimulus binding, which may preclude plant consciousness—or at least consciousness in the rich Umweltian sense discussed in previous chapters. The point, however, is that the evolution of proto-neural functions in plants—a clade that lacks motor systems and relies almost entirely on growth for movement—indicates the deep replicability of the action potentials, logical switching mechanisms, and molecular signaling networks that support sophisticated behavior.

In sum, it is highly possible that functional neurons originated more than once in the evolution of animal epigenesis, likely from multiple germ layers, by co-opting the basic signaling functions of secretory cells and recruiting low-weight molecules as neurotransmitters. If we conceive of nervous systems even more broadly to include molecular transduction networks that are designed to transmit information, such as those found in green plants, then the outcome is even more evolutionarily robust. Even if a homologous cell type—such as a monophyletic neuron—were implicated in the construction of nervous systems in all animal groups, this would not convert brain convergences into "Gouldian repetitions" (see chapter 5); for if there are broad constraints on the structure of bilaterian nervous systems, there is no reason to think they stem from any radically contingent properties of neurons themselves.

At the very minimum, then, we can say that the evolution of brains is likely robust if we presuppose the evolution of neurons and synaptic signaling mechanisms among the initial conditions of the replay. Whether brains are cosmically projectible, and if so, with what sort of frequency, depends on the evolutionary robustness of functional neurons as well as other major developmental and metabolic innovations that resulted in animal-grade complexity, which at present are not well understood. Of course, future findings may call all of this into question—that is the beauty of science. But for now, we have to make the most of our view from the frontier.

2.3 From Brains to Minds

Even if centralized nervous systems have evolved repeatedly and independently, this does not show that mind is a robustly replicable feature of the evolutionary process. This is because the existence of brains does not decisively establish the existence of Umweltian minds. We can be reasonably certain that the distributed nervous systems of cnidarians and echinoderms, and the primitive ganglion of bryozoans and nematodes, even if they accomplish some levels of information integration, do not engage in stimulus binding or give rise to a unified consciousness. But we are on more uncertain grounds when it comes to the centralized brains of arthropods and mollusks and the cognitive mechanisms they might support. Not all information processing and integration results in stimulus and semantic binding, mental representations of the body in space, attention, concepts, subjectivity, emotions, and so on. Because the inference from brains to minds is imperfect, the neuroanatomical evidence must be corroborated by observations of animal behavior. In the next chapter, we explore this critical source of evidence.

10 Finding Minds: Evidence from Behavior

We now have a reasonable grasp of where in the living world nervous systems of varying complexity can be found. And this distribution of nervous systems, when viewed in the light of evolutionary relationships among animal groups, can tell us something about how replicable brains might be. This is not sufficient, however, to determine the distribution and replicability of animal *minds*. To better interpret the neuroanatomical data, we need outwardly observable signatures of the cognitive functions that are closely linked to our intuitive notion of what it is to be a thinking being in a meaningful world. For this, we must look to behavior.

Our search for the behavioral signatures of mind will focus on invertebrates, for it is only here, in the deepest phylogenetic depths of animal evolution, that the replicability of Umweltian cognition and consciousness can be established. Because brain-bearing invertebrate lineages, like arthropods and cephalopod mollusks, are likely to share a brainless common ancestor both with vertebrates and with one another (as argued in chapter 9), any minds that are found to exist in these invertebrate animals would by implication have separate evolutionary origins. Although there is substantial evidence of convergent cognitive evolution both within mammals and between mammals and birds (discussed in the coda to part II), such intraphyletic patterns of convergence are built on a homologous vertebrate brain and Umweltian platform—and thus they fail to evince the levels of projectibility that could support a cosmic biology of mind.

Reasoning about invertebrate minds is trickier than reasoning about vertebrate minds, because invertebrates, as we have seen, lack homologous brain structures (such as the cortex and midbrain) that are associated with conscious states in vertebrates. The absence of vertebrate-specific brain structures obviously does not preclude other, independently evolved brain structures from giving rise to mentality in invertebrates. Though we can make tentative analogical-style inferences about animal minds based on convergent functional neuroanatomy,[1] discerning the functions of unfamiliar brain regions requires additional corroboration. Let us therefore consider the behavioral evidence for minds without spines.

1. Cephalopod Smarts

We saw in the previous chapter that coleoid cephalopod mollusks have complex brains that rival those of mammals, possessing a vertical lobe that serves as a center of higher-order information processing, signal integration, and executive control. What sorts of cognitive capacities do these brain structures give rise to and how are they manifested in coleoid behavior?

1.1 Measuring an Alien Mind

Experimental research on coleoid cognition and behavior is sparse. Unlike the wide-ranging, systematic, and carefully controlled work on arthropods (which will consume the bulk of our discussion), much of the evidence for cephalopod intelligence is anecdotal or gleaned from field observations that are open to multiple interpretations. For instance, there are reports of octopuses breaking out of their laboratory tanks during the night to steal a crustacean from an adjacent aquarium, only to return to their assigned tank before experimenters arrive in the morning. A cursory scouring of aquarium hobbyist blogs reveals similar reports of octopod antics with varying levels of hyperbole and credibility.

Such cheeky activities, even if they were empirically verified, are not far removed from the octopus's natural behavioral repertoire. Octopuses are central place foragers that often hunt in tide pools at night; this requires that they navigate obstacles and occasionally exit the water, only to return safely to their den with (or having already consumed) their bounty. Aquarium escape, foraging, and return behaviors are indicative of important types of cognition (more on these below), but they may not demonstrate the kind of flexibility that is indicative of "thinking" or "insight," let alone a brazen attempt to pull a fast one on experimenters and hobbyists. We might say the same for other octopod anecdotes, such as one recounted in a recent book by Peter Godfrey-Smith, in which a scuba enthusiast is led by an octopus, hand-in-tentacle, back to its den "as if he were being led across the sea floor by a very small eight-legged child."[2] It is tempting to see such a formative event (at least for the diver) as a veritable alien encounter—a communing of kindred intellect spirits across the chasms of evolutionary time. But procuring shiny new objects and transporting them back to their den is what octopuses do in the ordinary course of things.

A hallmark of sophisticated cognition and, perhaps, of the richness of mind, is the ability to produce flexible rather than stereotyped responses to challenges that animals do not ordinarily encounter in their natural habitat.[3] Cognition research on octopuses, and to a lesser extent on cuttlefish (very little has been done with squid), shows coleoids to be domain-general problem solvers and possibly even tool users. For instance, octopuses have a strong penchant for pulling, a behavior that may to some extent be incorporated into the autonomous neural routines of their arms. And yet they can learn in the laboratory to maneuver an L-shaped box through a square-sized hole by performing a subtle combination of pushing and pulling movements that is far removed from any standard behaviors they have to carry out.[4] They can also famously learn to unscrew jars and bottles to retrieve a prey item inside,[5] and to unscrew containers from the inside to escape their own confinement—complex motor sequences that have no natural analog.

There is also work that suggests octopuses may be capable of causal reasoning and tool use, though this work is based on field observation and not subject to experimental controls. For instance, researchers diving with octopuses near Bali have observed numerous cases of octopuses awkwardly carrying around coconut shell halves which they assemble as a defensive shelter only when needed, deploying an unnatural "stilting" gait to move across the sea floor with their makeshift shelter (see figure 10.1).[6]

This behavior differs in its underlying flexibility from instinctual defensive behaviors, such as the occupation of empty gastropod shells by hermit crabs or pom pom crabs carrying stinging anemones on their claws for defense, in several key respects: (1) the causal properties of the coconut shell and its utility in defense against predation are presumably learned; (2) the item is carried around in a cumbersome and unnatural manner, which entails incurring a short-term cost (reduced locomotion and increased energy expenditure) for a long-term benefit (predation avoidance); (3) the tool is deployed only in situations that call for it; and perhaps most important, (4) the octopus behavior smacks of insight, foresight (prospection), and perhaps even causal reasoning. The fact that multiple individuals were observed carrying around coconut shells in this way also raises the prospect of observational learning, although the researchers did not speculate as such.

Coleoids also exhibit spatial memory that they use in navigation. Octopuses, for instance, become familiar with the topography surrounding their den and negotiate complex foraging excursions. In the laboratory, this takes the form of successful maze learning and detour-taking, both by aid of visual landmarks and by remembering the location of a place or object despite temporarily losing visual contact with it.[7] This indicates that octopuses have goal and object

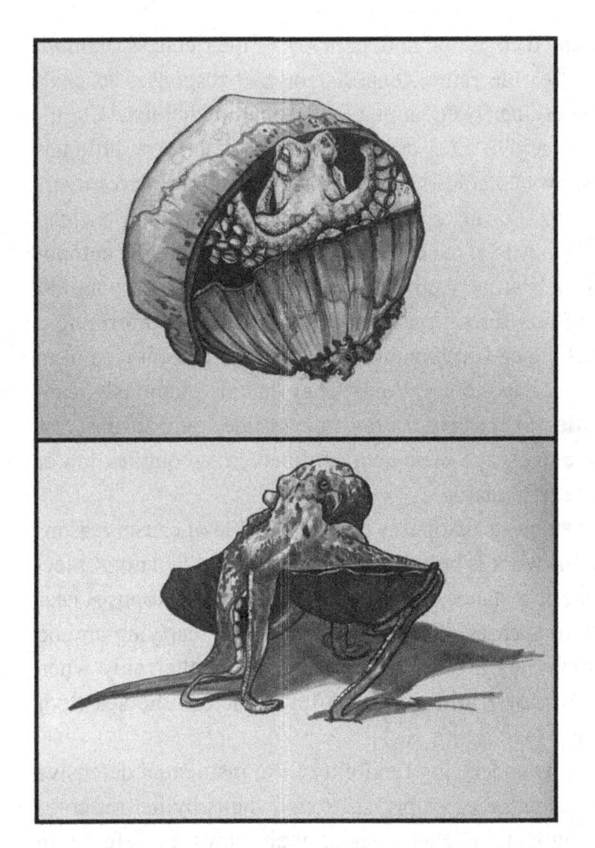

Figure 10.1
Veined octopus using coconut shell-half assembly as a shelter (*top*), and engaging in cumbersome stilt-walking with the tool, ready to deploy as needed (*bottom*)—a learned behavior that smacks of insight, planning, and causal reasoning. Redrawn from J. K. Finn, T. Tregenza, and M. D. Norman, "Defensive Tool Use in a Coconut-Carrying Octopus," *Current Biology* 19, no. 23 (2009): R1069–R1070.

permanence. They are also capable of conditional discrimination, and have long-term memory capacities that retain discrimination rules for weeks after the initial training period has ended. There is also recent evidence for episodic-like ("what/where/when") memory in cuttlefish (*Sepia*),[8] similar to the cognitive capacities that psychologists Nicola Clayton and Anthony Dickinson famously demonstrated in scrub jays.[9] More generally, octopuses are known to have an exploratory disposition that includes an affinity for novel objects, and they have been observed using their jet streams to "play" with unfamiliar objects in their aquarium.[10]

Cephalopod cognition researcher Jennifer Mather argues that coleoids demonstrate "imagination" by dangling a tentacle to lure a prey item, "causal reasoning" by using dynamic skin color displays to confuse their prey,[11] and "behavioral flexibility" by using human trash to adorn their dens. However, these behaviors have innate analogs in other animals to which we do not normally attribute proper imagination, causal reasoning, or unusually flexible behavior (such as anglerfish, sailfish, and bower birds, respectively). Experimental challenges that place cephalopods outside of their depth, so to speak, are more telling in these regards. Debate over the nature of coleoid consciousness continues. Some researchers, such as Sidney Carls-Diamante, argue that given the autonomy of its arms, the octopus may be comprised of several cognitive systems and multiple loci of consciousness.

From the perspective of many people who have spent time observing them in the ocean and in aquariums, octopuses show an inquisitiveness that seems to be lacking in most other creatures of the sea. Time will tell whether this common intuition can be translated into empirically tractable hypotheses. Meanwhile, research on cephalopod minds has begun to filter down into philosophy and public policy, with cephalopods even being included in some jurisdictions (such as the European Union) as an "honorary vertebrate" for the purposes of research ethics protocols. It may seem odd that an organism like the octopus, which has a relatively short lifespan and moves rapidly through developmental stages, would spend its brief lifetime learning so much about its world. This cognition-life-history conundrum will now be inflated to steroidal levels as we consider the stunning cognitive abilities of bees.

2. The Tiny Brain Revolution

In an episode of the American television series *The Twilight Zone* called "The Intruders" (1961), a lone woman in a rustic cabin hears a bang on her roof; shortly thereafter, she finds herself battling two miniature humanoid figures (each only a few inches tall) wearing pressure suits and wielding small radiation weapons. Terrified, she parries with them around the home until she manages to crush one of the creatures and toss it into the fireplace. She chases the second figure to the roof, where she finds a small spaceship and proceeds to destroy it with a hatchet. For the entire episode, there is no speaking at all, only shrieks of fear and pain, until the very end—when the tiny figure who had retreated to the small craft, with a now visible label "US Air Force Space Probe" blazoned on its side, frantically radios back to Earth that they have encountered a race of hostile giant humanoids and that the planet should be avoided at all costs. It becomes clear that the diminutive space aliens are in fact human astronauts visiting another planet.

Is it possible that *we* could be the backward, ignorant giants fearfully and gratuitously crushing intelligent aliens around us with nary a thought or care? Arthropods are the descendants of the very first lineage of life on Earth to evolve eyes, brains, and able bodies. Although arthropod cognition and behavior has been far more systematically investigated than that of cephalopods, this work has yet to make its way into mainstream philosophical thought and science policy. Indeed, the latter remain in the grips of traditional stereotypes about insects, such as the supposedly reflexive, instinctual, and inflexible nature of their behavior; and many continue to assume that tiny, pinhead-sized brains cannot possibly support sophisticated cognition or consciousness.

However, there is now a sizable body of experimental work documenting flexible learning capabilities in insects that, in some cases, rival those found in mammals and birds. These cognitive abilities range from conditional discrimination and concept formation to spatial cognition, planning, causal reasoning, and social learning. Indeed, a review of the insect cognition literature leaves one with the impression that bees are likely to outperform birds and mammals on many quintessential cognitive tasks, such as matching-to-sample discriminations and the cross-modal transfer of learned concepts—often necessitating fewer trials for success than is necessary to train up similar abilities in mammals (including primates!).[12]

Why has this rich and important body of comparative cognition research had little influence on philosophical thinking about the nature, evolution, and ethical significance of animal minds? Many have assumed that such absolutely small-brained animals could not be loci of sophisticated visual processing, learning, memory, consciousness, and emotion. However, this is an empirical question, not an a priori truth, and the empirical tide is now turning. As far as cutting-edge work on invertebrate cognition goes, arthropods are where the action is.

2.1 What Do Insects See?

One way to approach the question of arthropod minds is from the angle of image-forming perception. What do compound-eyed arthropods, such as insects, see? We might presume, given extensive homology in the vertebrate visual system, that the way humans see the world is unlikely to be fundamentally different from the way that other vertebrates see, at least in very broad strokes. Visual perception in invertebrates with convergent head/eye/brain complexes is another matter entirely. What is the world like, if anything, through the eyes and brain of a bee?

"Bug goggles" are a children's novelty toy that claim to simulate the visual world of arthropods. In effect, they break up the human visual field into numerous mini-fields that are supposed to represent ommatidia (or units) of the compound eye, each rendering a miniaturized panorama with human-typic levels of spatial resolution. In actuality, to simulate arthropod vision, bug goggles need to generate a single, unified, lower-resolution visual field that is assembled from light entering each individual ommatidium, broadcast across the retinal array, and then stitched together into a scene that is perceived without phenomenological gaps. Of course, one cannot blame the manufacturers of a plastic children's toy for not accurately simulating arthropod vision; indeed, creating a viable model of insect vision turns out to be exceedingly difficult and requires that we make certain theoretical commitments. Even if we fully understood arthropod optics, this does not tells us how visual information is processed and packaged in the animal's brain.

2.1.1 Holism versus Localism. How are certain animals, such as insects, capable of discriminating between objects? Do they perceive objects more or less as we do—with identifiable boundaries, motions, colors, and identities—or do they merely detect low-level features of objects and use these features as the basis of navigation, discrimination, and action? This is not simply an academic debate about the nature of visual representation—it is critical to questions about the convergent evolution of the Umwelt, and it goes to the very heart of what the world is like for these animals, if it is like anything at all.

Debates over "holism" that were prominent in early discussions of Gestalt psychology now play out in the form of a debate between "global perception" and "local feature detection." Some researchers have argued that visual information in insects is not assembled into complex representations, and thus is not "seen" in any meaningful way, at all. Instead, they claim, a cluster of unbound local features (such as edges, colors, etc.) are detected in visual processing streams that are confined to each ommatidium or to a small cluster of ommatidia.[13] If this is right, then low-level cues are never integrated into a single, seamless visual scene populated by meaningful objects, and thus there is no arthropod Umwelt. On this deflationary view, the behavioral feats of arthropods, no matter how sophisticated they appear, are merely reflexive responses to low-level cues; they do not involve holistic representations of objects in the world or of the organism's body within that object field.

If one subscribes to this deflationary view, then the answer to the question "What does the insect represent?" is something like "Local, unbound features of objects, not objects in their own right." And if this is true of arthropod visual representation, then the answer to the question "What does an insect see?" may very well be "Nothing at all," insofar as seeing implies a see-er, and the absence of stimulus and semantic binding implies the lack of Umweltian cognition and consciousness. What remains is a real-life zombie: an organism that can get around in a complex world in surprisingly sophisticated ways, even though it does not form complex representations or harbor an embodied subject of consciousness. This deflationary view of arthropod visual perception is complemented by a similarly deflationary interpretation of insect learning studies, according to which combinations of low-level cues in a scene, pattern, or object are learned and used to guide incentivized behavior, even though the insect never represents stimuli as patterns or objects in the way that vertebrates do.

As we will see, these deflationary interpretations do not hold up to empirical scrutiny. Before examining the behavioral evidence for global visual processing in bees, however, it is useful to consider the ecological tasks to which these animals are adaptively suited. In the wild, bees and wasps are faced with complex, visually guided, central-place foraging tasks. They must learn and remember the types of flowers that are likely to be nectar-rewarding and their locations relative to the hive. This is challenging, because flower resources are ephemeral, patchily distributed, variably visited by competitors, and often distantly located (frequently many miles away from the hive). Generalist wasps that colonize new locales must learn which insects are appropriate prey items; and properly exploiting other food sources like carrion and flower patches requires active navigational search and memory so as to support multiple trips to and from the nest. Visual information is crucial to solving these tasks; olfactory cues, which diffuse rapidly in air, are insufficiently precise to support the high-flying navigational feats and real-time target selections that are involved in bee and wasp foraging.

On the face of things, it would seem that simple visual cues—such as unbound edges, colors, or other local features proposed by the deflationary view as the explanation for arthropod visual discrimination—would hardly suffice for these complex, high-octane activities. Holistic representation would seem to be crucial for active visual foraging on the wing, for which object recognition needs to be stable across visual angles and motion disturbances. One might think that comparably low-resolution visual systems, such as that of arthropods, would be more likely to rely on elemental cues than on holistic representations; but as bee cognition researcher Aurore Avarguès-Weber and colleagues point out, low-resolution visual systems may actually lend themselves to global over localized information processing. This is because global processing provides more useful total information than does feature detection, and it allows the animal to glean ecologically relevant signatures without being swamped by noisy detail.

Moreover, the brains of flying insects (like bees and wasps) engaged in active navigation and foraging must compute and update their relative positions with respect to landmarks, reference points, and targets; some of these points of interest are not accessible at any given moment to the senses and hence must be retained in memory. They must also solve "reafference" problems, or confusions due to inputs generated by their own motion, in order to perform the navigation and discrimination feats that they accomplish on the wing every day of their short but productive lives. For these purposes, generating a single, panoramic, internal simulation (an idealized, dynamic, updating representation) of one's body in space that integrates all available sensory inputs may be more computationally efficient than deploying a battery of nonintegrated sensory modules, each of which must be linked up to action in the right way.[14] The Umwelt may be a computationally parsimonious way—perhaps *the* most parsimonious way—of rapidly navigating an informationally complex world where spatial representation is at a premium. Perhaps that explains its convergent evolution in distant branches of the tree of life.

2.1.2 Inverted Faces and Abstract Paintings. There is now strong evidence that some insects, especially those with hypertrophied mushroom bodies, are capable of binding visual and semantic properties to objects that retain their identity over space and time, as well as integrating information from olfactory, mechanosensory, and gustatory senses into what appears to be a unified, dynamic simulation of the surrounding world.[15] Evidence for global perception in arthropods comes first and foremost from pattern recognition studies that aim to control for local feature detection. The upshot of this work is that bee brains can generate learning rules based on the holistic configuration of low-level elements of a scene, bind affective valences to these global categorizations, and then apply these rules to novel stimuli.[16]

One such type of study tests whether insects can recognize faces even when experimenters control for local features. Paper wasps have been shown to recognize the faces of conspecifics,[17] and honeybees can learn to recognize human faces even when they are confronted with face recognition tasks that are designed to be challenging for human subjects.[18] More tellingly, when the images of human faces are turned 180 degrees, the ability of honeybees to recognize them markedly drops off, just as it does for humans. This "inversion effect," as it is called—wherein inverting the images in face recognition tasks results in far lower accuracy and longer reaction times—is generally taken to be diagnostic of "configural" (global or Gestalt-like) rather than featural processing. The logic is that because bees are subject to the inversion and scrambling effects in the very same way that humans are, this indicates that bees represent the stimuli in a similar way to humans: namely, as a Gestalt or holistic representation that "pops out" of the visual scene. Likewise, when pictures of real faces that contain all relevant low-level cues are presented in a scrambled arrangement, bees do not recognize them as the training stimulus.[19] Low-level cues, such as visual centers of gravity (empty regions of space), color, edges, and visual angles, have all been excluded as the basis of discrimination by manipulating highly schematized faces that control for these variables. These results cannot be explained in any obvious way by feature detection.

Even if bees represent *faces* holistically, one might wonder whether this ability is produced by a specialized subsystem that does not extend to object perception in general because it is designed specifically to attend to the faces or heads of conspecifics or predators. The data suggest otherwise. Gestalt-like perception in bees has been shown not only for faces, flower morphologies, and forest scenes,[20] but also for paintings and even artistic traditions—a surprising competency that has also been demonstrated in birds.[21] For instance, bees can learn to distinguish specific Monet paintings from specific Picasso paintings, an ability that is perhaps not too far removed from their vaunted pattern-recognition abilities. More impressively, however, they are capable of learning *artistic styles*, such as impressionism and cubism, and they can apply these style discrimination rules to novel paintings in the same tradition—even controlling for color, brightness, and other low-level features.[22]

Bees succeed at this task even controlling for elemental cues such as brightness, color, feature orientation, salience of edges, spatial frequency, and other properties that might subtly differ between artistic traditions. For example, because it is possible that bees could use complex color combinations as low-level cues for discriminations between paintings and painting styles, researchers tested whether the learned discrimination rules would transfer across greyscale images of the paintings. Not only did they successfully transfer, but greyscale discriminations of novel paintings were more successful, possibly because color patterns were interfering with holistic style judgments.

It is unlikely that simple elemental cues could account for these discriminations, given the extraordinary complexity of the individual paintings and the control of confounders. As with complex forested landscapes and visual landmarks, paintings are more effectively detected and remembered by virtue of Gestalt-like representations than by disarticulated low-level cues. Precisely how bees and birds, or humans for that matter, detect structural regularities across artistic traditions is not fully understood. Nevertheless, these results suggest that bees can represent very complex and ecologically alien patterns and apply those learning rules to novel tasks.

This raises the question of how readily insects, such as bees, take to global visual processing? Is the holistic representation of patterns and objects the predominant mode of arthropod visual perception, or must attention be modulated in the right ways to achieve these Gestalt-like effects? We know from experimental work that humans are disposed toward global information processing. That is, we tend to see the forest before the trees (which is presumably why the devil is always buried in the details). In contrast, it was long thought that insects would prefer local to global processing, even if they could be taught to discriminate using holistic representations. The idea was that local feature processing is a faster and more frugal cognitive strategy, and hence an evolutionarily preferable one for animals working with more limited computational power.

However, this view of insect visual perception is also being overhauled. Recent discrimination experiments that manipulate local and global variables in stimulus patterns (such as global shapes made up of smaller shapes) show that given an option, bees prefer global processing over local featural processing unless they are specifically trained to attend to local features.[23] It is unclear to what extent these observations extend to other arthropods with homologous or convergent visual systems. Regardless, what this work suggests is that whether lineages rely on local or global information processing hinges less on their absolute brain size or the acuity of their vison and more on the ecological design problems they need to solve.

2.1.3 Mimicry as a Natural Experiment in Visual Perception. There is another, overlooked source of evidence for holistic perception that has yet to be mined: patterns of signaling in nature. The evolution of mimicry and other forms of visual signaling constitutes a "natural experiment" in comparative cognition that can provide insights into the perception of the "intended" receivers of those signals. If there are forms of mimicry that are striking in their holistic content to humans, then it seems likely that the coevolving lineages at which these signals are aimed also perceive their world in a Gestalt-like way. Determining precisely which lineages are the intended receivers of mimicry signals is a difficult problem that we will bracket for the time being. Animals mimicking bees, ants, beetles, and millipedes routinely pass to naïve human observers as authentic individuals of their model, at least until closer inspection. Mimetic complexes are thought to evolve in a stepwise fashion, with each incremental modification conveying a slightly stronger signal until the entire suite is produced. This can involve many dozens of modifications to numerous aspects of the organism from morphology and behavior to olfactory and tactile elements.

Some remarkable cases of mimicry appear, at least on their face, to provide evidence for holistic visual perception in the receivers of those signals. For instance, several lepidopteran caterpillars have evolved to resemble venomous snakes. Henry Walter Bates described one such hawkmoth caterpillar as "the most extraordinary instance of imitation" he had ever seen—so startling in its physical and behavioral resemblance to a snake that it had reportedly frightened people in a local village. Caterpillars that mimic venomous snakes are equipped with a suite of features that include (inter alia) faux eyes on a triangular "head," "scales," a forked "tongue," and a rearing up and striking behavior (see figure 10.2).

Figure 10.2
Caterpillar snake mimic, *Hemeroplanes triptolemus*, indicating holistic signal perception. Photo courtesy of Andreas Kay.

Ant mimicry is among the most iterated forms of complex mimicry, involving numerous modifications that include (inter alia): forelimbs altered to mimic antennae (which are waved in ant-like fashion in the air), faux compound eyes (in the case of camera-eyed jumping spider ant mimics), the loss or reduction of wings (if the mimic originally had them), constricted abdominal regions, narrowing of the thorax to create an ant-like waist, an upward-angled rear segment, a large horizontally positioned head with illusorily large mandibles, elbowed and clubbed antennae, microstructural modifications to match the coloring, reflectance, and texture of ant bodies, and behavioral modifications such as engaging in a zig-zag/ stop-and-start running gate and foregoing non-ant-like motions like jumping and flying.[24]

This rich set of multimodal modifications presumably generates an "ant" percept in the receiver—a categorization that has some desirable effect from the mimic's standpoint, either in deterring a would-be predator (such as a bird or wasp) or allowing the sender to gain access to ants themselves or their resources. Ant mimicry is common in predatory terrestrial arthropods, such as spiders and mantids, though the target of this signaling is unclear and may involve a mix of defensive and aggressive motivations. In defensive cases, ant mimics capitalize on the toxicity and stinging reputations of true ants; in aggressive "wolf in sheep's clothing" scenarios, ant mimicry may act to deceive ants themselves in order to prey directly on ants, their larvae, or their vigilantly tended "domesticated" subjects, such as aphids.

One might want to infer from the iterated evolution of holistic mimetic complexes that the coevolutionary receivers of those signals are perceiving the world holistically. However, holistic-style mimicry could evolve even if the receivers do not perceive objects at all. For example, it could be that the more cues that are present in a given case of mimicry, the more likely it is that a receiver engaging in local feature processing, or some multitude of such receivers for whom different features are salient, will detect the signal. For some receivers, colors may be salient, whereas for others edges are more likely to be picked up. Perhaps, then, mimicry complexes evolve not because receivers form holistic representations but because the mimicry complexes contain salient and redundant low-level cues for a large variety of receivers. Of course, striking mimicry complexes can also be explained by the receivers perceiving objects in a gestalt-like fashion. The holism versus localism question thus seems to be underdetermined with respect to the mimicry data.

In addition, there is an apparent problem with patterns of mimicry that the holism view would have to satisfactorily resolve. If the strength of the signal (whether true or false) is directly correlated to its adaptiveness, and if better mimics send stronger signals, then mimics should tend toward suites or complexes that increase their resemblance to the model. But if this is true, then how can we account for the ubiquity of inferior mimetic signals that could be picked up merely through elemental feature detection, such as the simple eye spots on fish? The lack of a strong resemblance between many mimics and their models could support the deflationary view, insofar as it suggests that apparently weak signals are sufficient to fool the low-level feature detection mechanisms of the evolutionarily relevant receivers, be they vertebrates or arthropods.

However, the fact that mimicry is frequently imprecise can be attributed to a number of factors that have nothing to do with the perceptual representational capacities of the receivers. These include (1) the noxiousness of the model and the existence of alternative prey items (which affect the predator's foraging risk management), (2) the presence of other mimics (causing a breakdown in the mimetic signal), (3) the attempt to mimic multiple models simultaneously (with the result that no mimetic complex represents an adaptive peak), (4) the fact that a lineage is in the early phases of an adaptive optimization trajectory, (5) internal constraints or trade-offs that limit the moldability of the mimic, and (6) strategic evolutionary countermoves by the model away from the mimic, as the model suffers an evolutionary cost from the ubiquity of false signals (which creates an incentive for predators to unlearn or fail to learn the true signal).[25]

Taken together, these aspects of the evolution of mimicry can explain why, despite holistic representation on the part of their receivers, mimicry complexes are often less than striking. In addition, we should not take mimicry data in isolation. We have already seen experimental evidence that some visual discriminations in bees cannot be accounted for by local feature detection, and in addition that bees tend to prefer global over local processing. If the hymenopteran visual imaging systems studied thus far are representative of those possessed by the likely receivers of mimetic signals, then the existence of multifeature, multimodal mimetic signaling is likely to be the coevolutionary result of gestalt-like perception in the arthropod receivers of those signals (if there are any).

This brings us to the receiver problem. As noted earlier, to make any inferences about the representational capacities of the receivers of mimetic signals, we first need to determine who the receivers of the signals are; and identifying the receivers of mimetic signals requires, in turn, that we document the effects of a given case of mimicry on the foraging behaviors of potential predators or prey with which the senders coevolved. This turns out to be surprisingly difficult in practice, and investigations of this sort are sparse. It has long been presumed that many cases of mimetic signaling are aimed at thwarting visual predation by birds—an assumption that has driven much of the experimental work on the evolution of mimicry. The finding that birds are representing objects holistically is important but not all that surprising, given extensive homology in the vertebrate visual system. Gestalt-like mimetic signals aimed at fish and lizard receivers is somewhat more telling, though it, too, speaks to the basal nature of holistic visual perception in vertebrates.

Unfortunately, very little work has explored the extent to which mimetic signals are aimed specifically at *invertebrate* visual predators, such as dragonflies, wasps, mantids, and coleoid mollusks. There is evidence that dragonflies avoid dipteran (fly) bee mimics, and that mantids avoid mimics of toxic insects. The hoverfly bumblebee mimic not only avoids predatory attack by resembling a bee (see figure 10.3), but it also reportedly gains entry into bumblebee nests wherein it lays its eggs. Likewise, rove beetles have repeatedly evolved a complex suite of ant-like morphological and behavioral modifications that give them access to the nests of aggressive army ants; the mimics even join their army ant hosts on raids, foraging expeditions, and emigrations—when, that is, they are not busy consuming the army ant's brood.[26] The extent to which invertebrates are the receivers of mimetic signals is a question in need of further research and could prove important to the investigation of animal minds.

Figure 10.3
Hoverfly bumblebee mimic, *Volucella bombylans plumata*, photo by Charles Sharp, from Wiki Commons.

2.2 What Do Insects Think?

The Umweltian platform is the groundwork for a mental life upon which other cognitive mechanisms have been laid down. The high trainability of bees, using techniques pioneered by Nobel Prize-winning ethologist Karl von Frisch, allows researchers not only to design experiments that probe the ways that bees visually represent their world, but also to test for sophisticated cognitive abilities associated with intelligence, such as learning and problem solving. Cognitive competences like categorization, concept formation, planning, causal reasoning, and social learning offer flexible tools that can be applied to a wide range of life problems, including some that are far removed from a lineage's evolutionary history and encoded epistemic ecology. As we shall see in section 3, these generic tools of mind significantly enhance the goal-directed character of animal behavior.

2.2.1 Causal Reasoning and Social Learning. Like honeybees, bumblebees have proven adept at becoming competent with the ecologically foreign tasks that have been put to them in the laboratory. For instance, Lars Chittka's prolific bee cognition laboratory recently trained bumblebees to pull on a string to get a sucrose reward that was placed just out of reach under a transparent Plexiglas table. Although the videos of bumblebees performing this complex, spooling motor sequence are striking, the feat is somewhat less impressive than it may at first seem. Only two creative (or lucky!) bees out of nearly 300 spontaneously and without training figured out the solution; all other bees that learned to solve the task were trained in a stepwise fashion, with artificial flower rewards attached to a string placed at progressively more distant locations under the table until the full pulling sequence was learned.

It is unclear whether the bees came to understand the causal structure of the string-pulling solution. It is perhaps more likely that, like primates and birds that have learned to perform similar tasks, the bees solved the problem using perceptual or proprioceptive feedback and trial and error techniques, rather than employing causal reasoning. That said, the results should not be undersold: they demonstrate a degree of behavioral plasticity on a par with that observed in mammals and birds. Perhaps more noteworthy than the learning of the behavior itself is that this unnatural string-pulling skill was socially transmitted at high rates to other members of the colony who were able to view a demonstration of the behavioral innovation through a Plexiglas window.[27] After observing a demonstrator bee pull the string to obtain the reward, a whopping 60 percent of observers solved the string-pulling task without training and on the first try (recall that only two out of nearly 300 bees figured out the solution without training or demonstration).

The mechanisms that underwrite social learning and cultural transmission in bees may be fairly low level or even "hard wired," including some combination of local enhancement (attraction to specific locations that demonstrators have visited), stimulus enhancement (a learned association between flowers that conspecifics have recently visited and the extraction of a reward), and trial-and-error learning as enhanced location and stimuli are thoroughly explored. Social learning in bees may not, therefore, involve the sort of high-level cognitive mechanisms that underpin human-like cultural transmission, such as a theory of mind, causal reasoning, or even the deliberate copying of a complex sequence of behaviors with a specific end-goal in mind. Cultural learning in bees—as with more explicit "teaching" that has been observed in tandem-running ants[28] and socially learned mate choice in drosophila[29]—shows that social learning is a robust, multiply realizable phenomenon that is not limited to creatures capable of metacognition (i.e., reflecting on one's mental states or those of others).

Having said that, the underlying mechanisms of social transmission in bees are poorly understood, and there are two reasons to think that low-level cognitive mechanisms may not be the whole story. First, there is evidence that bees can reflect on their own epistemic states (such as their relative levels of certainty),[30] and thus metacognitive contributions to social learning in arthropods, though unlikely, are not definitively ruled out. Second, other recent experiments that have controlled for local and stimulus enhancement have found that bees may indeed be capable of causal reasoning and observational learning.

For instance, in an ingenious recent study, the Chittka laboratory taught bumblebees to bring a small yellow ball to a designated location on a blue platform in order to obtain a sucrose reward.[31] Experimenters used a plastic model bee to demonstrate how to successfully complete this unnatural task, and the vast majority of bees were able to solve it upon demonstration by the model. Next, to test the effects of social learning on skill acquisition, researchers trained bees on a similar task, except that three yellow balls were placed at varying distances from the target. In training the demonstrator bees, the researchers glued down the two yellow balls that were closest to the target so that the demonstrators learned to move only the furthest ball to the target in order to get the reward. Next, observer bees were placed in three conditions. In the first condition, a live demonstrator bee performed the task, moving the farthest yellow ball to the target whereupon a reward was produced; in the second "ghost" condition, a magnet moved the ball "magically" to the target and a reward was produced; and in the third condition, no demonstration of the task was provided. Observer bees in the first condition solved the task an impressive 99 percent of the time; observers in the second condition solved it 78 percent of the time; and observers in the last condition solved it only 7 percent of the time. More importantly, most of the successful observers moved the *closest* ball to the target, rather than the furthest one as shown by the demonstrator (the trained demonstrator had learned to ignore closer balls to the target because they were deliberately glued down in the training phase); this was true even when the closest ball was of a different color than the furthest ball used by the demonstrator.

This "shortcut" is striking for several reasons. One, it shows rather convincingly that observer bees were not mindlessly copying behaviors performed by the demonstrators or carrying out learned motor sequences in a blind, associative way. Instead, the causal shortcuts taken suggest that the bees understood the goal-oriented structure of the task before them and improved upon the demonstrator's tactics in light of the causal relations they observed. Such behavioral innovations even smack of means-end rationality. Two, unlike the string-pulling task, which required facing the target and acting on obstacles in its way, in the ball-rolling task the bees had to move *away* from the target stimulus in order to procure an object and transport it to the designated location to obtain a reward. This involved turning their backs to the target in order to drag the ball to the designated area, which suggests that the goal was not out of mind even when it was out of sight, hinting at planning, prospection, or imagination. Three, the ball-rolling experiment ruled out location enhancement and stimulus enhancement as the bases of social learning. Trained demonstrator bees interacted with balls located farthest from the target, whereas observers chose to perform the sequence with balls located in the area closest to target, showing that observers were not attracted to the location that the demonstrator conspecific had visited. Further, observer bees chose the ball closest to the target even if the ball was a different color than the demonstrator's ball, which suggests that the stimulus was not enhanced in a way that made success through trial and error more likely.

2.2.2 Abstract Concepts. Concept formation is another hallmark of "thinking." As we have already seen, bees are capable of categorizing natural objects in their environment, such as flowers, trees, forest landscapes and faces, as well as highly unnatural objects like contrived patterns and paintings.[32] This could be taken to imply that bees have conceptual ontologies, though these data are based largely on "matching-to-sample" discrimination tasks that are potentially vulnerable to deflationary "low-level" interpretations that deny insects perceive any objects or form any concepts at all. Discriminations based on *abstract* concepts, such as relations, more decisively rule out deflationary interpretations. Abstract relations between stimuli hold irrespective of the physical properties of the related objects, and so discriminations based on abstract *relata* cannot in any obvious way be explained in terms of local feature detection.[33]

Thus far, bees have been shown to master relational concepts that include above/below,[34] same/difference,[35] greater than/less than,[36] and bigger than/smaller than,[37] and they are capable of deploying several of these relational rules simultaneously.[38] Even more impressively, bees can transfer relational concepts across sensory modalities, so that, for example, the sameness/difference rules that are learned for visual patterns can be transferred to olfactory stimuli and vice versa. Cross-modal transfer is especially telling because there is no conceivable physical cue that could support the transfer of a relational rule between, say, visual and olfactory stimuli—sensory modalities that exploit entirely different energetic and representational bases.

For instance, in a standard matching-to-sample setup, free-flying bees enter a Y-maze from an opening in the middle of a sample pattern or color; they are then presented with a choice between two arms of the maze, one bearing a stimulus that matches the one displayed on the entrance to the maze, and one bearing a different stimulus. Bees are trained to choose the arm of the maze with either the same or different pattern or color (see figure 10.4).

If successful, the bees are rewarded with a sucrose solution (simulating nectar). Once they have mastered the sameness/difference rule, the bees are then tested to see if they can apply that abstract relational concept to visual stimuli they have never before encountered, such as novel patterns for the color-trained

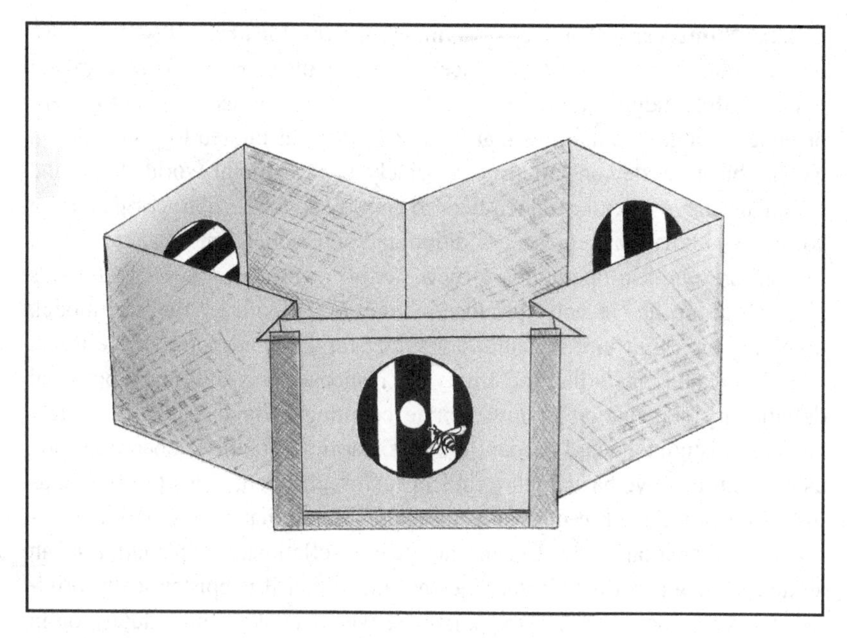

Figure 10.4
Y-maze training setup. Redrawn from A. Avarguès-Weber and M. Giurfa, "Conceptual Learning by Miniature Brains," *Proceedings of the Royal Society B 280*, no. 1772 (2013): 20131907.

bees and novel colors for the pattern-trained bees. Not only do bees succeed at this visual transfer task, but they are also capable of transferring these relational rules across sensory modalities—such as from visual to olfactory stimuli—even though these modalities share no common physical or representational cues.[39] More recently, Elizabeth Tibbetts and colleagues showed that paper wasps can master the logical operation of transitive inference, an ability associated with social dominance hierarchies. It bears stressing, once again, that these advanced, vertebrate-grade cognitive abilities are found in an animal with a brain no larger than a pinhead.

Do these results prove that bees can master abstract concepts, such as sameness/difference and above/below, in the higher-order cognitive sense that humans can? Although association-based explanations of these behaviors have not been ruled out,[40] the evidence is building that abstract concepts, as with holistic representation, is within the ken of animals with absolutely tiny brains. Even if deflationary models turn out to accurately describe the cognitive structure of abstract concept acquisition in insects, this would at a minimum show that bees have extraordinary learning capacities that more or less match higher cognitive learning in terms of identifying and exploiting complex causal structures in the world.

2.2.3 Numerosity, Time, and Planning for the Future. Bees have also demonstrated numerosity competences, including the capacity to make greater than/less than judgments up to the number four, as well as a concept of zero or none[41]—cognitive abilities that were long thought limited to primates and parrots but now demonstrated more widely in the animal world. Dyer and colleagues have recently trained bees to employ add and subtract rules using colored symbols that represent addition and subtraction, and to apply these learned arithmetical operations to new sample sizes and shapes.[42] Chittka's research group has recently produced a simple biologically realistic model, involving only four virtual neural units, that can replicate some of the insect counting results, including the empty set response, thus offering a potential deflationary explanation of invertebrate counting behaviors.[43] It is unclear whether this model could account for bees learning arithmetic operations and using them to solve basic math problems across different kinds of objects, as just discussed, though it of course raises the specter that an alternative deflationary explanation could. But the fact that a deflationary explanation of an animal behavior can be devised does not mean that it is epistemically privileged over higher-cognitive explanations, whether taken individually or in light of the totality of the evidence.[44]

Numerosity abilities are linked to the perception of time, which requires some form of internal counting. We saw in previous chapters that noncognitive counting/memory systems are widespread the living world, including in plants and unicellular organisms. We also saw that the implicit (cognitively encapsulated) ability to bind objects in time is a critical component of object perception. An *explicit* sense of time, on the other hand, is much rarer, though it too has been shown in insects: bees, for instance, are capable of attending to the future.[45]

Jumping spiders (salticids), another prominent model for arthropod cognition, have also been shown to count in discrete numbers up to three,[46] and it would not be surprising if they could also explicitly sense the passage of time. Salticids have been shown to plan routes that include elaborate detours,[47] they exhibit highly flexible and prey-specific hunting behaviors, and they are capable of reverse learning.[48] For instance, in a recent route-planning experiment designed by leading arachnid cognition researchers Fiona Cross and Robert Jackson, jumping spiders were placed on a platform from which they could see two walkways, only one of which led to a box with a desirable prey item.[49] Recall that jumping spiders have excellent vision—in fact, the best vision of any spider, courtesy of two powerful, forward-facing, camera-type eyes. To reach the prey, the spider had to leave the tower and take a detour away from the prey item to access the correct walkway, which by design goes out of view during the journey. After a period of surveying the situation from the tower, most of the spiders, without prompting, took the detour, chose the correct walkway, and retrieved the prey—even though none had any experience with the experimental setup or apparatus and indeed could have leapt out of it at any time. This complex navigational feat—one that is surely oversimplified compared to the actual spatial problems that jumping spiders face in the real world micro jungles they stalk—suggests a degree of planning and spatial cognition that meets our intuitive expectations about what minds are in the business of doing.

Some groups of salticids specialize in hunting other venomous spiders, including other salticids. These "Omar Littles" of the spider world (in reference to a character in the TV series *The Wire* who specialized in robbing drug dealers) exhibit extreme flexibility in their predation strategies.[50] They learn on the fly how to hunt dangerous prey that they have never encountered before, such as venomous web spiders and mantids, adapting to unfamiliar defenses, which they appear to probe for weaknesses.

Some salticid hunting techniques are eerily reminiscent of the stalking behavior of predatory cats like lions: they track down prey, carry out maneuvers to remain outside of the prey's field of view so as to not betray their position, and then pounce at the optimal time, often from behind or above after detouring to avoid detection.[51] Once a prey item with high escape ability or dangerous defenses is detected, the spiders orient and direct their attention toward the target, which they will typically approach from a flanking, rather than frontal, trajectory, often taking significant detours to do so. Their approach starts out in fast bursts, but slows to a crawl and finally to slow motion as they get close to the target. At this point they tuck away their pedipalps—fluttering chemosensory organs in the front of their head whose motion could give the game away. Various "movement masking" mechanisms are deployed, such as halting the approach when the prey item stops feeding or grooming or turns to face its stalker, then continuing the approach once the prey resumes feeding, grooming, or turns away.[52]

Jumping spiders have also been observed delaying the strike in favor of a closer approach if they are sufficiently camouflaged against their current background, thereby increasing the success rate of their attack.[53] In the case of dangerous prey, numerous venomous strikes are sometimes made, and the attacker waits (much like wolves, great white sharks, or komodo dragons) until the prey is sufficiently disabled so that it can safely be subdued.

We should not infer from these quintessential stalking behaviors that salticids have a theory of mind, or that they attribute belief states to prey, or that they can mentally simulate what other animals are seeing. If such metacognitive abilities are not necessary to explain the stalking behavior of predatory cats, then it would be gratuitous to impute them to spiders. Nor is anyone claiming that jumping spiders sit around ruminating on what kind of insect they might want to have for lunch. But this is not a deflationary conclusion of any sort. Much like the findings for bees, as discussed earlier, many of the flexible behaviors found in jumping spiders implicate working memory, attention, object permanency, spatial cognition, the crafting of complex behavior through learned experiences, and the ability to form representations that are indicative of planning, "model-based" thinking or even mental simulation.

2.2.4 Mental Maps and Spatial Awareness. Given the kinds of tasks that bees accomplish on a day-to-day basis, it would not be surprising if they were capable of complex spatial representations that include mental simulations of three-dimensional space. However, the evidence for so-called cognitive maps in insects is mixed, in part because theoretical work suggests that surprisingly complex navigational feats (such as shortcuts) can be achieved by "lower level" memory-based guidance systems that do not rise to the level of a mental map.[54]

These are not trivial cognitive capacities by any means, and some of them necessitate a basic awareness of the body in space (such as orientations vis-à-vis landmarks and targets). The Umwelt itself is essentially a spatial map, even if it is restricted at the lower limit to immediately perceptible objects in the enveloping phenomenal field. The point, however, is that none of these mechanisms require that the organism consult a proper mental map of their surrounds. Yet even if bees only learn spatially disconnected "route maps" to targets and do not integrate these into a single mental map of their surrounding world,[55] such abilities still require complex spatial representations and cannot be explained in purely associational terms.

Nevertheless, some research does provide tantalizing evidence of mental maps in bees, though some of this work has not been adequately replicated. It is well known that foraging bees that have located a productive food source engage in a "waggle dance" in the hive that specifies the coordinates (distance and direction) of the bounty in order to recruit other foragers to the site. In a brilliant series of studies, bee cognition researcher James Gould trained the same bee foragers to visit two feeding stations of equal distance and reward magnitudes, one positioned along the shore of a lake (a natural source) and one positioned in the center of the lake (an unnatural source).[56] The same dancers gave the coordinates of both sites, but attending foragers were not recruited to the ostensibly implausible lake site whereas they were vigorously recruited to the shore site. This could indicate that the bees attending the dance situated the instructed location in their own mental map of the surrounding world, realized the location was in the center of the lake, and wrote it off as an error.

In a follow-up study, researchers moved the hive and feeding stations overnight to a similar-looking location, except that it was a tree-lined field of grass rather than a tree-lined lake. Bees that had yet to be out of the hive since it had been relocated ignored the dance that instructed it to the feeding station that would, were the hive still in its original location, put the food source in the middle of the lake.[57] If these experiments could be replicated, they would suggest that the attending bee had a mental map of its surrounds in light of which it judged the instructed location to be implausible.

2.2.5 Metacognition. We have already noted evidence that bees can effectively "report" on their epistemic state of uncertainty. Given the furious pace of arthropod cognition research, additional meta-cognitive surprises are likely in store. Indeed, there is now at least one study showing that ants pass the mirror self-recognition test.[58]

In this experimental paradigm, developed by psychologist Gordon Gallup, an animal placed under anesthesia is marked on a part of the body that is not visible to it, often on top of its head. Upon regaining consciousness, the animal is placed in front of a mirror and observed. If it inspects the spot using the image in the mirror, then it is deemed to pass the test, ostensibly recognizing the image in the mirror as itself. After being marked with a visible blue dot on the top of their head by experimenters, ants placed in front of a mirror inspected the mark and attempted to remove it, whereas ants with a brown dot that matched their body color did not, providing the first evidence of self-recognition in an arthropod.

Animals can widely distinguish their own body movements, vocalizations, or scents from those of conspecifics; and even more basally, animal and plant immune systems distinguish endogenous elements from intruders. Indeed, the self-world boundary is arguably a fundamental property of life itself. But "self-recognition" in the mirror experiments is noteworthy because it indicates *psychological* self-representation.

If replicated, the ability of ants to pass the mark test raises three possibilities: (1) the mark test is an inadequate test for self-recognition; (2) ants are capable of self-recognition but do not have "self-awareness" in a thick narrativistic sense; or (3) ants are self-aware in the thick sense just noted. Option 3 seems rather unlikely, given the totality of ant behavior, whereas option 1 amounts to ad hoc goalpost moving, motivated by empirically undefended assumptions about the types of cognition that absolutely small-brained animals are capable of.[59] Option 2, by contrast, is not inherently implausible. It should not be assumed that forming a basic representation of the self is evolutionarily difficult to achieve, nor that it is has no ecological utility for "simpler" animals. One could easily imagine scenarios in which mirrored reflections are an informationally significant feature of the ant Umwelt. Self-representation could be useful, say, to diagnose and treat parasites that are otherwise out of view.

The extent to which the sophisticated cognitive abilities found in insect and arachnid test subjects can be projected to arthropods writ large is unknown.

Nevertheless, the ability of some invertebrates to replicate many of the complex tasks that "advanced" vertebrates are capable of learning calls into question prevailing assumptions about the relation between mind and brain. Lars Chittka and Jeremy Niven distill two key questions from insect cognition research: "How do insects generate such diversity and flexibility of behavior with so few neurons?" and "If so much can be achieved with relatively little neuronal hardware, what advantages are obtained with bigger brains?"[60] Both of these questions remain at the foreground of invertebrate cognition research.

2.3 What Do Insects Feel?

Thus far, we have been framing arthropod minds in wholly cognitive terms, such as memory, representation, association, and the like. Let us now switch gears to consider the second major way of thinking about the Umwelt: in terms of phenomenology. Consciousness is a more fraught investigation, of course, and some readers will, quite understandably, prefer to remain safely ensconced in the cognitive realm. Yet consciousness, which we will use synonymously with "sentience" and "subjectivity," is closely linked to, if not constituted by, cognitive structures which are themselves measurable. These include (*inter alia*) stimulus and semantic integration, attention, emotion, the open-ended capacity to form associations,[61] and other causal structures that are epistemically accessible to science.

2.3.1 Sentience. Biologist-philosopher duo Andrew Barron and Colin Klein propose,[62] following neuroscientist Bjorn Merker,[63] that the egocentric representation of the mobile animal in three-dimensional space is sufficient for subjectivity. We might fold in a fourth dimension that seems to be implicit in the Barron-Klein model, namely the integration of time—which, as we saw in chapter 8, is crucial for object identity, spatial navigation, and Umweltian cognition more broadly. Even the simple pursuit of a visual object involves the implicit perception of time, such as comparing current three-dimensional states of the world to previous ones in order to determine motion relative to the object and hence progress toward the goal.

Barron and Klein maintain that the egocentric representation of the body in space is achieved through the construction of a "unified model" or "integrated simulation" that is run in the organism's brain. By this, Barron and Klein mean something like a dynamic, updating model of the surrounding world: the organism's best guess about ecologically relevant states of the external environment and its own body within it. Their contention is that sentience is either equivalent to this simulation or arises out of it (their stance on the cognition-consciousness relation is unclear).

If this is right, then we can infer that arthropods represent an independently evolved locus (or loci, depending on how cognitively sophisticated the last common arthropod ancestor was) of sentience, given that (1) some insects and arachnids appear to form Umweltian representations, including a spatiotemporal awareness of the self (or body) in a field of affectively valenced objects, and (2) on the most plausible phylogeny of bilaterian brains, arthropods and vertebrates share a brainless and hence non-conscious (at least in the Umweltian sense) common ancestor (as discussed in chapter 9). Some annelids, such as the camera-eyed bristle worms (alciopids), are actively swimming visual predators that may also meet the conditions for Umwelt-like subjectivity and perhaps constitute a separate origin of consciousness—though to date little work has been done on bristle worm nervous systems, cognitive capacities, and behaviors.

The Umweltian ability to actively navigate and search a three-dimensional space implicates another cognitive process that is linked to animal sentience: attention. "Attention" refers to the selective focus of perceptual resources on some features of a scene rather than others. Attention is crucial for many search tasks, given the overwhelming amount of information that is available in the environment at any given moment, much of which amounts to ecological noise. Given constraints on information processing, we might expect attention-like mechanisms, which make some features of the environment more salient than others, to be a universal feature of minds wherever they evolve.

It is hard to make sense of the flexible learning and complex foraging behaviors of bees and other insects without the guidance of attentional processes. In a visual search, animals deploy a search image that focuses on certain salient characteristics, such as those associated with suitable prey items, foraging locations, competitors, or predators. Not surprisingly, both honeybees and bumblebees have been shown to use attentional mechanisms in foraging, navigation, and discrimination. As we saw earlier, bees can be trained to direct their attention to local rather than global features of objects, and vice versa. Studies of attention in honeybees show that visual search by these animals is comparatively slow and serial and thus more easily encumbered by distracting information (such as environmental clutter). To compensate for a lack of robust parallel processing, honeybees engage in a behavior called "peering," which involves a side-to-side scanning of a visual scene to obtain better spatial resolution. By contrast, in bumblebees, visual search is more mammal-like in that it is faster and carried out using parallel processes. Even the less-distractible bumblebees, however, are not capable of making complex scene discriminations on primate-level timescales, suggesting that there may be computational limits that come with absolutely small brains.[64]

There is no reason to think that attentional mechanisms are unique to bees among arthropods. The neural signatures of attentional tracking are found in the mushroom bodies of insects more broadly, such as in drosophila and cockroaches.[65] As for arachnids, it is difficult to imagine how jumping spiders are capable of stalking their prey, planning routes, and subduing deadly spiders in flexible ways without the filter of attentional mechanisms. The same is true of visual search in cephalopod mollusks. Octopuses, for instance, scan for, detect, navigate to, and remember rewarding foraging patches, they take an interest in novel objects, and they learn to make fine-grained prey discriminations—none of which could be accomplished without attention.

2.3.2 Emotional Intelligence. Normally when we talk about "emotional intelligence" in humans, we are referring to the ability to interpret the emotional states of oneself and others and to modulate one's affect in order to manage personal relationships and achieve one's goals. However, emotional capacity is a foundational type of intelligence in its own right. Emotions are central to adaptive action and an important part of what makes the world meaningful—what gives an organism a welfare of its own.

In short, emotions are action-tendencies that include a cognitive appraisal of some internally modeled state of affairs in the world and an accompanying somatic state or "feeling," which together generate appropriate action (such as approach or avoidance). Actions like visual search would arguably be impossible without affective valences being attached to perceived objects and outcomes. Neuroscientist Antonio Damasio argues that action is often guided by the "fainter image of an 'emotional' body state, without having to reenact it in the body proper."[66] Damasio conjectures that these affective markers, or "as-if" emotional simulations, are trained up over the course of ontogeny and permit rapid decision-making by bypassing the slow and energetically costly experience of a full-blown emotional state. These "as-if" feelings occur outside of attention or conscious awareness.

If Damasio is right about the causal relation between action, cognition, and emotion, then it would seem that the sophisticated learning, navigation, and discrimination behaviors exhibited by arthropods and cephalopods must involve affective capacities. Although emotion research on arthropods is still in its infancy, some recent work suggests that insects have full-blown emotion-like states, including positive and negative affect, that are mediated by dopamine—the same hormone that regulates emotion in humans and other vertebrates. For instance, Clint Perry and collaborators demonstrated that bumblebees, just like humans, tend to interpret ambiguous stimuli more optimistically after exposure to a pleasant stimulus, whereas they display pessimistic behaviors after exposure to a negatively valenced stimulus.[67]

Although the totality of the research discussed in this chapter indicates that insects are likely to experience some states as positive and others as aversive, relatively little is known about pain perception in invertebrates. Insects often do not respond to bodily damage in the same way that, say, mammals and birds do (such as by grooming the injured body parts). This should not be taken to imply that the world is not affectively meaningful to insects, however, or that they do not have a welfare of their own.[68] If the aversiveness of pain is an adaptive mechanism that motivates animals to avoid particular situations, then we might expect that any animal capable of instrumental learning through selective exposure to punishment or other negative stimuli—an ability amply demonstrated in insects—will also be capable of experiencing pain.

Having said all this, we are not yet in a position to definitively rule out the possibility, discussed earlier, that many or all arthropods are high-functioning philosophical zombies—that no subjective states accompany insect behaviors regardless of how sophisticated they might appear.[69] The same might be said for many nonhuman vertebrate animals. Although it is true that some surprisingly complex forms of information processing can take place outside of full conscious awareness, such as during sleepwalking and to a much lesser extent under general anesthesia, we do not see unconscious or even semiconscious downhill skiers or baseball outfielders—and yet comparably complex activities are expertly carried out by insects every day.

There are, moreover, no recognized cases of unconscious learning that involve attending to novel stimuli, observing the behavior of conspecifics, forming abstract concepts, and acquiring intricate motor sequences that have yet to be routinized and are far removed from a lineage's evolutionary history. Insofar as very low-level explanations of these documented behaviors have plausibly been excluded, the evidence supports an "inference to the best explanation" that some invertebrates have a mental life.[70] In light of the foregoing discussion, the burden of proof should shift to those who want to argue that arthropods are a diverse troupe of jointed-legged zombies—and this burden has not, in my view, been met.

In sum, when neuroanatomy, behavior, and evolution are taken together as mutually reinforcing lines of evidence, they set forth a powerful case for the presence of invertebrate minds.[71] The fact that sophisticated forms of cognition, and perhaps even consciousness, can be realized in animals with pinhead-sized brains—arthropods being proof of concept—should only increase our confidence in the cosmic character of mind and its evolution.

3. Umweltian Teleology and Macroevolutionary Ecology

Whatever one thinks of the relation between Umweltian cognition and consciousness, the capacity to situate the body in a three-dimensional field of objects is critical for the sophisticated goal-oriented behavior that marks intelligent life, be it on Earth or anywhere else.

3.1 Goal-Direction in an Umweltian Informational Space

A central if not defining feature of living things is that they are goal directed. Unlike inanimate physical systems, organisms are "teleologically" organized in that they appear to "aim" at certain ends, such as self-maintenance, reproduction, homeostasis, and so forth. The seminal account of goal-directedness in biology is due to the philosopher of science Ernst Nagel, who describes three features of goal-directed systems.[72] First is *plasticity*: the goal can be reached by the system following alternative paths or starting from different initial positions. Second is *persistence*: the system is maintained in its goal-directed behavior by compensating for disturbances that would otherwise prevent realization of the goal (provided these disturbances are not too great). And third is *orthogonality*: the variables coordinated must be independent of one another, in the sense that the value of either linked variable at any given moment must be nomically (read: physically) compatible with any value of the other variable (so as to rule out goal-directed behavior in purely physical systems, like candles and tornadoes).

Ernst Mayr defended an account of goal-direction according to which teleological (or what he preferred to call "teleonomic") behavior is behavior that results from the operations of an internal genetic program, which provides instructions for how to achieve the goal and how to respond to disturbances.[73] Rather than thinking of Mayr's proposal as an *account* of goal-directed behavior, it is probably better to think of it as an *explanation* of goal-directed behavior as defined in accordance with Nagel's property-systems criteria (or something in their vicinity). This allows us to remain neutral to the proximate mechanisms that realize goal-directed behavior, of which genetic programs, if they exist, may be simply one type. Belief-desire psychology, if *it* exists, may be another realizer of goal-direction.

Focusing on internal genetic programs also risks glossing over the structure of goal-directed systems. Hierarchical approaches to goal-direction, such as that proposed by theoretical biologist Daniel McShea, are more promising in this regard.[74] McShea understands goal-directed behavior as the movement of some entity within a larger "field" that contains and directs but does not rigidly determine the behavior of that entity. It is this wiggle room within a larger structure that makes plasticity and persistence possible, and the enveloping field ensures that the organism's behavior can be continually directed from any location within the field. Goal-directed systems, from torpedoes to tortoises, have this "upper-directed" structure. Like Nagel's account, McShea's theory of goal-directed behavior is agnostic to underlying mechanisms—the response properties of the enveloped entity could be caused by internal genetic programs, by the information-processing functions of sensory modalities, by agentic psychological capacities, or by some combination of these things—but the deep hierarchical structure of the system remains the same across all cases of goal-direction.

The potential payoff of McShea's account is broad, but its most natural application is in thinking about the structure of teleological systems in which physical relations are important. In physics, "fields" are physical quantities that assume different values at different locations in space. It is easy to see how chemical or light fields can envelop and guide but not rigidly determine behavior. In contrast, understanding internal homeostatic mechanisms as upper directed is somewhat belabored and requires postulating an abstract "phase space" (an issue we will not address here). Thinking of the organism as embedded within spatially enveloping information fields is helpful for understanding how basic goal-directed behavior was transformed by the evolution of the Umwelt.

Although McShea does not gloss fields in informational terms, for fields to do what they need to do on the hierarchical account—that is, to direct (but not determine) the behavior of the enveloped system—they must provide goal-relevant information to the organism. If they are to provide goal-relevant information, fields must be defined in relation to an organism's sensory, representational, and response capacities. Simple goal-directed behavior, such as that exhibited by bacteria, protists, plants, and animals with very rudimentary nervous systems, is guided by low-dimensional stimulus fields like basic chemical or light gradients. In contrast, sophisticated goal-oriented behavior is only achieved by integrating information from a wide range of spatiotemporally extended stimulus fields, increasing the persistence, plasticity, and sophistication of animal behavior.

The Umwelt is a nesting of the body (or subject) within a spatially extended field of objects bundled with physical properties and affective valences (see chapter 8). It is populated with meaningful entities, including things to be avoided (e.g., predators and obstacles), sought out (e.g., mates, conspecifics, and food), used in navigation (e.g., landmarks), and ignored (e.g., noise), enabling organisms to respond to fluctuations in their enveloping sensory fields with a degree of rapidity, flexibility, and precision that more diffuse, lower-dimensional stimulus fields do not support. With the aid of perceptual filters, attentional mechanisms, categorization abilities, and motivational states (drives or wants), the Umwelt generates goals and identifies ways of realizing these goals from a wide range of initial conditions and perturbations. It does this, for instance, by generating a stable visual field that can be used to direct foraging and navigation; by keeping a representation of a target in mind even when it is not immediately present to the senses; and by learning new associative and causal structures in the world that can be brought to bear on future problems.

The unity of the Umwelt is explanatorily crucial here. It entails the conjoined experience of numerous bound objects strewn over space and time, which are themselves integrated representations of multiple stimulus fields, such as shape, color, motion, identity, and valence. Detecting the motion of an object allows an organism to track it for pursuit or avoidance. Object permanency allows an animal to continue tracking an object even when it is partially occluded or temporarily out of view, as well as to project the object's future states, all of which may prove crucial to survival. Holistic forms of representation speed up detection, categorization, and affective appraisal, and with them goal-directed response. Some lineages have evolved the ability to act not only on concrete features of the bio-physical environment but also on abstract relations in the world, which further increases behavioral plasticity and persistence. Likewise, information about mental states adds yet another "informational field" that can direct and redirect an organism toward its ends. By enabling organisms to act on richer sources of information, Umweltian cognition and the cognitive bells and whistles that have been folded into it have greatly enhanced goal-directed behavior.

3.2 Active Predation and the Modern Evolutionary Arms Race

It is no accident that lineages boasting the greatest visual acuity—vertebrates, cephalopods, and arthropods—have independently evolved the most sophisticated nervous systems, motor capacities, and goal-directed behaviors in the living world.[75] Minds helped to construct the complex informational ecologies that image-forming eyes allowed. This, in turn, drove further expansions in cognitive-behavioral repertoires, which created additional problems and opportunities, which were solved and exploited through new cognitive, perceptual, proprioceptive, and motor innovations, and so on, resulting in a convergent complexification trend in at least three lineages of life.

Paleobiologists Mark McMenamin and Andrew Parker both have hypothesized that the evolution of eyes in the Cambrian seas kick-started the modern era of active predation (macroheterotrophy), spurring the rapid evolution of defensive countermeasures and triggering an arms race that would dominate the future history of animal life to the present day.[76] The Ediacaran fauna (encountered earlier in this book), which predated the true critters of the Cambrian, were soft-bodied, photoautotrophic animals that appear to have lacked any complex sensory apparatuses, heads, or swimming appendages that would be indicative of active lifeways. The Ediacaran world was, in essence, a two-dimensional world concentrated around microbial mats and occupied primarily with filter-feeding. Evidence for predation in the Ediacaran is virtually nonexistent, leading McMenamin to dub this period of multicellular evolution the "Garden of Ediacara." This is not to say that there was no heterotrophy at all—but to the extent that predation existed, it was limited to languidly following chemical gradients or fortuitously bumping into prey items.

Once eyes and, crucially, Umweltian cognition had coevolved in the base of the Cambrian, the listlessness of the primeval Garden was lost. As Dorian Sagan puts it in his forward to McMenamin's book, "the evil empire of carnophagy had begun … and sashimi was everywhere." The evolution of the first image-forming eye in arthropods enabled this clade to fill all active predatory niches during the early phases of animal evolution, as epitomized by the apex carnivore *Anomalocaris* whom we met near the outset of this book. The Earth's first major predators (such as the anomalocarids) are hard to come by in the fossil record, as they would not have had mineralized skeletons that are likely to be preserved. However, the signature of active predation is reflected in the near-simultaneous skeletonization of numerous animal groups (such as the elaborate spines of trilobites, wiwaxiids, and echinoderms, and the bore-resistant shells of brachiopods), as well as the emergence of complex aggressive and defensive behaviors (such as boring, burrowing, and masticating)—two key hallmarks of the Cambrian Explosion that are indicative of arms race escalations.[77] The emergence of active arthropod predators was rapidly followed by the evolution of the vertebrate camera eye and the rise of vertebrates to ecological prominence, and somewhat later by the mollusk eye and radiations of highly active ammonites and belemnoids in the Paleozoic seas.

Missing from Parker and McMenamin's "Light Switch" theory of the Cambrian explosion, however, is the crucial role of mind and meaning. As we saw in chapter 6, complex optical apparatuses exist in box jellyfish and even unicellular dinoflagellates to little or no neurocognitive effect. Image-forming organs thus seem to be a necessary, but not sufficient, condition for the evolution of brains and minds. As philosopher Michael Trestman rightly points out, Parker uses the term vision in a way that implies more than merely an image-forming optical apparatus, but he does so without acknowledging the critical role of cognition in the Cambrian story.[78] According to Trestman, the key was "cognitive embodiment," which allowed for coordinated control over active animal bodies. Therefore, it was not the image-forming eye per se but the eye combined with *Umwelt-grade cognition* that gave predators the search-and-destroy capabilities that likely precipitated a major ecological transition from the soft-bodied, inactive fauna of the Ediacaran to the great diversity of able-bodied and formidably defended bilaterian body plans that we see today.

Because of its contributions to goal-direction, the Umwelt is likely to be of great significance to the morphology, behavior, and relational structure of all complex living worlds.

Coda to Part II: Homage to *Homo Sapiens*

The evolutionary trajectory of this book might seem to be steamrolling straight into the inevitability of humanlike intelligence and technological prowess. This coda will throw a wrench into this optimistic juggernaut. The wrench in question is the radical contingency of human-grade cultural capacities, and it will help to resolve an old paradox that appears to have been given a new lease on life from the arguments canvassed in this book. If minds are robustly replicable outcomes of evolution, why is there no evidence for their cosmic existence?

1. Where Are They?

Over lunchtime conversation, the physicist Enrico Fermi asked why, if intelligent civilizations are so common in the universe, we have not a shred of evidence for their existence. "Fermi's paradox," as this question has become known, highlights a dissonance between the apparent observational entailments of a commonly held theory and our actual observations. The commonly held theory in question is the notion that intelligent life is widespread in the universe. The observational entailments of this theory are various empirically verifiable signatures of extraterrestrial intelligence discernable on Earth, in the solar system, and elsewhere in the galaxy (such as messages, artifacts, evidence of astroengineering, and visitations either by intelligent extraterrestrial organisms themselves or their autonomous machines). And these observational entailments are in tension with the "Great Silence" that descends upon us from the heavens.

In a recent book, Milan Ćirković, a proponent of the search for extraterrestrial intelligence (SETI), shows how Fermi's paradox arises from Copernican-style inferences (see chapter 1) about the cosmic typicality of terrestrial astronomy, geology, and biology as well as human-specific intelligences, technological capacities, and motivations.[1] Modern cosmology and astronomy have only strengthened the paradox, which Ćirković characterizes as one of the most profound and enduring problems of modern science. The present discussion lends considerable biological pressure to Fermi's paradox. Although part I argued that specific animal body plans may be radically contingent, part II showed that the evolution of mind transcends body plan constraints and is likely to be a robustly replicable feature of living worlds.

One obvious way of resolving the paradox is to reject the Copernican notion that human (or "peer-plus") grades of intelligence are cosmically typical. One can do this by targeting any of a number of jointly necessary conditions for the evolution of intelligent extraterrestrial life, such as those suggested by Peter Ward and Douglas Brownlee in their famous "Rare Earth" argument.[2] As the search for intelligent extraterrestrials continues, one great challenge before us is to identify biological focal points of contingency that could potentially dissolve the paradox. We have encountered several singular turning points in the history of life on Earth that present as good candidates, such as the evolution of eukaryote-grade cellular complexity and animal-grade morphological complexity—outcomes that are subject to observer selection effects and thus could account for the failure of Copernican principles.

In the homestretch of this book, we will highlight one important and underanalyzed condition for life to "progress" to a star-faring, or at least interstellar-communicating, civilization: the evolution human-grade technological capacities. Although this "step" was included as one of the fractions in the Drake equation (see chapter 1),[3] the formidable suite of contingencies that factor into the evolution of cumulative technological culture have been largely overlooked, and as a result the fraction of evolutionary histories that are expected to lead to robust technological civilizations has been overestimated.

The fact that cumulative technological culture—the stepwise improvement of technologies and the reliable transmission of these incremental improvements down the generations—was not achieved on Earth until so late in the evolutionary game bodes poorly for its cosmic replicability. In what follows, we will consider some explanations of this great delay and whether they can be marshaled into a tentative case for the radical contingency of robustly technological species. Navigating the fascinating landscape of human technological evolution would necessitate a book in its own right. What follows is just the dip of a theoretical toe into a vast and largely uncharted sea of empirically constrained speculation.

2. The Replicability of Higher Cognition

There is an enormous gap between the evolution of basic Umweltian minds, on the one hand, and the emergence of sophisticated technological cultures, on the other. Within this gap lie various "higher-cognitive" abilities on which the evolution of technological cultures is scaffolded. Precisely which higher cognitive abilities are implicated in the scaffolding of cumulative technology is unclear. However, causal reasoning, planning, imagination, means-ends rationality, metacognition, imitation, symbolism, self-awareness, and cognitive and cultural factors that underpin cooperation, communication, and social learning are all important parts of the story.

At the time of Stephen Jay Gould's writings on contingency, higher-cognitive properties were thought limited to human beings or at least to "higher" mammals like primates and cetaceans. Gould could be forgiven, therefore, for concluding that the evolution of humanlike cognitive properties on Earth were limited to mammals and thus hinged on the radically contingent dynamics of the dinosaur-mammal succession (discussed in chapter 2). Had a comet or asteroid never struck the Yucatán around 65 million years ago, or had it struck at a healthier time for dinosaurian ecosystems, the dinosaurs may have continued to dominate the cooler, dryer world of the Paleogene, precluding the diversification of mammals into their familiar modern orders (but see the discussion in chapter 2). Noting the lack of any detectable trend toward increasing brain complexity in dinosaurs, Gould surmised that the evolution of cognitive complexity may "lie outside the capabilities of reptilian design," and that therefore "we must assume that consciousness would not have evolved on our planet if a cosmic catastrophe had not claimed the dinosaurs as victims."[4]

The soundness of this claim depends, of course, on the sorts of cognitive complexity and consciousness that Gould is contemplating. As we have seen in part II, there is evidence for surprisingly sophisticated cognitive abilities and perhaps even consciousness in at least two phyla of invertebrates. Furthermore, it is evident that humanlike cognitive capacities are not, as Gould conjectured, beyond the capabilities of "reptilian" design. In fact, there are intelligent theropod dinosaurs living among us today: you might catch a glimpse of one silhouetted by the setting sun, or perhaps one is peering curiously over your shoulder as you read this. The dinosaurs of which we speak are, of course, birds. There is now a great deal of experimental evidence for complex cognition in birds, including causal reasoning, episodic-like memory, prospection, imagination, abstract concepts, language comprehension (both semantic and syntactic), numerosity, proto-moral behaviors, the multistage manufacture of tools, self-awareness, and metacognitive abilities such as the capacity to reflect on one's mental states.[5]

There is good reason to think that the impressive range of cognitive abilities exhibited by some living birds was also present, to some degree, in some of their extinct nonavian dinosaur relatives, especially the coelurosaurian theropods (which include raptors, tyrannosaurids, and compsognathids). Birds are more closely related to the feathered *Tyrannosaurus rex* than *T. rex* is to other nontheropodian dinosaurs (such as ceratopsians or apatosaurs) or even to other large theropods like *Allosaurus*. *T. rex* appears to have had a higher encephalization quotient (a measure of relative brain size) than living carnivores like dogs and cats.[6] This is perhaps not surprising, given that extinct theropods may have cooperated to solve more strategically complex foraging problems than those confronting modern birds—closer, perhaps, to lions and wolves hunting formidable prey in the grasslands and woodlands of our modern world. Maniraptorans, such as *Deinonychus* and other "raptors," had large, highly developed brains that were well matched for these tasks; indeed, the hypertrophy of the avian forebrain—which is analogous to the mammalian neocortex—began its evolutionary expansion deep in theropod evolution.[7]

All this suggests that the cognitive sophistication of birds was inherited to some extent from their extinct dinosaurian ancestors. Dinosaurs and mammals share a common ancestor that dates back to the Carboniferous, well before the great Permian catastrophe that would pave the way for the unlikely rise of the dinosaurs (see chapter 2). The convergent evolution of higher forms of cognition and consciousness in birds and mammals is thus indicative of a much deeper replicability of humanlike intelligence than Gould had entertained. Nevertheless, even cognitive convergence between dinosaurs and mammals is too phylogenetically "shallow" to support anything approaching cosmic levels of projectibility.

The fact that higher-cognitive convergence is confined to the vertebrate body plan could indicate that there are developmental limiting conditions on the evolution of very complex cognitive mechanisms that make these properties exceedingly rare among life worlds. Cognitive convergences between distant mammalian lineages, such as between whales and primates,[8] though striking, are even more restricted in their projectibility. In short, we cannot be confidant that iterations of higher cognition are free of developmental confounders if we only see patterns of cognitive convergence within a single body plan that is not itself subject to iteration.

3. Tool Use and Intelligence

Technology is closely (though not exclusively) associated with tool use, and tool use is a multiply-realizable behavior with a broad neural realization base. Defined in behavioral terms, tool use is a pervasively convergent phenomenon. However, only a small proportion of tool use among nonhuman animals is plausibly generated by higher-level cognition, such as planning, causal reasoning, insight, and working memory. Yet just as tool use per se does not necessarily indicate the presence of intelligence (in ants, for instance, it is likely "hard-wired"), the lack of tool use does not necessarily indicate the absence of intelligence. The dearth of tool use in intelligent clades could indicate that there are limiting conditions above and beyond higher cognition that restrict the evolution of technological cultures.

Consider, for instance, the general (though not categorical) lack of tool use in dolphins. Dolphins possess what is arguably the most complex mind in the extant animal world apart from that of humans, and this is reflected in the remarkable sophistication of dolphin brains, cognition, communication, culture, and society. Why is it, then, that humans—and not dolphins—are a space-faring species, even though toothed whales are a much older evolutionary lineage? Why do dolphins not dominate the oceans in the same way that humans dominate the terrestrial zone? Why is tool use in dolphins so rare, rather than a central axis of dolphin life?

Answers to these questions will be multifaceted, but ecological and developmental constraints on the evolution of tool use will be a critical part of the story. One major ecological constraint arises from the fact that the physics of tool use in water differ from that in air. Striking tools are less effective in water due to the energy dissipation that results from the higher density of water, which attenuates striking speeds. Buoyancy is another factor that constrains the striking efficacy and transportability of tools in water. And, of course, the manufacture of fire—a critical tool for external digestion, temperature regulation, and predator avoidance in humans—is forever inaccessible in the water medium. There are also numerous anatomical features that constrain the evolution of tool use in dolphins. Like fish, dolphins lack grasping appendages, and their rigid flippers and flukes have indispensable locomotory functions that preclude their modification into "hands" that are suited for tool manufacture and transport. Such transformations lie beyond the possibility space of dolphin body plan evolution.

In contrast, hominin bipedalism evolved in australopithecines for improved locomotory efficiency, freeing up the hands for subsequent technology-related modifications in *Homo*. There is now evidence that australopithecines—bipedal apes whose brain sizes were generally closer to those of chimpanzees than to those of early humans—were using tools manufactured through percussion more than 3 million years ago, well before the emergence of the first *Homo* species. What allowed these technological traditions to emerge in early hominins despite their chimp-sized brains?[9] Why did the distinctively human technology of *Homo erectus* exhibit bewilderingly little change on a global scale for over 1.5 million years?[10] If *erectus* populations were intelligent, intentional, foresighted, imaginative, symbolic, and proto or fully linguistic human beings who lived in highly cooperative societies with culturally scaffolded learning environments—as everything about their lifeways seems to suggest[11]—then why do we see a stunning lack of technological innovation for nearly the entire history of human evolution? Finally, what caused the great technology explosion that shattered this "Great Stagnance" just within the last 50,000 years? Many answers have been given to these questions, and all of them, or none of them, may be right. Our aim in the remaining pages is not to solve these enduring puzzles but merely to appreciate how ripe the landscape is for radical contingency.

4. Technology Made Human

The title of this section is intended as a double entendre. Not only does technology take on a uniquely cumulative character in modern humans, but in addition technological industry was a crucial driver of early human evolution. The transformation of *Homo* into an apex African predator[12] capable of migrating long distances and colonizing diverse environments represents a marked shift from the relatively simple "chimp-like" behavioral ecology of the australopithecines toward the highly cooperative lifeways of erectus-grade hominins. This shift could not have occurred without a technological industry that included tools for hunting, butchering, defense, and gathering; fire for cooking meats and hard-to-digest vegetables as well as for thermoregulation and keeping nocturnal predators at bay; primitive clothing which served as portable insulation; and basic sheltering techniques. Sustaining these industries, in turn, would have required the first "social technologies," including moral systems and culturally constructed learning environments.

Although the earliest direct evidence for the use of fire dates from 500,000 to 1.5 million years ago,[13] evolutionary anthropologist Richard Wrangham makes a persuasive case that the presence of cooking can be inferred from major changes in hominin skeletal anatomy and lifeways that indicate fire was in play at the evolutionary base of *erectus* around 1.7 million years ago.[14] These modifications include substantial reductions of teeth, jaws, and gut, an increase in hominin body size to modern human proportions, a switch to full-time life on the ground due to anatomical modifications for long-distance persistence hunting, and immigration into colder climes in Eurasia during glacial and interglacial periods despite having a body built for the tropics. If Wrangham's inference to the best explanation is right, then the reliable transmission of cooking, along with simple percussion-manufactured technologies (and probably gathering implements as well, which are unlikely to be preserved), were a crucial factor in the evolution of *erectus* from more primitive habilines. Technology quite literally made us human.

The deep conservation of *erectus* technology is bizarre from the standpoint of everything we know about these early humans. It may not only point to the very different character of human mind and culture in the Pleistocene, but also hint at the contingencies that may underlie the evolution of cumulative technological species. Within the last 100,000 years, stone tools and fire ignition technologies became more sophisticated, and projectile weapons, spearheads, and pulverizing tools were introduced. The "big bang" of cultural innovation occurred only within the last 40,000 years or so, with complex clothing, buttons, beads, jewelry, adornments, paintings, sculpture, religion, and music appearing on the scene, taking on the cultural morphology of human life as we know it.

Technological complexity was amplified again with the population boom and specialization of labor that followed in the wake of the Agricultural Revolution, and yet again with the mechanistic and methodological innovations of the Industrial and Scientific revolutions. Perhaps the next technological saltation will be driven by a revolution in artificial superintelligence.[15] Today, technology progresses at such a breakneck pace that older generations are compelled to acquire technological innovations from their descendants, reversing the intergenerational flow of cultural information that has characterized human societies for over a million years. In a break from nearly all of human history, technologies that are ubiquitous today bear essentially no resemblance to technologies that are hundreds or thousands, let alone millions, of years old. What forces were responsible for this sudden technological expansion?

5. Let the Parlor Games Begin

Cumulative culture was first posed as a central evolutionary explanandum by primatologist Michael Tomasello and colleagues in the early 1990s.[16] Their basic insight was that the cultural scaffolding of knowledge, skills, and material innovations enables descendants to build on the accumulated corpus of technical information that has been compiled and retained by ancestral generations. This scaffolding obviates the need for individuals to reinvent innovations in each generation through insight, trial-and-error learning, or luck, which in turn allows for the stepwise improvement of technologies that we take for granted today. In effect, cumulative technological culture extends the circle of co-operators beyond the grave to include distant generations of teachers and innovators. In contrast, the spread of technical innovations in nonhuman animals, such as chimpanzees, relies primarily on individual inventiveness, socially mediated trial-and-error learning, and imitation—vectors of technical innovation that are insufficient to produce what Tomasello and colleagues call the "cultural ratchet," which they argue is the key to human evolutionary success.[17]

Some of the factors that permit the spread of innovations in chimps are likely to figure in human cultural transmission as well. A central task of human evolutionary science, however, is to home in on the difference-making factors that contribute to the ratchet-like character of human culture. This project is ongoing and implicates a multitude of interwoven biocultural capacities, including symbolic language, communicative gestures, perspective-taking, mind reading, physical cognition, planning, insight, mental time travel, functional representations, cooperative motives, pedagogy, normative concepts and structures, and other traits that are not present, or not present to significant degrees, in nonhuman animals.

Part of the difficulty in making headway on this problem is that many of the implicated adaptations seem to have arisen well before the technological accelerations that we are trying to explain, and thus do not seem to be difference-making causes of cumulative culture, even if they are necessary ones. This temporal gap problem arises for spoken language,[18] symbolism, theory of mind, perspective-taking, joint intentionality,[19] mental simulation,[20] sophisticated causal reasoning,[21] functionalist orientations toward objects and artifacts,[22] normativity,[23] and other cognitive features that are apparently unique to humans. Other relevant traits, such as manual dexterity—including the precision-pinch and squeeze grip—precede the emergence of stone-tool manufacture altogether.[24]

The time lag problem is illustrated by two key traits that underpin the evolution of industry: language and morality. Language almost certainly coevolved with technology, enhancing the efficiency of interpersonal communication and with it the cultural ability to pass down innovations with ever greater precision and reliability. However, there is a large temporal gap between evolution of language and the very late emergence of cumulative technology in humans. As noted earlier, there is strong circumstantial evidence that *erectus* had language. There is even stronger evidence, based on comparative genomics,[25] that language was present in the last common ancestor of sapiens and Neanderthals, *Homo heidelbergensis*, who was a later variant of *erectus* that lived some 500,000 years ago. Both *erectus* and *heidelbergensis* were probably linguistic and to some extent symbolic, yet neither evolved cumulative culture. The earliest known cave paintings were painted not by Sapiens but by Neanderthals, who *ex hypothesi* lacked cumulative culture.[26] It seems more likely, then, that rather than language and symbolism ushering in behavioral modernity, cumulativity allowed language and symbolism to thrive in a way that they could not do before.

Normativity provides another crucial ingredient in the emergence of cumulative technology. It is hard to imagine anything approaching the levels of cooperation achieved by early humans, let alone the ultra-cooperation of modern human societies, without social norms to structure, coordinate, and incentivize cooperative behavior. The prevailing view is that morality was selected in early human groups (likely in the late Pleistocene) for managing intragroup conflicts that would otherwise impede collective action. Morality did this by reducing free-riding through the administration of punishment, by inculcating altruistic attitudes that enabled individuals to resist the temptation to act selfishly, and by enforcing an antihierarchical ethos that prevented dominant individuals from monopolizing the fruits of cooperation.[27]

The rewards of this new-found cooperation were substantial and far exceeded what could have been achieved alone or in groups with poor coordination. Moral systems were likely present in *heidelbergensis* at least 400,000 years ago, and probably earlier as indicated by levels of cooperation in *erectus*, which appears to have included highly deliberative and collaborative seafaring voyages that established populations on isolated Asian islands like Java and Flores. Both language and morality were crucial for constructing the learning environments in which technological traditions could be reliably sustained and improved upon. Yet both language and morality predated cumulative culture by hundreds of thousands if not more than a million years.

In his book *The Evolved Apprentice*, Kim Sterelny distinguishes between two types of cultural cumulativity that characterize behavioral modernity.[28] The first is the sheer volume and diversity of material artifacts that begin to be forged from a wider array of material substrates; the second is the stepwise improvement of technologies through incremental modifications that are retained and transmitted down the generations. Sterelny attributes the first feature to the increasing bandwidth of cultural learning, or the magnitude of information that culture is capable of transmitting; he attributes the second feature to the increasing fidelity of cultural transmission. The high-volume, high-fidelity features of human culture were made possible, Sterelny argues, by the creation of apprenticeship institutions. These include the explicit teaching of crafts and design as well as the organization of a learning space in which apprentices could learn by exploring and experimenting with the materials and by-products of the design process. Over time, the coevolution of numerous cognitive and social factors, such as social attuning mechanisms (e.g., joint intentionality), copying biases (e.g., a tendency toward conformity), larger populations (in which innovations are more likely to arise and be preserved), and more elaborate normative categories (e.g., the concepts of "teacher" and "student"), would enhance the cumulative capacities of culture.

If cumulative technology requires cooperatively scaffolded learning environments, and if these cooperative structures are underpinned not only by language but also by normative judgments and concepts, then the evolution of cumulative technology requires the evolution of normativity. Like language, normativity appears to have very low evolvability due to both its cognitive demandingness and the limited ecological conditions under which altruism can evolve. Modeling and ethnographic work have shown that the norms underpinning cooperation will not be sufficiently adhered to and free-riding strategies will tend to invade unless moral norms are coercively enforced through institutions of moralizing punishment.[29] If punishment is expensive to carry out, then the evolution of third-party punishment appears to pose a higher-order altruism problem that only group-level selection can solve. If this is right, then we are faced, in effect, with an evolutionary catch-22: competition among cohesive cultural groups is a necessary condition for the evolution of punishment and hence for morality to get off the ground, but institutionalized punishment is a necessary condition for the coherence and stability of cultural groups. Perhaps these two aspects could evolve in tandem, but we can begin to see why it is so difficult to evolve the normative structures that underpin cumulative technology.

6. A Silence Less Eerie

In short, cumulative culture hinges on the confluence of a large number of anatomical, cognitive, and social adaptations, each of which is exceedingly rare in the animal world if not exclusively possessed by humans. It is plausible to think that all or many of these components must be in place for cumulative culture to take off, even if we are uncertain as to why culture became cumulative when it did. Most of the adaptations implicated in cumulative culture have plausible selective functions in their own right; only later were they coopted in the service of cultural cumulativity. If this complex exaptation picture is correct, then it could explain the great time lag between the origins of key components of culture and the emergence of cultural cumulativity.

How might the evolution of some components of cumulative culture have affected the evolution of others? This is an important question for our present purposes because coevolutionary feedback loops can make certain outcomes more evolutionarily robust. We have seen that image-forming eyes, brains, and able bodies coevolved to produce the first mindful animals on Earth. Do similar coevolutionary feedback processes make the evolution of cumulative culture more likely? There have almost certainly been feedback effects in human biocultural evolution, most obviously between cognition on the one hand and the complexity of human cooperative societies and technological industries on the other. The evolution of language, mind reading, causal reasoning, and pedagogy-related adaptations would have facilitated the cultural construction of learning environments in which more sophisticated technical skills and norms could be acquired, which in turn would have become potent new sources of selection pressures for enhanced cognitive functions, and so on.

Feedback loops of this sort presumably drove the rapid expansion of the human brain and the swift complexification of human societies and industries that we see over the middle to late Pleistocene. It is certainly conceivable that other post-*erectus* human clades, such as the Neanderthals or Denisovans, might have independently evolved cumulative culture had they not been nudged into extinction by their more robustly cultural cousins, the Sapiens. However, the Neanderthals and Denisovans shared (via common ancestry) a large number of the singular adaptations that underlie cumulative culture in Sapiens, and so any hypothetical iterations of cumulative culture in these hominin clades would amount to Gouldian repetitions rather than true convergence—and repetitions with a very shallow replicability depth at that.

Even if feedback dynamics played a role in the evolution of cultural cumulativity, they do not rescue technological species from the jaws of radical contingency. There is no reason to think that there is any single, let alone robustly replicable, innovation that could account for the emergence of a complexly configured exaptation like cumulative culture. Rather than a Sherlock Holmes murder mystery that can be pinned unequivocally on a single culprit, the evolution of cumulative culture is more like the film *Murder on the Orient Express* (1974), in which each of twelve suspects turns out to have stabbed the victim in a vastly unlikely conspiracy. If there is a "Great Filter"[30] in the emergence of intelligent civilizations that dissolves the Fermi paradox, it may very well be at this juncture, in the evolution of cumulative technological species.

7. Summary of Part II

The upshot of part II is that mind in the rich Umweltian sense is likely to emerge across deep replays of the tape of animal life on Earth. The bolder proposition that the Umwelt is a ubiquitous feature of the living cosmos remains highly plausible but cannot be established beyond a reasonable doubt at this time. This is because there are lingering uncertainties about the contingent or replicable nature of some key evolutionary transitions and innovations in the history of life about which there is still great uncertainty. Nevertheless, a strong case can be made, based on patterns of convergence, for the striking notion that minds of a particular sort are written into the living fabric of the universe. Even if minds turn out to be relatively uncommon, wherever they *do* arise, they are likely to assume an Umweltian character—not because we cannot imagine minds any other way, but because evolution does not (or cannot) build minds any other way. Wherever minds are found, they are likely to have deep similarities to our own. This conclusion lends considerable pressure to Fermi's paradox. However, if the unique suite of cognitive capacities that underpins cumulative culture in behaviorally modern humans is radically contingent, this would explain the very late emergence of robust technology in the history of life on Earth and the puzzling lack of evidence for the ubiquity of other older, wiser we's out there in the universe.

Conclusion: The Morals of Macroevolution

In the film *Contact* (1997), based on a novel by Carl Sagan, SETI researcher Dr. Ellie Arroway, played brilliantly by Jodie Foster, is whisked off by an alien-designed transporter device on an eighteen-hour journey through the galaxy. At one poignant moment, Arroway emerges from a wormhole to witness a remarkable stellar event that she attempts to describe scientifically, in third-person terms, only to be overcome by the subjective aesthetic of the experience. "They should have sent a poet," she famously exclaims.

The story of macroevolution is just as poetic and profound as any journey through the physical universe. This remains true whether bodies and minds as we know them are contingent accidents of history or robustly replicable outcomes written into the laws of life. Either way, the picture painted by macroevolutionary science provides us with a deep source of self-knowledge, placing humanity in an unbroken chain of reproduction that extends all the way back to the origin of life on Earth, and forging connections, through laws or through chance, to life as it may exist elsewhere in the cosmos.

At the same time that natural history enriches our self-understanding, it also enables us to transcend our selves. During transcendent experiences, the self fades from view, allowing one to feel "connected" to some hidden grandeur that had been there all along but had remained out of sight. Psychologist Jonathan Haidt argues that transcendence is an adaptive mechanism that promoted the evolution of human cooperation by modulating group identity. In my view, transcendence is not so much a relic of human evolution as it is a primitive feature of consciousness itself. If the awareness of the body in space and time is a critical component of consciousness, then there is an important sense in which richer forms of conscious experience flow from a better understanding of one's place in the cosmos. This understanding can be simple—relating, for example, to an animal's position relative to its nest or other meaningful objects in its immediate microcosmos—or it can involve a more abstract representation of the broader enveloping world, even one of open-ended spatiotemporal depth that reposes on the theoretical knowledge of scaffolded science.

Transcendence is not the sole province of mystics and shamans. It is an important part of many people's lives, and it is central to the value of science. The iconic "Earthrise" image, taken on Christmas Eve 1968 by the crew of Apollo 8, shows the bright blue Earth, churning with oceans and clouds, rising over the barren gray surface of the moon. Logically, of course, there was nothing surprising about the view; and yet, like Sagan's Arroway, words failed the astronauts of Apollo 8 as they were faced with nature's transformative grandeur. For at that moment, they not only *knew* their place in the universe, but they *experienced* it in a different and deeper way. The event was more revelation than confirmation.

As the frontiers of scientific knowledge shrink, either renewed confidence or increasing doubt will swirl around the conclusions of this book. Convergence, if it proves to be a potent force in evolution, could provide a cosmic connectedness—for it is a story about the immense chasms that life will cross to reach the same place. If convergence is an ode to similarity and destiny, contingency is a celebration of the diversity and uniqueness that results from an evolutionary process that performs more like improvisational jazz or a Jerry Garcia guitar riff than it does like the highly structured composures of classical form. Radical contingency, should *it* prevail, would compel us to confront the infinite isolation of a desolate universe, drawing us closer to the only cosmic companions we will ever have: the meaningful minds with whom we share this pale blue dot.

Notes

Chapter 1

1. C. de Duve, "Life as a Cosmic Imperative?," *Philosophical Transactions of the Royal Society of London A: Mathematical, Physical and Engineering Sciences* 369, no. 1936 (2011): 620–623.

2. C. R. Woese, "On the Evolution of Cells," *Proceedings of the National Academy of Sciences of the United States of America* 99, no. 13 (2002): 8742–8747.

3. C. Mariscal and W. F. Doolittle, "Eukaryotes First: How Could That Be?," *Philosophical Transactions of the Royal Society of London B: Biological Sciences* 370 (2015), https://doi.org/10.1098/rstb.2014.0322.

4. D. W. McShea and W. Hordijk, "Complexity by Subtraction," *Evolutionary Biology* 40, no. 4 (2013): 504–520.

5. D. W. McShea, "Three Trends in the History of Life: An Evolutionary Syndrome," *Evolutionary Biology* 43, no. 4 (2016): 531–542.

6. W. F. Doolittle and O. Zhaxybayeva, "What Is a Prokaryote?," in *The Prokaryotes: Prokaryotic Biology and Symbiotic Associations*, ed. E. Rosenberg, E.F. DeLong, S. Lory, E. Stackebrandt, and F. Thompson, 21–37 (Springer, 2013).

7. M. P. Callahan, K. E. Smith, H. J.Cleaves II, et al., "Carbonaceous Meteorites Contain a Wide Range of Extraterrestrial Cucleobases," *Proceedings of the National Academy of Sciences of the United States of America* 108, no. 34 (2011): 13995–13998.

8. A. P. Nutman, V. C. Bennett, and C. R. L. Friend, "Rapid Emergence of Life Shown by Discovery of 3,700-Million-Year-Old Microbial Structures," *Nature* 537 (2016): 535–538.

9. M. S. Dodd, D. Papineau, and T. Grenne, "Evidence for Early Life in Earth's Oldest Hydrothermal Vent Precipitates," *Nature* 543 (2017): 60–64.

10. N. Lane, *The Vital Question: Energy, Evolution, and the Origins of Complex Life* (Norton, 2016).

11. S. B. Hedges, "The Origin and Evolution of Model Organisms," *Nature Reviews Genetics* 3 (2002): 838–849.

12. O. Abramov and S. J. Mojzsis, "Microbial Habitability of the Hadean Earth during the Late Heavy Bombardment," *Nature* 459 (2009): 419–422.

13. C. Sagan, "The Abundance of Life-Bearing Planets," *Bioastronomy News* 7, no. 4 (1995): 1–4.

14. J. C. Tarter, P. R. Bakus, R. L. Mancinelli, et al., "A Reappraisal of the Habitability of Planets around M Dwarf Stars," *Astrobiology* 7, no. 1 (2007): 30–65.

15. E. V. Koonin, "On the Origin of Cells and Viruses," *Annals of the New York Academy of Sciences* 1178 (2009): 47–64.

16. E. V. Koonin and W. Martin, "On the Origin of Genomes and Cells within Inorganic Compartments," *Trends in Genetics* 21 (2005): 647–654.

17. F. H. C. Crick, "The Origin of the Genetic Code," *Journal of Molecular Biology* 38 (1968): 367–379.

18. D. Penny, "An Interpretive Review of the Origin of Life Research," *Biology and Philosophy* 20 (2005): 633–671.

19. M. L. Rosenzweig and R. D. McCord, "Incumbent Replacement: Evidence for Long-Term Evolutionary Progress," *Paleobiology* 17, no. 3 (1991): 202–213.

20. D. Erwin, "A Public Goods Approach to Major Evolutionary Innovations," *Geobiology* 13 (2015): 308–315.

21. K. Vetsigian, C. Woese, and N. Goldenfeld, "Collective Evolution and the Genetic Code," *Proceedings of the National Academy of Sciences of the United States of America* 103 (2006): 10696–10701.

22. D. M. Raup and J. W. Valentine, "Multiple Origins of Life," *Proceedings of the National Academy of Sciences of the United States of America* 80, no. 10 (1983): 2981–2984.

23. C. H. Lineweaver and D. Grether, "What Fraction of Sun-Like Stars Have Planets?," *Astrophysics* 598 (2003): 1350–1360.

24. M. Ćirković, *The Great Silence: Science and Philosophy of Fermi's Paradox* (Oxford University Press, 2018).

25. A. Vilenkin, "Predictions from Quantum Cosmology," *Physical Review Letters* 74, no. 6 (1995): 846–849.

26. See Ćirković, *The Great Silence*.

27. B. Carter, "The Anthropic Principle and Its Implications for Biological Evolution," *Philosophical Transactions of the Royal Society of London A: Mathematical, Physical and Engineering Sciences* 310, no. 1512 (1983): 347–363.

28. J. Leconte, F. Forget, B. Charnay, et al. "Increased Insolation Threshold for Runaway Greenhouse Processes on Earth-like Planets." *Nature* 504, no. 7479 (2013): 268–271.

29. See, for example, P. A. Wilson, "Carter on Anthropic Principle Predictions," *British Journal for the Philosophy of Science* 45 (1994): 241–253; M. M. Ćirković, B. Vukotić, and I. Dragićević, "Galactic 'Punctuated Equilibrium': How to Undermine Carter's Anthropic Argument in Astrobiology," *Astrobiology* 9, no. 5 (2009): 491–501.

30. See Tarter et al., "A Reappraisal of the Habitability of Planets," 30–65.

31. G. Basalla, *Civilized Life in the Universe: Scientists on Intelligent Extraterrestrials* (Oxford University Press, 2006), 9.

32. H. Reichenbach, *Experience and Prediction* (University of Chicago Press, 1938); see also W. Salmon, "The Uniformity of Nature," *Philosophy and Phenomenological Research* 14, no. 1 (1953): 39–48.

33. E. Sober, *The Nature of Selection* (MIT Press, 1984).

34. R. Brandon, *Adaptation and Environment* (Princeton University Press, 1990).

35. P. Kitcher, "Darwin's Achievement," in *Reason and Rationality in Science*, ed. N. Rescher, 127–189 (University Press of America, 1985).

36. R. N. Brandon, *Adaptation and Environment* (Princeton University Press, 1990).

37. See M. M. Ćirković, "Evolutionary Contingency and SETI Revisited," *Biology and Philosophy* 29, no. 4 (2014): 539–557.

38. G. G. Simpson, "The Nonprevalence of Humanoids," *Science* 143, no. 3608 (1962): 769–775.

39. E. Mayr, "The Search for Intelligence," *Science* 259, no. 5101 (1993): 1522–1523.

40. See, for example, P. D. Ward and D. Brownlee, *Rare Earth: Why Complex Life Is Uncommon in the Universe* (Springer, 2000).

41. D. Raup, *Science and Alien Intelligence* (Cambridge University Press, 1985).

42. J. Diamond, "Alone in a Crowded Universe," *Natural History* 99, no. 6 (1990): 30–34.

43. S. J. Gould, "SETI and the Wisdom of Casey Stengel," in *The Flamingo's Smile: Reflections in Natural History* (Norton, 1985), 403–413.

Chapter 2

1. "Franz Ferdinand, Whose Assassination Sparked a World War," *New York Times*, June 28, 2016, https://www.nytimes.com/interactive/projects/cp/obituaries/archives/archduke-franz-ferdinand-world-war.

2. R. N. Lebow, *Archduke Franz Ferdinand Lives! A World without World War I* (St. Martin's Press, 2014), 11.

3. David McCullough, *1776* (Simon & Schuster, 2005), 294.

4. S. J. Gould, *The Structure of Evolutionary Theory* (Belknap Press of Harvard University Press, 2002), 47.

5. See, for example, P. Tetlock, R. Lebow, and N. Parker, *Unmaking the West: "What-If" Scenarios That Rewrite World History* (University of Michigan Press, 2006).

6. A. Roberts, *The Storm of War: A New History of the Second World War* (HarperCollins, 2011).

7. J. Diamond, *Guns, Germs, and Steel: The Fates of Human Societies* (Norton, 1996).

8. K. Sterelny, "Contingency and History," *Philosophy of Science* 83, no. 4 (2016): 521–539.

9. Tetlock et al., *Unmaking the West.*

10. S. J. Gould, *Wonderful Life: The Burgess Shale and the Nature of History* (Norton, 1989).

11. Gould, *Wonderful Life*, 290.

12. D. Fu et al., "The Qingjiang Biota—A Burgess Shale–Type Fossil Lagerstätte from the Early Cambrian of South China," *Science* 363, no. 6433 (2019): 1338–1342.

13. J. Yang, J. Ortega-Hernández, S. Gerber, et al., "A Superarmored Lobopodian from the Cambrian of China and Early Disparity in the Evolution of Onychophora," *Proceedings of the National Academy of Sciences of the United States of America* 112, no. 28 (2015): 8678–8683.

14. J. B. Caron and C. Aria, "Cambrian Suspension-Feeding Lobopodians and the Early Radiation of Panarthropods," *BMC Evolutionary Biology* 17, no. 1 (2017): 29, https://doi.org/10.1186/s12862-016-0858-y.

15. A. H. Knoll, *Life on a Young Planet: The First Three Billion Years of Evolution on Earth* (Princeton University Press, 2005).

16. Gould, *Wonderful Life*, 322–323.

17. Gould, *Wonderful Life*, 1160–1161.

18. See R. Powell, "Convergent Evolution and the Limits of Natural Selection," *European Journal for the Philosophy of Science* 2, no. 3 (2012): 355–373.

19. D. H. Erwin, *Extinction: How Life on Earth Nearly Ended 250 Million Years Ago* (Princeton University Press, 2006).

20. S. D. Burgess and S. A. Bowring, "High-Precision Geochronology Confirms Voluminous Magmatism before, during, and after Earth's Most Severe Extinction," *Science Advances* 1, no. 7 (2015): e1500470.

21. J. J. Sepkoski, "A Factor Analytic Description of the Phanerozoic Marine Fossil Record," *Paleobiology* 7 (1981): 36–53.

22. P. M. Hull, S. A. F. Darroch, and D. H. Erwin, "Rarity in Mass Extinctions and the Future of Ecosystems," *Nature* 528, no. 7582 (2015): 345–351.

23. D. Jablonski, "Mass Extinctions and Macroevolution," *Paleobiology* 31, no. S2 (2005): 192–210.

24. For a state of the art review of species selection, see D. Jablonski, "Species Selection: Theory and Data," *Annual Review of Ecology, Evolution, and Systematics* 39 (2008): 501–524.

25. Jablonski, "Species Selection," 501–524.

26. D. M. Raup, *Extinction: Bad Genes or Bad Luck?* (Norton, 1991).

27. S. M. Stanley, *Macroevolution* (Freeman, 1979).

28. J. J. Sepkoski, "A Kinetic Model of Phanerozoic Taxonomic Diversity. III. Post-Paleozoic Families and Mass Extinctions," *Paleobiology* 10 (1984): 246–267.

29. R. B. J. Benson, P. D. Mannion, R.J. Butler, et al., "Cretaceous Tetrapod Fossil Record Sampling and Faunal Turnover: Implications for Biogeography and the Rise of Modern Clades," *Palaeogeography, Palaeoclimatology, Palaeoecology* 372 (2013): 88–107.

30. M. J. Benton, "On the Nonprevalence of Competitive Replacement in the Evolution of Tetrapods," in *Evolutionary Paleobiology*, ed. J. W. Valentine, 185–210 (University of Chicago Press, 1996); M. J. Benton, "Large-Scale Replacements in the History of Life," *Nature* 302 (1983): 16–17.

31. Gould, *Wonderful Life*, 317.

32. W. Alvarez, *T. rex and the Crater of Doom* (Princeton University Press, 1997).

33. J. Alroy, "The Fossil Record of North American Mammals: Evidence for a Paleocene Evolutionary Radiation," Systematic Biology 48 (1999): 107–118.

34. M. A. O'Leary, J. I. Bloch, J. J. Fynn, et al., "The Placental Mammal Ancestor and the Post–K-Pg Radiation of Placentals," *Science* 339, no. 6120 (2013): 662–667.

35. G. P. Wilson, A. R. Evans, I. J. Corfe, et al., "Adaptive Radiation of Multituberculate Mammals before the Extinction of Dinosaurs," *Nature* 483 (2012): 457–460.

36. M. S. Springer, W. J. Murphy, E. Eizirik, and S. J. O'Brien, "Placental Mammal Diversification and the Cretaceous–Tertiary Boundary," *Proceedings of the National Academy of Sciences of the United States of America* 100, no. 3 (2003): 1056–1061.

37. D. M. Grossnickle and E. Newham, "Therian Mammals Experience an Ecomorphological Radiation during the Late Cretaceous and Selective Extinction at the K–Pg Boundary," *Proceedings of the Royal Society B: Biological Sciences* 283, no. 1832 (2016), https://doi.org/10.1098/rspb.2016.0256.

38. S. L. Brusatte, R. J. Butler, P. M. Barrett, et al., "The Extinction of the Dinosaurs," *Biological Reviews* 90, no. 2 (2015): 628–642.

39. M. Sakamoto, M. J. Benton, and C. Venditti, "Dinosaurs in Decline Tens of Millions of Years before Their Final Extinction," *Proceedings of the National Academy of Sciences of the United States of America* 113, no. 18 (2016): 5036–5040.

40. S. J. Gould and C. B. Calloway, "Clams and Brachiopods—Ships That Pass in the Night," *Paleobiology* 6 (1980): 383–396.

41. M. J. Benton, "The Late Triassic Tetrapod Extinction Events," in *The Beginning of the Age of Dinosaurs: Faunal Change across the Triassic-Jurassic Boundary*, ed. K. Padian, 303–320 (Cambridge University Press, 1986).

42. R. A. DePalma, et al., "A Seismically Induced Onshore Surge Deposit at the KPg Boundary, North Dakota," *Proceedings of the National Academy of Sciences* 116, no. 17 (2019): 8190–8199.

43. G. J. Slater, "Phylogenetic Evidence for a Shift in the Mode of Mammalian Body Size Evolution at the Cretaceous-Palaeogene Boundary," *Methods in Ecology and Evolution* 4, no. 8(2013): 734–744.

44. S. L. Brusatte, M. J. Benton, M. Ruta, and G. T. Lloyd, "Superiority, Competition, and Opportunism in the Evolutionary Radiation of Dinosaurs," *Science* 321, no. 5895 (2008): 1485–1488.

45. S. J. Gould and R. C. Lewontin, "The Spandrels of San Marco and the Panglossian Paradigm: A Critique of the Adaptationist Programme," *Proceedings of the Royal Society of London B: Biological Sciences* 205, no. 1161 (1979): 585.

46. Gould, *Structure of Evolutionary Theory*, 880.

47. J. C. Schank and W. C. Wimsatt, "Generative Entrenchment and Evolution," *Philosophy of Science* 2 (1986): 33–60.

48. R. C. Lewontin, "Adaptation," *Scientific American* 239 (1978): 212–230.

49. On the relation between selection and stasis, see R. Brandon and D. McShea, *The Zero Force Evolutionary Law* (University of Chicago Press, 2011).

50. E. H. Davidson, *The Regulatory Genome: Gene Regulatory Networks in Development and Evolution* (Academic Press, 2010); I. Peter and E. H. Davidson, *Genomic Control Process: Development and Evolution* (Academic Press, 2015).

51. E. H. Davidson and D. H. Erwin, "Gene Regulatory Networks and the Evolution of Animal Body Plans," *Science* 311 (2006): 796–800.

52. M. Thattai and A. van Oudenaarden, "Intrinsic Noise in Gene Regulatory Networks," *Proceedings of the National Academy of Sciences of the United States of America* 98, no. 15 (2001): 8614–8619.

53. Davidson, *Regulatory Genome*.

54. Gould, *Wonderful Life*, 216.

55. D. H. Erwin, "Evolutionary Uniformitarianism," *Developmental Biology* 357, no. 1 (2011): 32.

56. M. Foote, "Evolution of Morphological Diversity," *Annual Review of Ecology and Systematics* 28 (1997): 129–152.

57. J. Maclaurin and K. Sterelny, *What Is Biodiversity?* (University of Chicago Press, 2008).

58. G. E. Budd and S. Jensen, "A Critical Reappraisal of the Fossil Record of the Bilaterian Phyla," *Biological Reviews* 75, no. 2 (2000): 253–295.

59. R. P. S. Jefferies, "The Origin of Chordates—a Methodological Essay," in *The Origin of Major Invertebrate Groups*, ed. M. R. House, 443–477 (Academic Press, 1979).

60. Budd and Jensen, "Critical Reappraisal," 256.

61. Yang et al., "Superarmored Lobopodian," 8678–8683.

62. For the mollusk-affinity hypothesis, see M. R. Smith, "Ontogeny, Morphology and Taxonomy of the Soft-Bodied Cambrian 'Mollusc' Wiwaxia," *Palaeontology* 57, no. 1 (2014): 215–229.

63. See I. Bobrovskiy, J. M. Hope, A. Ivantsov, et al., "Ancient Steroids Establish the Ediacaran Fossil *Dickinsonia* as One of the Earliest Animals," *Science* 361, no. 6408 (2018): 1246–1249.

64. K. Brysse, "From Weird Wonders to Stem Lineages: The Second Reclassification of the Burgess Shale Fauna," *Studies in History and Philosophy of Biological and Biomedical Sciences* 39, no. 3 (2008): 298–313.

65. D.-G. Shu, H.-L. Luo, S. Conway Morris, et al., "Lower Cambrian Vertebrates from South China," *Nature* 402 (1999): 42–46.

66. S. C. Morris and J. B. Caron, "A Primitive Fish from the Cambrian of North America," *Nature* 512, no. 7515 (2014): 419–422.

67. D. H. Erwin, "Wonderful Life Revisited: Chance and Contingency in the Ediacaran-Cambrian Radiation," in *Chance in Evolution*, ed. G. Ramsey and C. Pence, 277–298 (University of Chicago Press, 2016).

Chapter 3

1. J. Beatty, "Replaying Life's Tape," *Journal of Philosophy* 103, no. 7 (2006): 336–362.

2. Beatty, "Replaying Life's Tape," 345.

3. S. J. Gould, *Wonderful Life: The Burgess Shale and the Nature of History* (Norton, 1989), 283.

4. S. P. Laplace, "Essai philosophique sur les probabilité" (1820), quoted in E. Nagel, *The Structure of Science: Problems in the Logic of Scientific Explanation* (Harcourt, Brace and World, 1961), 281–282.

5. Y. Ben-Menahem, "Historical Contingency," *Ratio* 10, no. 2 (1997): 99–107.

6. Gould, *Wonderful Life*, 51; S. J. Gould, *The Structure of Evolutionary Theory* (Belknap Press of Harvard University Press, 2002), 1333.

7. Gould, *Structure of Evolutionary Theory*, 1333.

8. D. Turner, "Gould's Replay Revisited," *Biology and Philosophy* 26 (2011): 69.

9. Beatty, "Replaying Life's Tape," 345. See also R. Inkpen and D. Turner, "The Topography of Historical Contingency," *Journal of the Philosophy of History* 6, no. 1 (2012): 1–19.

10. R. N. Brandon, "The Principle of Drift: Biology's First Law," *Journal of Philosophy* 103 (2006): 319–335.

11. E. Desjardins, "Historicity and Experimental Evolution," *Biology and Philosophy* 26 (2011): 339–364.

12. E. Szathmary, "Path Dependence and Historical Contingency in Biology," in *Understanding Change: Models, Methodologies, and Metaphors*, ed. Andreas Wimmer and Reinhart Kossler, 140–157 (Macmillan, 2006), 141.

13. S. J. Gould, "Dollo on Dollo's Law: Irreversibility and the Status of Evolutionary Laws," *Journal of the History of Biology* 3, no. 2 (1970): 189–212.

14. R. Powell, "Is Convergence More than an Analogy? Homoplasy and its Implications for Macroevolutionary Predictability," *Biology & Philosophy* 22, no. 4 (2007): 565–578; R. Powell, "Contingency and Convergence in Macroevolution: A Reply to Beatty," *Journal of Philosophy* 106, no. 7 (2009): 390–404; R. Powell, "Convergent Evolution and the Limits of Natural Selection," *European Journal for the Philosophy of Science* 2, no. 3 (2012): 355–373; R. Powell and C. Mariscal, "Convergent Evolution as Natural Experiment: The Tape of Life Reconsidered," *Journal of the Royal Society Interface Focus* 5, no. 6 (2015): 1–13.

15. Gould, *Wonderful Life*, 290 (emphasis added).

16. Gould, *Wonderful Life*, 283 (emphasis added).

17. Gould, *Wonderful Life*, 283 (emphasis added).

18. Gould, *Structure of Evolutionary Theory*, 55–56.

19. D. Brownlee and P. Ward, *Rare Earth: Why Complex Life Is Uncommon in the Universe* (Copernicus Springer-Verlag, 1999).

20. S. Conway Morris, *Life's Solution: Inevitable Humans in a Lonely Universe* (Cambridge University Press, 2003), 309.

21. Gould, *Structure of Evolutionary Theory*, 696, 1055, 1227, 1333.

22. J. Beatty, "The Evolutionary Contingency Thesis," in *Concepts, Theories and Rationality in the Biological Sciences*, ed. J. G. Lennox and G. Wolters, 45–81 (University of Pittsburgh Press, 1995).

23. A. Rosenberg, "The Supervenience of Biological Concepts," *Philosophy of Science* 45 (1978): 368–386.

24. A. Rosenberg, "How Is Biological Explanation Possible?," *British Journal for the Philosophy of Science* 52, no. 4 (2001): 735–760.

25. L. van Valen, "A New Evolutionary Law," *Evolutionary Theory* 1 (1973): 1–30.

26. B. van Fraassen, *Laws and Symmetry* (Clarendon Press, 1989), 27.

27. R. N. Brandon, *Adaptation and Environment* (Princeton University Press, 1995).

28. R. N. Brandon, "The Principle of Drift: Biology's First Law," *Journal of Philosophy* 103 (2006): 319–335.

29. R. N. Brandon, "Natural Selection," in *Stanford Encyclopedia of Philosophy*, ed. E. N. Zalta (Metaphysics Research Lab, Center for the Study of Language and Information, Stanford University, 2014), https://plato.stanford.edu/entries/natural-selection/.

30. R. N. Brandon and D. McShea, *Biology's First Law: The Tendency for Diversity and Complexity to Increase in Evolutionary Systems* (University of Chicago Press, 2012).

31. F. A. Smith, et al., "The Evolution of Maximum Body size of Terrestrial Mammals," *Science* 330, no. 6008 (2010): 1216–1219.

32. D. McShea, "The Hierarchical Structure of Organisms: A Scale and Documentation of a Trend in the Maximum," *Paleobiology* 27 (2001): 405–423.

33. E. Sober, "Two Outbreaks of Lawlessness in Recent Philosophy of Biology," *Philosophy of Science* 64 (1997): S458–S467.

34. Gould, *Wonderful Life*, 233–234.

35. See, for instance, S. J. Gould and R. C. Lewontin, "The Spandrels of San Marco and the Panglossian Paradigm: A Critique of the Adaptationist Programme," *Proceedings of the Royal Society of London B: Biological Sciences* 205, no. 1161 (1979): 581–598.

Chapter 4

1. M. J. S. Rudwick, *Worlds before Adam: The Reconstruction of Geohistory in the Age of Reform* (University of Chicago Press, 2008).

2. C. Lyell, *Principles of Geology* (J. Murray, 1830), 123.

3. D. N. Stamos, *Darwin and the Nature of Species* (SUNY Press, 2008).

4. M. Travisano, J. A. Mongold, A. F. Bennett, and R. E. Lenski, "Experimental Tests of the Roles of Adaptation, Chance, and History in Evolution," *Science* 267, no. 5194 (1995): 87–90.

5. P. R. Grant and B. R. Grant, "Unpredictable Evolution in a 30-Year Study of Darwin's Finches," *Science* 296, no. 5568 (2002): 707–711.

6. E. E. Goldberg and B. Igić, "On Phylogenetic Tests of Irreversible Evolution," *Evolution* 62, no. 11 (2008): 2727–2741.

7. Z. D. Blount, "History's Windings in a Flask: Microbial Experiments in Evolutionary Contingency," in *Chance in Evolution*, ed. G. Ramsey and C. Pence, 244–263 (University of Chicago Press, 2015).

8. M. Foote, "Contingency and Convergence: Review of *The Crucible of Creation* by Simon Conway Morris," *Science* 280, no. 5372 (1998): 2068–2069.

9. S. J. Gould, *Wonderful Life: The Burgess Shale and the Nature of History* (Norton, 1989).

10. S. Conway Morris, *Life's Solution: Inevitable Humans in a Lonely Universe* (Cambridge University Press, 2003).

11. G. McGhee, *Convergent Evolution: Limited Forms Most Beautiful* (MIT Press, 2013).

12. M. Mori, "The Uncanny Valley," *Energy* 7, no. 4 (1970): 33–35.

13. Conway Morris, *Life's Solution.*

14. McGhee, *Convergent Evolution.*

15. G. J. Vermeij, *Evolution and Escalation: An Ecology History of Life* (Princeton University Press, 2006).

16. J. B. Losos, *Improbable Destinies: Fate, Chance, and the Future of Evolution* (Riverhead Books, 2017).

17. L. D. Martin and T. J. Meehan, "Extinction May Not Be Forever," *Naturwissenschaften* 92, no. 1 (2005): 1–19.

18. D. C. Dennett, *Darwin's Dangerous Idea: Evolution and the Meaning of Life* (Simon & Schuster, 1995).

19. S. Conway Morris, *The Crucible of Creation: The Burgess Shale and the Rise of Animals* (Oxford University Press, 1998), 205.

20. Dennett, *Darwin's Dangerous Idea*, 306.

21. Dennett, *Darwin's Dangerous Idea*, 308.

22. G. McGhee, "Convergent Evolution: A Periodic Table of Life?," in *The Deep Structure of Biology*, ed. S. Conway Morris, 17–31 (Templeton Foundation Press, 2008).

23. J. Beatty, "Replaying Life's Tape," *Journal of Philosophy* 103, no. 7 (2006): 336–362.

24. M. M. Ćirković, "Woodpeckers and Diamonds: Some Aspects of Evolutionary Convergence in Astrobiology," *Astrobiology* 18, no. 5 (2018): 491–502.

25. G. J. Slater and B. Van Valkenburgh, “Long in the Tooth: Evolution of Sabertooth Cat Cranial Shape,” *Paleobiology* 34, no. 3 (2008): 403–419.

26. R. C. Lewontin, “Adaptation,” *Scientific American* 239 (1978): 212–230.

27. See, for instance, A. M. Currie, *Rock, Bone & Ruin: An Optimist's Guide to the Historical Sciences* (MIT Press, 2018).

28. T. Lingham-Soliar, “Convergence in Thunniform Anatomy in Lamnid Sharks and Jurassic Ichthyosaurs,” *Integrative and Comparative Biology* 56, no. 6 (2016): 1323–1336.

29. A. M. Currie, “Convergence as Evidence,” *British Journal for the Philosophy of Science* 64 (2013): 763–786.

30. C. T. Stayton, “Is Convergence Surprising? An Examination of the Frequency of Convergence in Simulated Datasets,” *Journal of Theoretical Biology* 252, no. 1 (2008): 1–14.

31. McGhee, *Convergent Evolution*.

32. McGhee, *Convergent Evolution*, 271.

33. Z. D. Blount, “History's Windings in a Flask: Microbial Experiments in Evolutionary Contingency,” in *Chance in Evolution*, ed. G. Ramsey and C. Pence, 244–263 (University of Chicago Press, 2015), 246; Gould, *Wonderful Life*; S. J. Gould, *The Structure of Evolutionary Theory* (Belknap Press of Harvard University Press, 2002).

34. Gould, *Structure of Evolutionary Theory*, 1122–1123.

35. J. Maynard Smith, R. Burian, S. Kauffman, et al., “Developmental Constraints and Evolution: A Perspective from the Mountain Lake Conference on Development and Evolution,” *Quarterly Review of Biology* 60, no. 3 (1985): 265–287.

36. R. Amundson, “Two Concepts of Constraint: Adaptationism and the Challenge from Developmental Biology,” *Philosophy of Science* 61 (1994): 556–578.

37. S. J. Gould and R. C. Lewontin, “The Spandrels of San Marco and the Panglossian Paradigm: A Critique of the Adaptationist Programme,” *Proceedings of the Royal Society of London B: Biological Sciences* 205, no. 1161 (1979): 585, p. 594.

38. See, for example, G. P. Wagner, “The Influence of Variation and of Developmental Constraints on the Rate of Multivariate Phenotypic Evolution,” *Journal of Evolutionary Biology* 1 (1988): 45–66.

39. S. Kauffman, *At Home in the Universe: The Search for the Laws of Self-Organization and Complexity* (Oxford University Press, 1996).

40. A. Louis, “Contingency, Convergence and Hyper-Astronomical Numbers in Biological Evolution,” *Studies in History and Philosophy of Biological and Biomedical Sciences* 58 (2016): 107–116.

41. S. J. Gould, “A Developmental Constraint in *Cerion*, with Comments on the Definition and Interpretation of Constraint in Evolution,” *Evolution* 43 (1989): 516–539.

42. Gould, *Structure of Evolutionary Theory*, 1134.

43. E. Desjardins, “Reflections on Path Dependence and Irreversibility: Lessons from Evolutionary Biology,” *Philosophy of Science* 78, no. 5 (2011): 724–738.

44. R. Powell and C. Mariscal, “Convergent Evolution as Natural Experiment: The Tape of Life Reconsidered,” *Journal of the Royal Society Interface Focus* 5, no. 6 (2015): 1–13.

45. McGhee, *Convergent Evolution*, 271.

46. Gould, *Wonderful Life*, 289–290.

47. C. Haufe, “Gould's Laws,” *Philosophy of Science* 82 (2015): 1–20.

48. Gould and Lewontin, “Spandrels of San Marco,” 585.

Chapter 5

1. G. G. Simpson, “The Nonprevalence of Humanoids,” *Science* 143, no. 3608 (1962): 769–775.

2. P. D. Ward and D. Brownlee, *Rare Earth: Why Complex Life Is Uncommon in the Universe* (Springer, 2000).

3. S. Okasha, “Experiment, Observation and the Confirmation of Laws,” *Analysis* 71 (2011): 222–232.

4. R.N. Brandon, “Does Biology Have Laws? The Experimental Evidence,” *Philosophy of Science* 64 (1997): S444–S457.

5. J. M. Diamond, “Overview: Laboratory Experiments, Field Experiments, and Natural Experiments,” in *Community Ecology*, ed. J. Diamond and T. J. Case, 3–22 (Harper & Row, 1986).

6. M. J. Benton, Z. Csiki, D. Grigorescu, et al., “Dinosaurs and the Island Rule: The Dwarfed Dinosaurs from Haţeg Island,” *Palaeogeography, Palaeoclimatology, Palaeoecology* 293, no. 3–4 (2010): 438–454.

7. See, for example, J. B. Losos, T. R. Jackman, A. Larson, et al., “Contingency and Determinism in Replicated Adaptive Radiations of Island Lizards,” *Science* 279 (1998): 2115–2118.

8. J. B. Losos, *Lizards in an Evolutionary Tree: Ecology and Adaptive Radiation of Anoles* (University of California Press, 2009).

9. See, for example, M. S. Morgan, “Nature’s Experiments and Natural Experiments in the Social Sciences,” *Philosophy of Social Science* 43 (2013): 341–357; J. Beatty, “Replaying Life’s Tape,” *Journal of Philosophy* 103, no. 7 (2006): 336–362.

10. R. Sansom, “Constraining the Adaptationism Debate,” *Biology and Philosophy* 18 (2003): 493–512.

11. See O. Haas and G. G. Simpson, “Analysis of Some Phylogenetic Terms, with Attempts at Redefinition,” *Proceedings of the American Philosophical Society* 90, no. 5 (1946): 319–349.

12. E. O. Wiley and B. S. Lieberman, *Phylogenetics: Theory and Practice of Phylogenetic Systematics* (Wiley-Blackwell, 2011).

13. G. Ramsey and A. Peterson, “Sameness in Biology,” *Philosophy of Science* 79 (2012): 255–275.

14. See, for example, R. Powell, “Is Convergence More than an Analogy? Homoplasy and its Implications for Macroevolutionary Predictability,” *Biology & Philosophy* 22, no. 4 (2007): 565–578; T. Pearce, “Convergence and Parallelism in Evolution: A Neo-Gouldian Account,” *British Journal for the Philosophy of Science* 63 (2011): 429–448; A. M. Currie, “Venomous Dinosaurs and Rearfanged Snakes: Homology and Homoplasy Characterized,” *Erkenntnis* 79 (2014): 701–727.

15. G. McGhee, *Convergent Evolution: Limited Forms Most Beautiful* (MIT Press, 2013).

16. A. G. Dyer, S. Boyd-Gerny, S. McLoughlin, et al., “Parallel Evolution of Angiosperm Colour Signals: Common Evolutionary Pressures Linked to Hymenopteran Vision,” *Proceedings of the Royal Society B: Biological Sciences* 279, no. 1742 (2012): 3606–3615.

17. J. D. Mclver and G. Stonedahl, “Myrmecomorphy: Morphological and Behavioral Mimicry of Ants,” *Annual Review of Entomology* 38, no. 1 (1993): 351–377.

18. “Holobionts” are assemblages of species that form proper ecological units. See L. Margulis, *Symbiosis as a Source of Evolutionary Innovation* (MIT Press, 1991).

19. See R. Amundson, *The Changing Role of the Embryo* (Cambridge University Press, 2005).

20. See, for instance, S. H. Orzack and E. Sober, “Optimality Models and the Test of Adaptationism,” *American Naturalist* 143 (1994): 361–380.

21. Losos, *Lizards in an Evolutionary Tree,* 362.

22. J. Arendt and D. Reznick, “Convergence and Parallelism Reconsidered: What Have We Learned about the Genetics of Adaptation?,” *Trends in Ecology and Evolution* 23, no. 1 (2007): 26–32.

23. S. Conway Morris, *Life’s Solution: Inevitable Humans in a Lonely Universe* (Cambridge University Press, 2003), 435n1.

24. G. G. Simpson, “The Principles of Classification and a Classification of Mammals,” *Bulletin of the American Museum of Natural History* 85 (1945), http://hdl.handle.net/2246/1104.

25. Quoted in O. Haas and G. G. Simpson, “Analysis of Some Phylogenetic Terms with Attempts at Redefinition,” *Proceedings of the American Philosophical Society* 90, no. 5 (1946): 319–349.

26. Haas and Simpson, “Analysis of Some Phylogenetic Terms.”

27. C. Darwin, *On the Origin of Species by Means of Natural Selection* (Appleton and Company, 1882), 375.

28. G. Ramsey and A. Peterson, "Sameness in Biology," *Philosophy of Science* 79 (2012): 255–275.

29. S. B. Carroll, *Endless Forms Most Beautiful: The New Science of Evo Devo* (W.W. Norton, 2005).

30. W. J. Gehring and K. Ikeo, "*Pax 6*: Mastering Eye Morphogenesis and Eye Evolution," *Trends in Genetics* 15, no. 9 (1999): 371–377.

31. C. R. Hitchcock, "Screening-Off and Visibility to Selection," *Biology and Philosophy* 12 (1997): 521–529.

32. E. Sober, "Screening-Off and the Units of Selection," *Philosophy of Science* 59 (1992): 142–152.

33. K. Waters, "Causes That Make a Difference," *Journal of Philosophy* 104 (2007): 551–579.

34. J. Woodward, *Making Things Happen: A Theory of Causal Explanation* (Oxford University Press, 2003).

35. Beatty, "Replaying Life's Tape," 336–362.

36. M. D. Shapiro, M. E. Marks, C. L. Peichel, et al., "Genetic and Developmental Basis of Evolutionary Pelvic Reduction in Threespine Sticklebacks," *Nature* 428, no. 6984 (2004): 717–723.

37. M. P. Harris, S. M. Hasso, M. W. Ferguson, and J. F. Fallon, "The Development of Archosaurian First Generation Teeth in a Chicken Mutant," *Current Biology* 16, no. 4 (2006): 371–377.

38. Z. Yang, B. Jiang, M. E. McNamara, et al., "Pterosaur Integumentary Structures with Complex Feather-like Branching," *Nature Ecology and Evolution* 3, no. 1 (2019): 24–30.

39. A. Louchart et al., "Structure and Growth Pattern of Pseudoteeth in *Pelagornis mauretanicus* (Aves, Odontopterygiformes, Pelagornithidae)," *PloS One* 8, no. 11(2013): e80372.

40. F. Jacob, "Evolution and Tinkering," *Science* 196, no. 4295 (1977): 1161–1166.

41. M. L. Rosenzweig and R. D. McCord, "Incumbent Replacement: Evidence for Long-Term Evolutionary Progress," *Paleobiology* 17, no. 3 (1991): 202–213.

Chapter 6

1. M. Ruse, *Monad to Man: The Concept of Progress in Evolutionary Biology* (Harvard University Press, 1996).

2. G. A. Gelman, *The Essential Child: Origins of Essentialism in Everyday Thought* (Oxford University Press, 2003).

3. E. Sober, "Evolution, Population Thinking, and Essentialism," *Philosophy of Science* 47, no. 3 (1980): 350–383.

4. See M. Ghiselin, "A Radical Solution to the Species Problem," *Systematic Zoology* 23 (1974): 536–544.

5. See, for example, D. Hull, "On Human Nature," *Proceedings of the Philosophy of Science Association* 2 (1986): 3–13.

6. Ruse, *Monad to Man*, 501.

7. P. Boyer and C. Ramble, "Cognitive Templates for Religious Concepts: Cross-Cultural Evidence for Recall of Counter-intuitive Representations," *Cognitive Science* 25 (2001): 535–564.

8. C. Sagan, *The Demon-Haunted World: Science as a Candle in the Dark* (Ballantine Books, 2000).

9. R. Powell and C. Mariscal, "Convergent Evolution as Natural Experiment: The Tape of Life Reconsidered," *Journal of the Royal Society Interface Focus* 5, no. 6 (2015): 1–13.

10. S. J. Gould, *The Structure of Evolutionary Theory* (Belknap Press of Harvard University Press, 2002), 1212.

11. G. McGhee, *Convergent Evolution: Limited Forms Most Beautiful* (MIT Press, 2013), 272.

12. A. Feduccia, *The Origin and Evolution of Birds* (Yale University Press, 1999), 204–205.

13. J. Vinther, M. Stein, N. R. Longrich, and D. A. T. Harper, "A Suspension-Feeding Anomalocarid from the Early Cambrian," *Nature* 507, no. 7493 (2014): 496–499.

14. K. V. Ewart, Q. Lin, and C. L. Hew, "Structure, Function and Evolution of Antifreeze Proteins," *Cellular and Molecular Life Sciences* 55, no. 2 (1999): 271–283.

15. R. R. Hoy and D. Robert, "Tympanal Hearing in Insects," *Annual Review of Entomology* 41 (1996): 433–450.

16. P.S. Shamble et al, "Airborne Acoustic Perception by a JumpingSpider," *Current Biology* 26, no. 21 (2016): 2913–2920.

17. L. Miersch, W. Hanke, S. Wieskotten, et al., "Flow Sensing by Pinniped Whiskers," *Philosophical Transactions of the Royal Society B: Biological Sciences* 366, no. 1581 (2011): 3077–3084.

18. M. Gagliano, S. Mancuso, and D. Robert, "Towards Understanding Plant Bioacoustics," *Trends in Plant Science* 17, no. 6 (2012): 323–325.

19. The multiple realizability of function as a source of lawlessness in biology is discussed in chapter 3.

20. J. M. Donley, C. A. Sepulveda, P. Konstantinidis, et al., "Convergent Evolution in Mechanical Design of Lamnid Sharks and Tunas," *Nature* 429, no. 6987 (2004): 61–65.

21. J. B. Losos, *Improbable Destinies: Fate, Chance, and the Future of Evolution* (Riverhead Books, 2017).

22. L. D. Martin and T. J. Meehan, "Extinction May Not Be Forever," *Naturwissenschaften* 92 (2005): 1–19.

23. Feduccia, *Origin and Evolution of Birds*.

24. M. F. Land and D. E. Nilsson, *Animal Eyes* (Oxford University Press, 2012).

25. M. F. Land and R. D. Fernald, "The Evolution of Eyes," *Annual Review of Neuroscience* 15 (1992): 1–29.

26. B. S. Leander, "A Hierarchical View of Convergent Evolution in Microbial Eukaryotes," *Journal of Eukaryotic Microbiology* 55, no. 2 (2008): 59–68.

27. D. Arendt, "Evolution of Eyes and Photoreceptor Cell Types," *International Journal of Developmental Biology* 47 (2003): 563–571; W. J. Gehring, "Historical Perspective on the Development and Evolution of Eyes and Photoreceptors," *International Journal of Developmental Biology* 48 (2004): 707–717.

28. T. D. Lamb, "Evolution of Vertebrate Retinal Photoreception," *Philosophical Transactions of the Royal Society of London B: Biological Sciences* 364, no. 1531 (2009): 2911–2924.

29. Z. Kozmik, J. Ruzickova, K. Jonasova, et al., "Assembly of the Cnidarian Camera-Type Eye from Vertebrate-like Components," *Proceedings of the National Academy of Sciences of the United States of America* 105, no. 26 (2008): 8989–8993.

30. T. Cavalier-Smith, "Origin of Animal Multicellularity: Precursors, Causes, Consequences—the Choanoflagellate/Sponge Transition, Neurogenesis and the Cambrian Explosion," *Philosophical Transactions of the Royal Society B: Biological Sciences* 372, no. 1713 (2017): 20150476; D. H. Erwin and J. W. Valentine, *The Cambrian Explosion: The Construction of Animal Biodiversity* (Roberts and Co., 2013).

31. W. J. Gehring, "New Perspectives on Eye Development and the Evolution of Eyes and Photoreceptors," *Journal of Heredity* 96, no. 3 (2005): 171–184.

Coda to Part I

1. J. Maynard Smith and E. Szathmary, *The Major Transitions in Evolution* (Oxford University Press, 1997).

2. For critiques, see M. A. O'Malley and R. Powell, "Major Problems in Evolutionary Transitions: How a Metabolic Perspective Can Enrich Our Understanding of Macroevolution," *Biology and Philosophy* 31 (2016): 159–169; D. W. McShea and C. Simpson, "The Miscellaneous Transitions in Evolution," in *The Major Transitions in Evolution Revisited*, ed. B. Calcott and K. Sterelny, 17–33 (MIT Press, 2011).

3. See, for example, A. Booth and W. F. Doolittle, "Eukaryogenesis, How Special Really?," *Proceedings of the National Academy of Sciences of the United States of America* 112 (2015): 10278–10285; S. J. Gould, *Full House: The Spread of Excellence from Plato to Darwin* (Harmony Books, 1997); R. Powell and M. O'Malley, "Metabolic and Microbial Perspectives on the Evolution of Evolution," *Journal of Experimental Zoology B*, in press.

4. D. McShea, "The Hierarchical Structure of Organisms: A Scale and Documentation of a Trend in the Maximum," *Paleobiology* 27 (2001): 405–423.

5. E. Szathmary, "Toward Major Evolutionary Transitions Theory 2.0," *Proceedings of the National Academy of Sciences of the United States of America* 112 (2015): 10104–10111.

6. J. G. Umen, "Green Algae and the Origins of Multicellularity in the Plant Kingdom," *Cold Spring Harbor Perspectives in Biology* 6, no. 11 (2014): a016170.

7. M. D. Herron and R. Michod, "Evolution of Complexity in the Volvocine Algae: Transitions in Individuality through Darwin's Eye," *Evolution* 62, no. 2 (2008): 436–451.

8. N. A. Lyons and R. Kolter, "On the Evolution of Bacterial Multicellularity," *Current Opinion in Microbiology* 24 (2015): 21–28.

9. T. Cavalier-Smith, "Origin of Animal Multicellularity: Precursors, Causes, Consequences—the Choanoflagellate/Sponge Transition, Neurogenesis and the Cambrian Explosion," *Philosophical Transactions of the Royal Society of London B: Biological Sciences* 372, no. 1713 (2017): 20150476.

10. A. Seilacher, D. Grazhdankin, and A. Legouta, "Ediacaran Biota: The Dawn of Animal Life in the Shadow of Giant Protists," *Paleontological Research* 7, no. 1 (2003): 43–54.

11. See I. Bobrovskiy, J. M. Hope, A. Ivantsov, et al., "Ancient Steroids Establish the Ediacaran Fossil *Dickinsonia* as One of the Earliest Animals," *Science* 361 (2018): 1246–1249; M. L. Droser and J. G. Gehling, "The Advent of Animals: The View from the Ediacaran," *Proceedings of the National Academy of Sciences of the United States of America* 112, no. 16 (2015): 4865–4870.

12. B. Holldobler and E. O. Wilson, *The Ants* (Harvard University Press, 1990).

13. M. A. Nowak, C. E. Tarnita, and E. O. Wilson, "The Evolution of Eusociality," *Nature* 466, no. 7310 (2010): 1057–1062.

14. See N. Lane, *The Vital Question: Energy, Evolution, and the Origins of Complex Life* (Norton, 2016).

15. H. Gest, "Evolutionary Roots of the Citric Acid Cycle in Prokaryotes," *Biochemical Society Symposium* 54 (1987): 3–16.

16. A. Wagner, "Metabolic Networks and Their Evolution," in *Evolutionary Systems Biology*, Advances in Experimental Medicine and Biology, vol. 751, ed. O. Soyer, 29–52 (Springer, 2012).

17. For a brief discussion, see Powell and O'Malley, "Metabolic and Microbial Perspectives on the Evolution of Evolution."

18. S. Conway Morris, *Life's Solution: Inevitable Humans in a Lonely Universe* (Cambridge University Press, 2003), 310.

Chapter 7

1. B. S. Hansson and M. C. Stensmyr, "Evolution of Insect Olfaction," *Neuron* 72, no. 5 (2011): 698–711.

2. M. S. Lewicki, B. A. Olshausen, A. Surlykke, and C. F. Moss, "Scene Analysis in the Natural Environment," *Frontiers in Psychology* 5 (2014): 199. A detailed discussion of phenomenology and the delineation of sensory modalities is contained in chapter 8.

3. T. W. Cronin, S. Johnsen, N. J. Marshall, and E. J. Warrant, *Visual Ecology* (Princeton University Press, 2014).

4. A. Wilde and C. W, Mullineaux, "Light-Controlled Motility in Prokaryotes and the Problem of Directional Light Perception," *FEMS Microbiology Reviews* 41, no. 6 (2017): 900–922.

5. For a recent review, see G. Jékely, "Evolution of Phototaxis," *Philosophical Transactions of the Royal Society of London B: Biological Sciences* 364, no. 1531 (2009): 2795–2808.

6. N. D. Larusso, B. E. Ruttenberg, A. K. Singh, and T. H. Oakley, "Type II Opsins: Evolutionary Origin by Internal Domain Duplication?," *Journal of Molecular Evolution* 66 (2008): 417–423.

7. See T. H. Oakley, "On Homology of Arthropod Compound Eyes," *Integrative and Comparative Biology* 43, no. 4 (2003): 522–530.

8. A. Packard, "Cephalopods and Fish: The Limits of Convergence," *Biological Reviews* 47 (1972): 241–307; A. M. Sweeney, S. H. Haddock, and S. Johnsen, "Comparative Visual Acuity of Coleoid Cephalopods," *Integrative and Comparative Biology* 47, no. 6 (2007): 808–814.

9. G. S. Gavelis, S. Hayakawa, R. A. White, et al., "Eye-like Ocelloids Are Built from Different Endosymbiotically Acquired Components," *Nature* 523, no. 7559 (2015): 204–207.

10. D. E. Nilsson, L. Gislén, M. M. Coates, et al., "Advanced Optics in a Jellyfish Eye," *Nature* 435, no. 7039 (2005): 201–205.

11. M. F. Land and D. E. Nilsson, *Animal Eyes* (Oxford University Press, 2012).

12. D. E. Nilsson and S. Pelger, "A Pessimistic Estimate of the Time Required for an Eye to Evolve," *Proceedings of the Royal Society of London B: Biological Sciences* 256 (1994): 53–58.

13. Land and Nilsson, *Animal Eyes,* 46.

14. T. Kimchi and J. Terkel, "Seeing and Not Seeing," *Current Opinion in Neurobiology* 12, no. 6 (2002): 728–734.

15. B. N. Schenkman and M. E. Nilsson, "Human Echolocation: Blind and Sighted Persons' Ability to Detect Sounds Recorded in the Presence of a Reflecting Object," *Perception* 39 (2010): 483–501; L. Thaler, S. R. Arnott, and M. A. Goodale, "Neural Correlates of Natural Human Echolocation in Early and Late Blind Echolocation Experts," *PLoS One* 6, no. 5 (2011): e20162.

16. J. A. Simmons, "The Resolution of Target Range by Echolocating Bats," *Journal of the Acoustic Society of America* 54 (1973): 157–173; J. Simmons, "Perception of Echo Phase Information in Bat Sonar," *Science* 204 (1979): 1336–1338.

17. G. Jones, "Echolocation," *Current Biology* 15, no. 13 (2005): R484–R488.

18 C. O'Callaghan, "Object Perception: Vision and Audition," *Philosophy Compass* 3, no. 4 (2008): 803–829.

19. For a review, see Y. Yovel, M. O. Franz, P. Stilz, and H. U. Schnitzler, "Complex Echo Classification by Echo-locating Bats: A Review," *Journal of Comparative Physiology A: Neuroethology, Sensory, Neural, and Behavioral Physiology* 197, no. 5 (2011): 475–490.

20. C. F. Moss and A. Surlykke, "Auditory Scene Analysis by Echolocation in Bats," *Journal of the Acoustical Society of America* 110, no. 4 (2001): 2207–2226.

21. M. Geva-Sagiv, L. Las, Y. Yovel, and N. Ulanovsky, "Spatial Cognition in Bats and Rats: From Sensory Acquisition to Multiscale Maps and Navigation," *Nature Reviews Neuroscience* 16, no. 2 (2015): 94–108.

22. D. Genzel and L. Wiegrebe, "Size Does Not Matter: Size-Invariant Echo-Acoustic Object Classification," *Journal of Comparative Physiology A: Neuroethology, Sensory, Neural, and Behavioral Physiology* 199, no. 2 (2013): 159–168.

23. Geva-Sagiv et al, "Spatial Cognition," 94–108.

24. H. Harley, E. A. Putman, and H. L. Roitblat, "Bottlenose Dolphins Perceive Object Features through Echolocation," *Nature* 424 (2003): 667–669; A. A. Pack, L. M. Herman, M. Hoffmann-Kuhnt, and B. K. Branstetter, "The Object Behind the Echo: Dolphins (*Tursiops truncatus*) Perceive Object Shape Globally through Echolocation," *Behavioral Processes* 58, no. 1–2 (2002): 1–26.

25. A. A. Pack and L. M. Herman, "Sensory Integration in the Bottlenosed Dolphin: Immediate Recognition of Complex Shapes across the Senses of Echolocation and Vision," *Journal of Acoustic Society of America* 98 (1995): 722–733.

26. For a review, see E. I. Knudsen, "The Hearing of the Barn Owl," *Scientific American* 245, no. 6 (1981): 113–125.

27. K. C. Catania and F. E. Remple, "Asymptotic Prey Profitability Drives Star-Nosed Moles to the Foraging Speed Limit," *Nature* 433 (2005): 519–522.

28. L. Miersch, W. Hanke, S. Wieskotten, et al., "Flow Sensing by Pinniped Whiskers," *Philosophical Transactions of the Royal Society of London B: Biological Sciences* 366, no. 1581 (2011): 3077–3084.

29. K. Kral and M. Poteser, "Motion Parallax as a Source of Distance Information in Locusts and Mantids," *Journal of Insect Behavior* 10, no. 1 (1997): 145–163; M. Lehrer, M. V. Srinivasan, S. W. Zhang, and G. A. Horridge, "Motion Cues Provide the Bee's Visual World with a Third Dimension," *Nature* 332, no. 6162 (1988): 356–357.

30. D. S. Jacobs and A. Bastian, *Predator–Prey Interactions: Co-evolution between Bats and their Prey* (Springer, 2017).

31. M. B. Fenton, "Questions, Ideas and Tools: Lessons from Bat Echolocation," *Animal Behaviour* 85 (2013): 869–879.

32. For instance, P. J. Miller, M. P. Johnson, and P. L. Tyack, "Sperm Whale Behaviour Indicates the Use of Echolocation Click Buzzes 'Creaks' in Prey Capture," *Proceedings of the Royal Society of London B: Biological Sciences* 271, no. 1554 (2004): 2239–2247.

33. C. Chiu, W. Xian, and C. F. Moss, "Flying in Silence: Echolocating Bats Cease Vocalizing to Avoid Sonar Jamming," *Proceedings of the National Academy of Sciences of the United States of America* 105, no. 35 (2008): 13116–13121.

34. H. T. Arita and M. B. Fenton, "Flight and Echolocation in the Ecology and Evolution of Bats," *Trends in Ecology and Evolution* 12 (1997): 53–58.

35. J. R. Speakman and P. A. Racey, "No Cost of Echolocation for Bats in Flight," *Nature* 350, no. 6317 (1991): 421–423.

36. M. Hagedorn and W. Heiligenberg, "Court and Spark: Electric Signals in the Courtship and Mating of Gymnotoid Fish," *Animal Behaviour* 33, no. 1 (1985): 254–265.

37. M. E. Arnegard and B. A. Carlson, "Electric Organ Discharge Patterns during Group Hunting by a Mormyrid Fish," *Proceedings of the Royal Society of London B: Biological Sciences* 272, no. 1570 (2005): 1305–1314.

38. J. A. Kaufman, A. Hladik, and P. Pasquet, "On the Expensive-Tissue Hypothesis: Independent Support from Highly Encephalized Fish," *Current Anthropology* 44, no. 5 (2003): 705–707.

39. G. von der Emde and S. Fetz, "Distance, Shape and More: Recognition of Object Features during Active Electrolocation in a Weakly Electric Fish," *Journal of Experimental Biology* 210, no. 17 (2007): 3082–3095; G. von der Emde, K. Behr, B. Bouton, et al., "3-Dimensional Scene Perception during Active Electrolocation in a Weakly Electric Pulse Fish," *Frontiers in Behavioral Neuroscience* 4 (2010): 26.

40. G. von der Emde and S. Schwarz, "Three-Dimensional Analysis of Object Properties during Active Electrolocation in Mormyrid Weakly Electric Fishes (*Gnathonemus petersii*)," *Philosophical Transactions of the Royal Society of London B: Biological Sciences* 355, no. 1401 (2000): 1143–1146.

41. G. von der Emde, S. Schwarz, L. Gomez, et al., "Electric Fish Measure Distance in the Dark," *Nature* 395, no. 6705 (1998): 890–894.

42. G. von der Emde, "Active Electrolocation of Objects in Weakly Electric Fish," *Journal of Experimental Biology* 202, no. 10 (1999): 1205–1215; G. von der Emde and S. Schwarz, "Imaging of Objects through Active Electrolocation in *Gnathonemus petersii*," *Journal of Physiology–Paris* 96, no. 5 (2002): 431–444.

43. C. Graff, G. Kaminski, M. Gresty, and T. Ohlmann, "Fish Perform Spatial Pattern Recognition and Abstraction by Exclusive Use of Active Electrolocation," *Currrent Biology* 14, no. 9 (2004): 818–823.

44. G. von der Emde, "Distance and Shape: Perception of the 3-Dimensional World by Weakly Electric Fish," *Journal of Physiology* 98 (2004): 67–80; Von der Emde et al., "Electric Fish Measure Distance," 890–894.

45. S. Schumacher, T. Burt de Perera, J. Thenert, and G. von der Emde, "Cross-Modal Object Recognition and Dynamic Weighting of Sensory Inputs in a Fish," *Proceedings of the National Academy of Sciences of the United States of America* 113, no. 27 (2016): 7638–7643.

46. G. S. Berns, P. F. Cook, S. Foxley, et al., "Diffusion Tensor Imaging of Dolphin Brains Reveals Direct Auditory Pathway to Temporal Lobe," *Proceedings of the Royal Society of London B: Biological Sciences* 282, no. 1811 (2015), https://doi.org/10.1098/rspb.2015.1203.

47. J. Parker, G. Tsagkogeorga, J. A. Cotton, et al., "Genome-wide Signatures of Convergent Evolution in Echolocating Mammals," *Nature* 502 (2013): 228–236.

Chapter 8

1. T. Bayne, *The Unity of Consciousness* (Oxford University Press, 2011). See also T. Bayne and D. Chalmers, "What Is the Unity of Consciousness?," in *The Unity of Consciousness: Binding, Integration and Dissociation*, ed. A. Cleeremans, 23–58 (Oxford University Press, 2003).

2. J. von Uexküll, *Umwelt und Innenwelt der Tiere*, 2nd ed. (Springer, 1921).

3. Von Uexküll, *Umwelt und Innenwelt*, 45.

4. Von Uexküll, *Umwelt und Innenwelt*, 167.

5. See M. Tønnessen, "Umwelt Transitions: Uexküll and Environmental Change," *Biosemiotics* 2, no. 1 (2009): 47–64.

6. See, for example, P. Godfrey-Smith, "Signs and Symbolic Behavior," *Biological Theory* 9, no. 1 (2014): 78–88.

7. See, for example, C. T. Bergstrom and M. Rosvall, "The Transmission Sense of Information," *Biology and Philosophy* 26, no. 2 (2011): 159–176; N. Shea, "Representation in the Genome and in Other Inheritance Systems," *Biology & Philosophy* 22, no. 3 (2007): 313–331.

8. C. Brentari, *Jakob von Uexküll: The Discovery of the Umwelt between Biosemiotics and Theoretical Biology* (Springer, 2015).

9. Tønnessen, "Umwelt Transitions," 47–64.

10. S. N. Salthe, "Theoretical Biology as an Anticipatory Text: The Relevance of Uexkull to Current Issues in Evolutionary Systems," *Semiotica* 134, no. 1/4 (2001): 359–380.

11. T. Nagel, "What Is It Like to Be a Bat?," *Philosophical Review* 83, no. 4 (1974): 435–450.

12. D. Chalmers, "Can We Construct a Science of Consciousness?," in *The Cognitive Neurosciences*, vol. 3, ed. M. Gazzaniga, 1111–1119 (MIT Press, 2004).

13. T. Metzinger, *The Ego Tunnel: The Science of the Soul and the Myth of the Self* (Basic Books, 2009).

14. See, for example, B. Merker, "Consciousness without a Cerebral Cortex: A Challenge for Neuroscience and Medicine," *Behavioral and Brain Sciences* 30, no. 1 (2007): 63–81, and its accompanying commentaries.

15. See, for example, D. B. Edelman and A. K. Seth, "Animal Consciousness: A Synthetic Approach," *Trends in Neurosciences* 32, no. 9 (2009): 476–484; see also G. M. Edelman, *The Remembered Present* (Basic Books, 1989).

16. For a defense of the higher-order representational view, see P. Carruthers, "Brute Experience," *Journal of Philosophy* 86, no. 5 (1989): 258–269.

17. P. Godfrey-Smith, "Mind, Matter, and Metabolism," *Journal of Philosophy* 113, no. 10 (2016): 495.

18. G. M. Edelman and G. Tononi, *A Universe of Consciousness* (Basic Books, 2000), xiii.

19. S. J. Shettleworth, *Cognition, Evolution, and Behavior*, 2nd ed. (Oxford University Press, 2010), 4.

20. A. Trewavas, *Plant Behaviour and Intelligence* (Oxford University Press, 2015).

21. Godfrey-Smith, "Mind, Matter, and Metabolism," 481–506.

22. See, for example, M. van Duijn, "Phylogenetic Origins of Biological Cognition: Convergent Patterns in the Early Evolution of Learning," *Journal of the Royal Society Interface Focus* 7 (2017): 20160158.

23. F. A. Keijer, "Evolutionary Convergence and Biologically Embodied Cognition," *Journal of the Royal Society Interface Focus* 7 (2017): 20160123.

24. G. Tononi, "An Information Integration Theory of Consciousness," *BMC Neuroscience* 5, no. 1 (2004): 42; G. Tononi, "The Integrated Information Theory of Consciousness: An Updated Account," *Archives italiennes de biologie* 150, no. 2/3 (2011): 56–90; C. Koch, M. Massimini, M. Boly, and G. Tononi, "Neural Correlates of Consciousness: Progress and Problems," *Nature Reviews Neuroscience* 17, no. 5 (2016): 307–321.

25. C. E. Shannon, "The Mathematical Theory of Communication," in C. E. Shannon, *The Mathematical Theory of Communication* (University of Illinois Press, 1949), https://www.itsoc.org/about/shannon-1948.

26. O. Sacks, *An Anthropologist on Mars* (Picador, 1995).

27. G. Tononi and C. Koch, "Consciousness: Here, There and Everywhere?," *Philosophical Transactions of the Royal Society of London B: Biological Sciences* 370 (2015): 20140167.

28. E. Schechter, *Self-Consciousness and "Split" Brains: The Mind's I* (Oxford University Press, 2018).

29. T. Nagel, "Brain Bisection and the Unity of Consciousness," *Synthese* 22, no. 3–4 (1971): 396–413.

30. G. M. Edelman and G. Tononi, *A Universe of Consciousness* (Basic Books, 2000), 15.

31. J. Quilty-Dunn, "Iconicity and the Format of Perception," *Journal of Consciousness Studies* 23, no. 3–4 (2016): 255–263.

32. Z. Z. Bronfman, S. Ginsburg, and E. Jablonka, "The Transition to Minimal Consciousness through the Evolution of Associative Learning," *Frontiers in Psychology* 7 (2016): 1954.

33. A. Revonsuo, "Binding and the Phenomenal Unity of Consciousness," *Consciousness and Cognition* 8, no. 2 (1999): 173–185.

34. O. Sacks, *The Man Who Mistook His Wife for a Hat* (Picador, 2009).

35. S. Baron-Cohen and J. E. Harrison, *Synaesthesia: Classic and Contemporary Readings* (Blackwell, 1997).

36. C. O'Callaghan, "Seeing What You Hear: Cross-Modal Illusions and Perception," *Philosophical Issues* 18, no. 1 (2008): 316–338.

37. On neural synchrony: L. Goldfarb and A. Treisman, "Counting Multidimensional Objects: Implications for the Neural-Synchrony Theory," *Psychological Science* 24, no. 3 (2013): 266–271. On reentry: S. Bouvier and A. Treisman, "Visual Feature Binding Requires Reentry," *Psychological Science* 21, no. 2 (2010): 200–204.

38. Quoted in D. L. Cheney and R. M. Seyfarth, *Baboon Metaphysics: The Evolution of a Social Mind* (University of Chicago Press, 2008).

39. L. Weiskrantz, *Blindsight: A Case Study Spanning 35 Years and New Developments* (Oxford University Press, 2009).

40. For a defense of the "zombie bee" position, see S. Allen-Hermanson, "Insects and the Problem of Simple Minds: Are Bees Natural Zombies?," *Journal of Philosophy* 105, no. 8 (2008): 389–415.

41. B. Baars, *A Cognitive Theory of Consciousness* (Cambridge University Press, 1988).

42. Problems encountered in attempts to construct a sensory modality space are discussed at length in R. Gray, "Is There a Space of Sensory Modalities?," *Erkenntnis* 78, no. 6 (2013): 1259–1273.

43. B. L. Keeley, "Making Sense of the Senses: Individuating Modalities in Humans and Other Animals," *Journal of Philosophy* 99, no. 1 (2002): 5–28.

Chapter 9

1. H. Putnam, "The Nature of Mental States," in *Mind, Language and Reality: Philosophical Papers*, vol. 2, 249–240 (Cambridge University Press, 1967); but see T. Polger, "Evaluating the Evidence for Multiple Realization," *Synthese* 167, no. 3 (2009): 457–472.

2. T. W. Polger, "Realization and the Metaphysics of Mind," *Australasian Journal of Philosophy* 85, no. 2 (2007): 233–259.

3. S. Conway Morris, "Evolution: Like Any Other Science It Is Predictable," *Philosophical Transactions of the Royal Society of London B: Biological Sciences* 365, no. 1537 (2010): 133–145; S. Conway Morris, *The Runes of Evolution: How the Universe Became Self-Aware* (Templeton Press, 2015).

4. M. Giurfa, "Cognition with Few Neurons: Higher-Order Learning in Insects," *Trends in Neurosciences* 36, no. 5 (2013): 285–294.

5. S. M. Farris, "Evolution of Complex Higher Brain Centers and Behaviors: Behavioral Correlates of Mushroom Body Elaboration in Insects," *Brain Behavior and Evolution* 82 (2013): 9–18.

6. On possible justifications for the parsimony preference in cladistics, see E. Sober, *Reconstructing the Past: Parsimony, Evolution, and Inference* (MIT Press, 1991).

7. B. Galliot, M. Quiquand, L. Ghila, et al., "Origins of Neurogenesis, a Cnidarian View," *Developmental Biology* 332, no. 1 (2009): 2–24.

8. D. H. Erwin and J. W. Valentine, *The Cambrian Explosion: The Construction of Animal Biodiversity* (Roberts and Co., 2013).

9. R. G. Northcutt, "Evolution of Centralized Nervous Systems: Two Schools of Evolutionary Thought," *Proceedings of the National Academy of Sciences of the United States of America* 109, Suppl. 1 (2012): 10626–10633.

10. F. Hirth, L. Kammermeier, E. Frei, et al., "An Urbilaterian Origin of the Tripartite Brain: Developmental Genetic Insights from *Drosophila*." *Development* 130, no. 11 (2003): 2365–2373.

11. D. Acampora, V. Avantaggiato, F. Tuorto, et al., "Murine Otx1 and *Drosophila* otd Genes Share Conserved Genetic Functions Required in Invertebrate and Vertebrate Brain Development," *Development* 125 (1998): 1691–1702.

12. T. Cavalier-Smith, "Origin of Animal Multicellularity: Precursors, Causes, Consequences—the Choanoflagellate/Sponge Transition, Neurogenesis and the Cambrian Explosion," *Philosophical Transactions of the Royal Society of London B: Biological Sciences* 372, no. 1713 (2017): 20150476.

13. G. H. Wolff and N. J. Strausfeld, "Genealogical Correspondence of a Forebrain Centre Implies an Executive Brain in the Protostome–Deuterostome Bilaterian Ancestor," *Philosophical Transactions of the Royal Society of London B: Biological Sciences* 371, no. 1685 (2016): 20150055.

14. T. Shomrat, A. L. Turchetti-Maia, N. Stern-Mentch, et al., "The Vertical Lobe of Cephalopods: An Attractive Brain Structure for Understanding the Evolution of Advanced Learning and Memory Systems," *Journal of Comparative Physiology A: Neuroethology, Sensory, Neural, and Behavioral Physiology* 201, no. 9 (2015): 947–956.

15. J. Z. Young, *The Anatomy of the Nervous System of Octopus vulgaris* (Clarendon Press, 1971).

16. For a recent review, see J. Mather, "What Is in an Octopus's Mind?," *Animal Sentience* 4, no. 26 (2019): 1.

17. P. N. Lee, P. Callaerts, H. G. de Couet, M. Q. Martindale, et al., "Cephalopod *Hox* Genes and the Origin of Morphological Novelties," *Nature* 424 (2003): 1061–1065.

18. Shomrat et al., "Vertical Lobe," 947–956.

19. C. B. Albertin, O. Simakov, T. Mitros, et al., "The Octopus Genome and the Evolution of Cephalopod Neural and Morphological Novelties," *Nature* 524, no. 7564 (2015): 220–224.

20. E. J. Steele, S. Al-Mufti, K.A. Augustyn, et al., "Cause of Cambrian Explosion—Terrestrial or Cosmic?," *Progress in Biophysics and Molecular Biology* 136 (2018): 3–23.

21. L. L. Moroz and A. B. Kohn, "Independent Origins of Neurons and Synapses: Insights from Ctenophores," *Philosophical Transactions of the Royal Society of London B: Biological Sciences* 371 (2016): 20150041.

22. C. R. Smarandache-Wellmann, "Arthropod Neurons and Nervous System," *Current Biology* 26, no. 20 (2016): R960–R965.

23. W. B. Kristan Jr., "Early Evolution of Neurons," *Current Biology* 26, no. 20 (2016): R949–R954.

24. R. Powell and N. Shea, "Homology across Inheritance Systems," *Biology and Philosophy* 29, no. 6 (2014): 781–806; V.L. Roth, "Homologies and Hierarchies: Problems Solved and Unresolved," *Evolutionary Biology* 4 (1991): 167–194.

25. N. V. Whelan, K. M. Kocot, T. P. Moroz, et al. "Ctenophore Relationships and Their Placement as the Sister Group to All Other Animals," *Nature Ecology and Evolution* 1, no. 11 (2017): 1737–1746.

26. R. Feuda, M. Dohrmann, W. Pett, et al., "Improved Modeling of Compositional Heterogeneity Supports Sponges as Sister to All Other Animals," *Current Biology* 27, no. 24 (2017): 3864–3870. e4; Cavalier-Smith, "Origin of Animal Multicellularity," 20150476.

27. L. L. Moroz, "The Genealogy of Genealogy of Neurons," *Communicative and Integrative Biology* 7, no. 6 (2014): e993269.

28. W. B. Kristan Jr., "Early Evolution of Neurons," *Current Biology* 26, no. 20 (2016): R949–R954, p. R954.

29. Y. Forterre, J. M. Skotheim, J. Dumais, and L. Mahadevan, "How the Venus Flytrap Snaps," *Nature* 433, no. 7024 (2005): 421–425.

30. A. Trewavas, *Plant Behaviour and Intelligence* (Oxford University Press, 2014).

Chapter 10

1. See P. Low, "The Cambridge Declaration on Consciousness," Francis Crick Memorial Conference (Cambridge, United Kingdom, 2012), http://fcmconference.org/img/CambridgeDeclarationOnConsciousness.pdf.

2. P. Godfrey-Smith, *Other Minds: The Octopus, the Sea, and the Deep Origins of Consciousness* (Farrar, Straus and Giroux, 2017).

3. I. Mikhalevich, R. Powell, and C. Logan, "Is Behavioural Flexibility Evidence of Cognitive Complexity? How Evolution Can Inform Comparative Cognition," *Journal of the Royal Society Interface Focus* 7, no. 3 (2017): 20160121; J. L. Gould, "Animal Cognition," *Current Biology* 14 (2004): R372–R375.

4. J. N. Richter, B. Hochner, and M. J. Kuba, "Pull or Push? Octopuses Solve a Puzzle Problem," *PloS One* 11, no. 3 (2016): e0152048.

5. C. Bertapelle, G. Polese, and A. Di Cosmo, "Enriched Environment Increases PCNA and PARP1 Levels in *Octopus vulgaris* Central Nervous System: First Evidence of Adult Neurogenesis in Lophotrochozoa," *Journal of Experimental Zoology Part B: Molecular and Developmental Evolution* 328, no. 4 (2017): 347–359.

6. J. K. Finn, T. Tregenza, and M. D. Norman, "Defensive Tool Use in a Coconut-Carrying Octopus," *Current Biology* 19, no. 23 (2009): R1069–R1070.

7. C. Alves, J. G. Boal, and L. Dickel, "Short-Distance Navigation in Cephalopods: A Review and Synthesis," *Cognitive Processing* 9, no. 4 (2008): 239–247.

8. C. Jozet-Alves, M. Bertin, and N. S. Clayton, "Evidence of Episodic-like Memory in Cuttlefish," *Current Biology* 23, no. 23 (2013): R1033–R1035.

9. N. S. Clayton and A. Dickinson, "Episodic-like Memory during Cache Recovery by Scrub Jays," *Nature* 395, no. 6699 (1998): 272–274.

10. M. J. Kuba, R. A. Byrne, D. V. Meisel, and J. A. Mather, "When Do Octopuses Play? Effects of Repeated Testing, Object Type, Age, and Food Deprivation on Object Play in *Octopus vulgaris*," *Journal of Comparative Psychology* 120, no. 3 (2006): 184–190.

11. For a recent review, see J. A. Mather and L. Dickel, "Cephalopod Complex Cognition," *Current Opinion in Behavioral Sciences* 16 (2017): 131–137, p. 135.

12. See, for example, A. Avarguès-Weber, A. G. Dyer, M. Combe, and M. Giurfa, "Simultaneous Mastering of Two Abstract Concepts by the Miniature Brain of Bees," *Proceedings of the National Academy of Sciences of the United States of America* 109, no. 19 (2012): 7481–7486.

13. See, for example, A. Horridge, "The Spatial Resolutions of the Apposition Compound Eye and Its Neuro-sensory Feature Detectors: Observation versus Theory," *Journal of Insect Physiology* 51, no. 3 (2005): 243–266; A. Horridge, "What Does an Insect See?," *Journal of Experimental Biology* 212, no. 17 (2009): 2721–2729.

14. A. B. Barron and C. Klein, "What Insects Can Tell Us about the Origins of Consciousness," *Proceedings of the National Academy of Sciences of the United States of America* 113, no. 18 (2016): 4900–4908.

15. M. V. Srinivasan, S. W. Zhang, and H. Zhu, "Honeybees Link Sights to Smells," *Nature* 396, no. 6712 (1998): 637–638; L.-Z. Zhang, S.-W. Zhang, Z.-L. Wang, et al., "Cross-Modal Interaction between Visual and Olfactory Learning in *Apis cerana*," *Journal of Comparative Physiology A: Neuroethology, Sensory, Neural, and Behavioral Physiology* 200, no. 10 (2014): 899–909.

16. S. Stach, J. Benard, and M. Giurfa, "Local-Feature Assembling in Visual Pattern Recognition and Generalization in Honeybees," *Nature* 429, no. 6993 (2004): 758–761.

17. M. J. Sheehan and E. A. Tibbetts, "Specialized Face Learning Is Associated with Individual Recognition in Paper Wasps," *Science* 334, no. 6060 (2011): 1272–1275.

18. A. G. Dyer, C. Neumeyer, and L. Chittka, "Honeybee (*Apis mellifera*) Vision Can Discriminate between and Recognise Images of Human Faces," *Journal of Experimental Biology* 208, no. 24 (2005): 4709–4714.

19. A. Avarguès-Weber, G. Portelli, J. Benard, et al., "Configural Processing Enables Discrimination and Categorization of Face-like Stimuli in Honeybees," *Journal of Experimental Biology* 213, no. 4 (2010): 593–601.

20. A. G. Dyer, M. G. Rosa, and D. H. Reser, "Honeybees Can Recognise Images of Complex Natural Scenes for Use as Potential Landmarks," *Journal of Experimental Biology* 211 (2008): 1180–1186.

21. S. Watanabe, J. Sakamoto, and M. Wakita, "Pigeons' Discrimination of Paintings by Monet and Picasso," *Journal of the Experimental Analysis of Behavior* 63, no. 2 (1995): 165–174.

22. W. Wu, A. M. Moreno, J. M. Tangen, and J. Reinhard, "Honeybees Can Discriminate between Monet and Picasso Paintings," *Journal of Comparative Physiology A: Neuroethology, Sensory, Neural, and Behavioral Physiology* 199, no. 1 (2013): 45–55.

23. A. G. Dyer, S. R. Howard, and J. E. Garcia, "Through the Eyes of a Bee: Seeing the World as a Whole," *Animal Studies Journal* 5, no. 1 (2016): 97–109.

24. J. D. McIver and G. Stonedahl, "Myrmecomorphy: Morphological and Behavioral Mimicry of Ants," *Annual Review of Entomology* 38, no. 1 (1993): 351–377; M. Maruyama and J. Parker, "Deep-Time Convergence in Rove Beetle Symbionts of Army Ants," *Current Biology* 27, no. 6 (2017): 920–926.

25. G. D. Ruxton, T. N. Sherratt, and M. P. Speed, *Avoiding Attack: The Evolutionary Ecology of Crypsis, Warning Signals and Mimicry* (Oxford University Press, 2004).

26. M. Maruyama and J. Parker, "Deep-Time Convergence in Rove Beetle Symbionts of Army Ants," *Current Biology* 27, no. 6 (2017): 920–926.

27. S. Alem, C. J. Perry, X. Zhu, et al., "Associative Mechanisms Allow for Social Learning and Cultural Transmission of String Pulling in an Insect," *PLoS Biology* 14, no. 10 (2016): e1002564.

28. N. R. Franks and T. Richardson, "Teaching in Tandem-Running Ants," *Nature* 439, no. 7073 (2006): 153.

29. E. Danchin et al., "Cultural Flies: Conformist Social Learning in Fruitflies Predicts Long-Lasting Matechoice Traditions," *Science* 362(2018): 1025–1030.

30. C. J. Perry and A. B. Barron, "Honey Bees Selectively Avoid Difficult Choices," *Proceedings of the National Academy of Sciences of the United States of America* 110 (2013): 19155–19159.

31. O. J. Loukola, C. J. Perry, L. Coscos, and L. Chittka, "Bumblebees Show Cognitive Flexibility by Improving on an Observed Complex Behavior," *Science* 355, no. 6327 (2017): 833–836.

32. S. Zhang, M. V. Srnivasan, H. Zhu, and J. Wong, "Grouping of Visual Objects by Honeybees," *Journal of Experimental Biology* 207, no. 19 (2004): 3289–3298.

33. A. Avarguès-Weber and M. Giurfa, "Conceptual Learning by Miniature Brains," *Proceedings of the Royal Society of London B: Biological Sciences* 280, no. 1772 (2013): 20131907.

34. A. Avarguès-Weber, A. G. Dyer, and M. Giurfa, "Conceptualization of Above and Below Relationships by an Insect," *Proceedings of the Royal Society of London B: Biological Sciences* 278 (2011): 898–905.

35. M. Giurfa, S. Zhang, A. Jennet, et al., "The Concepts of 'Sameness' and 'Difference' in an Insect," *Nature* 410, no. 6831 (2001): 930–933.

36. S. R. Howard, A. Avarguès-Weber, J. E. Garcia, et al., "Numerical Ordering of Zero in Honey Bees," *Science* 360, no. 6393 (2018): 1124–1126.

37. A. Avarguès-Weber, D. d'Amaro, M. Metzler, and A.G. Dyer, "Conceptualization of Relative Size by Honeybees," *Frontiers in Behavioral Neuroscience* 8 (2014): 80.

38. Avarguès-Weber et al., "Simultaneous Mastering," 7481–7486.

39. Giurfa et al., "Concepts of 'Sameness' and 'Difference' in an Insect."

40. For an association-based model of sameness/difference, see A. J. Cope et al, "Abstract Concept Learning in a Simple Neural Network Inspired by the Insect Brain," *PLoS Computational Biology* 14 (9): e1006435. It is unclear whether this model can explain cross-modal transfer results, however. For a discussion of association-based explanations of above/below discriminations, see V. Vasas and L. Chittka, "Insect-Inspired Sequential Inspection Strategy Enables an Artificial Network of Four Neurons to Estimate Numerosity," *iScience* 11(2019): 85-92.

41. Howard, Avarguès-Weber, Garcia, et al., "Numerical Ordering of Zero in Honey Bees."

42. S. R. Howard, "Numerical Cognition in Honeybees Enables Addition and Subtraction," *Science Advances* 5, no. 2 (2019): eaav0961.

43. Vasas and Chittka, "Insect-Inspired Sequential Inspection Strategy."

44. For a discussion of the role of simplicity and parsimony in adjudicating hypotheses in animal cognition science, see I. Mikhlaevich (published as Meketa), "A Critique of the Principle of Cognitive Simplicity in Comparative Cognition," *Biology & Philosophy* 29, no. 5 (2014): 731–745; S. Fitzpatrick, "Doing Away with Morgan's Canon," *Mind & Language* 23 (2008): 224–246.

45. P. Skorupski and L. Chittka, "Animal Cognition: An Insect's Sense of Time?," *Current Biology* 16, no. 19 (2006): R851–R853.

46. F. R. Cross and R. R. Jackson, "Representation of Different Exact Numbers of Prey by a Spider-Eating Predator," *Interface Focus* 7, no. 3 (2017): 20160035.

47. F. R. Cross and R. R. Jackson, "The Execution of Planned Detours by Spider-Eating Predators," *Journal of Experimental Analysis of Behavior* 105, no. 1 (2016): 194–210; M. S. Tarsitano and R. R. Jackson, "Araneophagic Jumping Spiders Discriminate between Detour Routes That Do and Do Not Lead to Prey," *Animal Behaviour* 53 (1997): 257–266.

48. J. Liedtke and J. M. Schneider, "Association and Reversal Learning Abilities in a Jumping Spider," *Behavioural Processes* 103 (2014): 192–198.

49. Cross and Jackson, "Execution of Planned Detours," 194–210.

50. R. R. Jackson and F. R. Cross, "A Cognitive Perspective on Aggressive Mimicry," *Journal of Zoology* 290, no. 3 (2013): 161–171.

51. D. P. Harland and R. R. Jackson, "Eight-Legged Cats and How They See: A Review of Recent Research on Jumping Spiders (Araneae: Salticidae)," *Cimbebasia* 16 (2000): 231–240.

52. M. Bartos, "Hunting Prey with Different Escape Potentials—Alternative Predatory Tactics in a Dune Dwelling Salticid," *Journal of Arachnology* 35, no. 3 (2007): 499–508.

53. A. Bear and O. Hasson, "The Predatory Response of a Stalking Spider, *Plexippus paykulli*, to Camouflage and Prey Type," *Animal Behaviour* 54, no. 4 (1997): 993–998.

54. See, for instance, M. Collett, L. Chittka, and T. S. Collett, "Spatial Memory in Insect Navigation," *Current Biology* 23, no. 17 (2013): R789–R800; K. Cheng, "How to Navigate without Maps: The Power of Taxon-like Navigation in Ants," *Comparative Cognition and Behavior Reviews* 7 (2012): 1–22.

55. F. C. Dyer, "Bees Acquire Route-Based Memories but Not Cognitive Maps in a Familiar Landscape," *Animal Behaviour* 41, no. 2 (1991): 239–246.

56. J. L. Gould, "The Locale Map of Honey Bees: Do Insects Have Cognitive Maps?," *Science* 232, no. 4752 (1986): 861–863.

57. J. L. Gould and C. G. Gould, *The Honey Bee* (W. H. Freeman, 1988).

58. M. C. Cammaerts and R. Cammaerts, "Are Ants (Hymenoptera, Formicidae) Capable of Self-recognition?," *Journal of Science* 5, no. 7 (2015): 521–532.

59. Mikhalevich (published as Meketa), "Critique of the Principle of Cognitive Simplicity."

60. L. Chittka and J. Niven, "Are Bigger Brains Better?," *Current Biology* 19, no. 21 (2009): R996.

61. Z. Z. Bronfman, S. Ginsburg, and E. Jablonka, "The Transition to Minimal Consciousness through the Evolution of Associative Learning," *Frontiers in Psychology* 7 (2016): 1954.

62. Barron and Klein, "What Insects Can Tell Us."

63. B. Merker, "Consciousness without a Cerebral Cortex: A Challenge for Neuroscience and Medicine," *Behavioral and Brain Sciences* 30, no. 1 (2007): 63–81.

64. L. Morawetz and J. Spaethe, "Visual Attention in a Complex Search Task Differs between Honeybees and Bumblebees," *Journal of Experimental Biology* 215, no. 14 (2012): 2515–2523.

65. For a review, see R. J. Greenspan and B. Van Swinderen, "Cognitive Consonance: Complex Brain Functions in the Fruit Fly and Its Relatives," *Trends in Neurosciences* 27, no. 12 (2004): 707–711.

66. A. R. Damasio, *Descartes' Error: Emotion, Reason, and the Human Brain* (Putnam, 1994), 155.

67. C. J. Perry, L. Baciadonna, and L. Chittka, "Unexpected Rewards Induce Dopamine-Dependent Positive Emotion-like State Changes in Bumblebees," *Science* 353, no. 6307 (2016): 1529–1531.

68. I. Mikhalevich and R. Powell, "Minds without Spines: Toward a More Inclusive Animal Ethics" (unpublished manuscript, 2019).

69. See, for instance, S. Allen-Hermanson, "Insects and the Problem of Simple Minds: Are Bees Natural Zombies?," *Journal of Philosophy* 105, no. 8 (2008): 389–415.

70. See C. Allen and M. Bekoff, *Species of Mind* (MIT Press, 1997); C. Allen and M. Trestman, "Animal Consciousness," in *Stanford Encyclopedia of Philosophy,* ed. E. N. Zalta (Stanford University, 1995/2016), https://plato.stanford.edu/entries/consciousness-animal/.

71. For a discussion of this convergent regularity and its role in cognition inferences, see Mikhalevich et al., "Behavioural Flexibility Evidence," 20160121.

72. E. Nagel, "Goal-Directed Processes in Biology," *Journal of Philosophy* 74, no. 5 (1977): 261–279.

73. E. Mayr, "The Idea of Teleology," *Journal of the History of Ideas* 53 (1992): 117–135.

74. D. W. McShea, "Upper-Directed Systems: A New Approach to Teleology in Biology," *Biology and Philosophy* 27, no. 5 (2012): 663–684.

75. I make a preliminary case for this in my master's thesis: R. Powell, "Extinction and Reiteration in the History of Life: A Macroevolutionary Dialectic" (Duke University, 2009).

76. A. S. McMenamin, *The Garden of Ediacara: Discovering the First Complex Life* (Columbia University Press, 1998); A. Parker, *In the Blink of an Eye* (Basic Books, 2004).

77. G. J. Vermeij, *Evolution and Escalation: An Ecological History of Life* (Princeton University Press, 1987).

78. M. Trestman, "The Cambrian Explosion and the Origins of Embodied Cognition," *Biological Theory* 8, no. 1 (2013): 80–92.

Coda to Part II

1. M. Ćirković, *The Great Silence: Science and Philosophy of Fermi's Paradox* (Oxford University Press, 2018).

2. P. D. Ward and D. Brownlee, *Rare Earth: Why Complex Life Is Uncommon in the Universe* (Springer, 2000).

3. F. D. Drake, *Intelligent Life in Space* (Macmillan, 1962).

4. S. J. Gould, *Wonderful Life: The Burgess Shale and the Nature of History.* (Norton, 1989), 319.

5. For corvids, see N. J. Emery and N. S. Clayton, "The Mentality of Crows: Convergent Evolution of Intelligence in Corvids and Apes," *Science* 306, no. 5703 (2004): 1903–1907; for parrots, see I. M. Pepperberg, *The Alex Studies: Cognitive and Communicative Abilities of Grey Parrots* (Harvard University Press, 2009).

6. S. Brusatte, "Tyrannosaur Paleobiology: New Research on Ancient Exemplar Organisms," *Science* 329, no. 5998 (2010): 1481–1485.

7. A. M. Balanoff, G. S. Bever, T. B. Rowe, and M. A. Norell, "Evolutionary Origins of the Avian Brain," *Nature* 501, no. 7465 (2013): 93–96.

8. For a review, see L. Marino, R. C. Connor, R. E. Fordyce, et al., "Cetaceans Have Complex Brains for Complex Cognition," *PLoS Biology* 5, no. 5 (2007): e139.

9. S. Harmand, J. E. Lewis, C. S. Feibel, et al., "3.3-Million-Year-Old Stone Tools from Lomekwi 3, West Turkana, Kenya," *Nature* 521 (2015): 310–315.

10. J. McNabb, "Hominids and the Early-Middle Pleistocene Transition: Evolution, Culture and Climate in Africa and Europe," in *Early-Middle Pleistocene Transitions: The Land-Ocean Evidence*, ed. M. J. Head and P. L. Gibbard, 287–304 (Geological Society, 2005).

11. D. L. Everett, *How Language Began: The Story of Humanity's Greatest Invention* (Liveright, 2017).

12. M. Domínguez-Rodrigo, "Hunting and Scavenging by Early Humans: The State of the Debate," *Journal of World Prehistory* 16, no. 1 (2002): 1–54.

13. J. A. J. Gowlett, "The Discovery of Fire by Humans: A Long and Convoluted Process," *Philosophical Transactions of the Royal Society of London B: Biological Sciences* 371, no. 1696 (2016), https://doi.org/10.1098/rstb.2015.0164.

14. R. W. Wrangham, *Catching Fire: How Cooking Made Us Human* (Basic Books, 2009).

15. N. Bostrom, *Superintelligence: Paths, Dangers, Strategies* (Oxford University Press, 2016).

16. M. Tomasello, A. C. Kruger, and H. H. Ratner, "Cultural Learning," *Behavioral and Brain Sciences* 16 (1993): 495–511. For a recent similar view, see V. Kumar and R. Campbell, "How Morality Evolves" (unpublished manuscript).

17. C. Tennie, J. Call, and M. Tomasello, "Ratcheting Up the Ratchet: On the Evolution of Cumulative Culture," *Philosophical Transactions of the Royal Society of London B: Biological Sciences* 364, no. 1528 (2009): 2405–2415.

18. J. J. Shea, *Stone Tools in Human Evolution: Behavioral Differences among Technological Primates* (Cambridge University Press, 2017).

19. M. Tomasello, *A Natural History of Human Thinking* (Harvard University Press, 2014); M. Tomasello, *Why We Cooperate* (MIT Press, 2009).

20. T. Suddendorf and M. C. Corballis, "The Evolution of Foresight: What Is Mental Time Travel, and Is It Unique to Humans?," *Behavioral and Brain Sciences* 30, no. 3 (2007): 299–313.

21. J. Tooby and I. DeVore, “The Reconstruction of Hominid Behavioral Evolution through Strategic Modeling,” in *Primate Models of Hominid Behavior*, ed. W. G. Kinzey, 183–237 (SUNY Press, 1987).

22. D. Kelemen and S. Carey, “The Essence of Artifacts: Developing the Design Stance,” in *Creations of the Mind: Theories of Artifacts and their Representation*, ed. S. Laurence and E. Margolis, 212–230 (Oxford University Press, 2007).

23. R. Joyce, *The Evolution of Morality* (MIT Press, 2006).

24. M. M. Skinner, N. B. Stephens, Z. J. Tsegai, et al., “Human-like Hand Use in *Australopithecus africanus*,” *Science* 347, no. 6220 (2015): 395–399.

25. Krause, J., C. Lalueza-Fox, L. Orlando, et al., “The Derived FOXP2 Variant of Modern Humans Was Shared with Neandertals,” *Current Biology* 17, no. 21 (2007): 1908–1912.

26. E. Marris, “Neanderthal Artists Made Oldest-Known Cave Paintings,” *Nature*, February 22, 2018, https://www.nature.com/articles/d41586-018-02357-8.

27. On evolutionary accounts of morality, see R. Powell and A. Buchanan, *The Evolution of Moral Progress* (Oxford University Press, 2018); M. Tomasello, *A Natural History of Human Morality* (Harvard University Press, 2016); J. Haidt, *The Righteous Mind* (Pantheon, 2012); R. Joyce, *The Evolution of Morality* (MIT Press, 2007); C. Boehm, *Hierarchy in the Forest: The Evolution of Egalitarian Behavior* (Harvard University Press, 2001); Kumar and Campbell, “How Morality Evolves.”

28. K. Sterelny, *The Evolved Apprentice: How Evolution Made Humans Unique* (MIT Press, 2012).

29. S. Mathew and R. Boyd, “Punishment Sustains Large-Scale Cooperation in Prestate Warfare,” *Proceedings of the National Academy of Sciences of the United States of America* 108, no. 28 (2011): 11375–11380; J. Henrich, R. McElreath, A. Barr, et al., “Costly Punishment Across Human Societies,” *Science* 312, no. 5781 (2006): 1767–1770.

30. R. Hanson, “The Great Filter—Are We Almost Past It?,” September 15,1998, http://hanson.gmu.edu/greatfilter.html.

Index

Printed in the United States
by Baker & Taylor Publisher Services